Universitext

Universitext

Universitext is a series of textbooks that presents material from a wide variety of mathematical disciplines at master's level and beyond. The books, often well class-tested by their author, may have an informal, personal, even experimental approach to their subject matter. Some of the most successful and established books in the series have evolved through several editions, always following the evolution of teaching curricula, into very polished texts.

Thus as research topics trickle down into graduate-level teaching, first textbooks written for new, cutting-edge courses may make their way into *Universitext*.

For further volumes:
www.springer.com/series/223

Anton Deitmar

Automorphic Forms

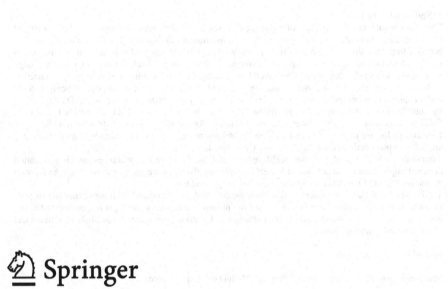

Springer

Anton Deitmar
Inst. Mathematik
Universität Tübingen
Tübingen, Baden-Württemberg
Germany

Translation from German language edition: Automorphe Formen by Anton Deitmar
© 2010, Springer Berlin Heidelberg
Springer Berlin Heidelberg is a part of Springer Science+Business Media
All Rights Reserved

ISSN 0172-5939   ISSN 2191-6675 (electronic)
Universitext
ISBN 978-1-4471-4434-2   ISBN 978-1-4471-4435-9 (eBook)
DOI 10.1007/978-1-4471-4435-9
Springer London Heidelberg New York Dordrecht

Library of Congress Control Number: 2012947923

Mathematics Subject Classification (2010): 11F12, 11F70, 11-01, 11F37

Printed on acid-free paper

Springer is part of Springer Science+Business Media (www.springer.com)

Introduction

This book is an introduction to the theory of automorphic forms. Starting with classical modular forms, it leads to representation theory of the adelic GL(2) and corresponding L-functions. Classical modular forms, which are introduced in the beginning of the book, will serve as the principal example until the very end, where it is verified that the classical and representation-theoretic approaches lead to the same L-functions. Modular forms are defined as holomorphic functions on the upper half plane, satisfying a particular transformation law under linear fractional maps with integer coefficients. We then lift functions on the upper half plane to the group $SL_2(\mathbb{R})$, a step that allows the introduction of representation-theoretic methods to the theory of automorphic forms. Finally, ground rings are extended to adelic rings, which means that number-theoretical questions are built into the structure and can be treated by means of analysis and representation theory.

For this book, readers should have some knowledge of algebra and complex analysis. They should be acquainted with group actions and the basic theory of rings. Further, they should be able to apply the residue theorem in complex analysis. Additionally, knowledge of measure and integration theory is useful but not necessary. One needs basic notions of this theory, like that of a σ-algebra and measure and some key results like the theorem of dominated convergence or the completeness of L^p-spaces. For the convenience of the reader, we have collected these facts in an appendix.

The present book focuses on the interrelation between automorphic forms and L-functions. To increase accessibility, we have tried to obtain the central results with a minimum of theory. This has the side effect that the presentation is not of the utmost generality; therefore the interested reader is given a guide to the literature.

In Chap. 1 the classical approach to modular forms via doubly periodic functions is presented. The Weierstrass $\wp$-function leads to Eisenstein series and thus to modular forms. The modular group and its modular forms are the themes of Chap. 2, which concludes with the presentation of L-functions. According to Dieudonné, there have been two revolutions in the theory of automorphic forms: the *intervention of Lie groups* and the *intervention of adeles*. Lie groups intervene in Chap. 3, and adeles in the rest of the book. We try to maintain continuity of presentation by continually referring back to the example of classical modular forms. Chapters 4 and 5 pave the way for Tate's thesis, which is introduced in Chap. 6. We present it in a simplified form over the rationals instead of an arbitrary number field. This is more than adequate for our purposes, as it brings out the central ideas better. In Chap. 7 automorphic forms on the group of invertible 2×2 matrices with adelic entries are investigated, and Chap. 8 we transfer the ideas of Tate's thesis to this setting and perform the analytic continuation of L-functions. For classical cusp forms we finally show that the classical and representation-theoretic approaches give the same L-functions.

For proofreading, pointing out errors, and many useful comments I thank Ralf Beckmann, Eberhard Freitag, Stefan Kühnlein, Judith Ludwig, Frank Monheim and Martin Raum.

Contents

Notation

We write $\mathbb{N} = \{1, 2, 3, \ldots\}$ for the set of natural numbers and $\mathbb{N}_0 = \{0, 1, 2, \ldots\}$ for the set of natural numbers with zero, as well as $\mathbb{Z}$, $\mathbb{Q}$, $\mathbb{R}$ and $\mathbb{C}$ for the sets of integer, rational, real, and complex numbers, respectively.

If A is a subset of a set X, we write $\mathbf{1}_A : X \to \mathbb{C}$ for the *indicator function* of A, i.e.

$$\mathbf{1}_A(x) = \begin{cases} 1 & \text{if } x \in A, \\ 0 & \text{if } x \notin A. \end{cases}$$

A *ring* is always considered to be commutative with unit. If R is a ring, we denote by $R^\times$ its group of invertible elements.

Chapter 1
Doubly Periodic Functions

In this chapter we present meromorphic functions on the complex plane which are periodic in two different directions, hence the wording 'doubly periodic'. These are constructed using infinite sums. The dependence of these sums on the periods leads us to the notion of a modular form.

1.1 Definition and First Properties

We recall the notion of a meromorphic function. Let D be an open subset of the complex plane $\mathbb{C}$. A *meromorphic function* f on D is a holomorphic function $f : D \smallsetminus P \to \mathbb{C}$, where $P \subset D$ is a countable subset and the function f has poles at the points of P.

The set of poles P can be empty, so every holomorphic function is an example of a meromorphic function. An accumulation point of poles is always an essential singularity. As we do not allow essential singularities, this means that the set P has no accumulation points inside D, so poles can accumulate only on the boundary of D.

Let $\widehat{\mathbb{C}} = \mathbb{C} \cup \{\infty\}$ be the one-point compactification of the complex plane, also called the *Riemann sphere* (see Exercise 1.1). Let f be meromorphic on D and let P be its set of poles. We extend f to a map $f : D \to \widehat{\mathbb{C}}$, by setting $f(p) = \infty$ for every $p \in P$.

A meromorphic function can thus be viewed as an everywhere defined, $\widehat{\mathbb{C}}$-valued map.

For a point $p \in D$ and a meromorphic function f on D, there exists exactly one integer $r \in \mathbb{Z}$ such that $f(z) = h(z)(z - p)^r$, where h is a function that is holomorphic and non-vanishing at p. This integer r is called the *order* of f at p. For this we write

$$r = \mathrm{ord}_p f.$$

Note: the order of f at p is positive if p is a zero of f, and negative if p is a pole of f.

A. Deitmar, *Automorphic Forms*, Universitext,
DOI 10.1007/978-1-4471-4435-9_1, © Springer-Verlag London 2013

Definition 1.1.1 A *lattice* in $\mathbb{C}$ is a subgroup Λ of the additive group $(\mathbb{C}, +)$ of the form

$$\Lambda = \Lambda(a, b) = \mathbb{Z}a \oplus \mathbb{Z}b = \{ka + lb : k, l \in \mathbb{Z}\},$$

where $a, b \in \mathbb{C}$ are supposed to be linearly independent over $\mathbb{R}$. In this case one says that the lattice is generated by a and b, or that a, b is a $\mathbb{Z}$-*basis* of the lattice.

A lattice has many sublattices; for example, $\Lambda(na, mb)$ is a sublattice of $\Lambda(a, b)$ for any $n, m \in \mathbb{N}$. A subgroup $\Sigma \subset \Lambda$ is a sublattice if and only if the quotient group Λ / Σ is finite (see Exercise 1.3). For instance, one has

$$\Lambda(a, b) / \Lambda(ma, nb) \cong \mathbb{Z}/m\mathbb{Z} \times \mathbb{Z}/n\mathbb{Z}.$$

Definition 1.1.2 Let Λ be a lattice in $\mathbb{C}$. A meromorphic function f on $\mathbb{C}$ is said to be *periodic* with respect to the lattice Λ if

$$f(z + \lambda) = f(z)$$

for every $z \in \mathbb{C}$ and every $\lambda \in \Lambda$. If f is periodic with respect to Λ, then it is so with respect to every sublattice. A function f is called *doubly periodic* if there exists a lattice Λ with respect to which f is periodic (see Exercise 1.2).

Proposition 1.1.3 *A doubly periodic function which is holomorphic is necessarily constant.*

Proof Let f be holomorphic and doubly periodic. Then there is a lattice $\Lambda = \Lambda(a, b)$ with $f(z + \lambda) = f(z)$ for every $\lambda \in \Lambda$. Let

$$\mathcal{F} = \mathcal{F}(a, b) = \{ta + sb : 0 \le s, t < 1\}.$$

The set $\mathcal{F}$ is a bounded subset of $\mathbb{C}$, so its closure $\overline{\mathcal{F}}$ is compact. The set $\mathcal{F}$ is called a *fundamental mesh* for the lattice Λ.

Two points $z, w \in \mathbb{C}$ are said to be *congruent modulo* Λ if $z - w \in \Lambda$.

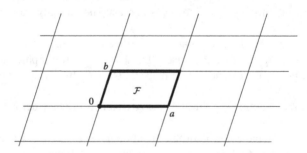

To conclude the proof of the proposition, we need a lemma.

Lemma 1.1.4 *Let $\mathcal{F}$ be a fundamental mesh for the lattice $\Lambda \subset \mathbb{C}$. Then $\mathbb{C} = \mathcal{F} + \Lambda$, or more precisely, for every $z \in \mathbb{C}$ there is exactly one $\lambda \in \Lambda$ such that $z + \lambda \in \mathcal{F}$. Equivalently, we can say that for every $z \in \mathbb{C}$ there is exactly one $w \in \mathcal{F}$ such that $z - w \in \Lambda$.*

Proof of the Lemma Let a, b be the $\mathbb{Z}$-basis of Λ, which defines the fundamental mesh $\mathcal{F}$, so $\mathcal{F} = \mathcal{F}(a, b)$. Since a and b are linearly independent over $\mathbb{R}$, they form a basis of $\mathbb{C}$ as an $\mathbb{R}$-vector space. Thus, for a given $z \in \mathbb{C}$ there are uniquely determined $r, v \in \mathbb{R}$ with $z = ra + vb$. There are uniquely determined $m, n \in \mathbb{Z}$ and $t, s \in [0, 1)$ such that

$$r = m + t \quad \text{and} \quad v = n + s.$$

This implies

$$z = ra + vb = \underbrace{ma + nb}_{\in \Lambda} + \underbrace{ta + sb}_{\in \mathcal{F}}$$

and this representation is unique.   □

We now show the proposition. As the function f is holomorphic, it is continuous, so $f(\overline{\mathcal{F}})$ is compact, hence bounded. For an arbitrary $z \in \mathbb{C}$ there is, by the lemma, a $\lambda \in \Lambda$ with $z + \lambda \in \mathcal{F}$, so $f(z) = f(z + \lambda) \in f(\mathcal{F})$, which means that the function f is bounded, hence constant by Liouville's theorem.   □

Proposition 1.1.5 *Let $\mathcal{F}$ be a fundamental mesh of a lattice $\Lambda \subset \mathbb{C}$ and let f be an Λ-periodic meromorphic function. Then there is $w \in \mathbb{C}$ such that f has no pole on the boundary of the translated mesh $\mathcal{F}_w = \mathcal{F} + w$. For every such w one has*

$$\int_{\partial \mathcal{F}_w} f(z)\, dz = 0,$$

where $\partial \mathcal{F}_w$ is the positively oriented boundary of $\mathcal{F}_w$.

Proof If f had poles on the boundary of $\mathcal{F}_w$ for every w, then f would have uncountably many poles, contradicting the meromorphicity of the function f. We can therefore choose w in such a way that no poles of f are located on the boundary of $\mathcal{F}_w$.

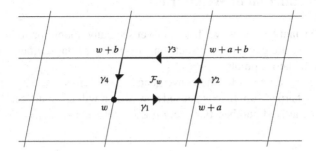

The path of integration $\partial \mathcal{F}_w$ is composed of the paths $\gamma_1, \gamma_2, \gamma_3, \gamma_4$ as in the picture. The path γ_3 is the same as γ_1, only translated by $b \in \Lambda$ and running in the reverse direction. The function f does not change when one translates the argument by b and the change of direction amounts to a change of sign in the integral. Therefore we get

$$\int_{\gamma_1} f(z)\,dz + \int_{\gamma_3} f(z)\,dz = 0 \quad \text{and similarly} \quad \int_{\gamma_2} f(z)\,dz + \int_{\gamma_4} f(z)\,dz = 0,$$

which together give $\int_{\partial \mathcal{F}_w} f(z)\,dz = 0$ as claimed.   □

Proposition 1.1.6 *Let $f \neq 0$ be a meromorphic function, periodic with respect to the lattice $\Lambda \subset \mathbb{C}$ and let $\mathcal{F}$ be a fundamental mesh for the lattice Λ. For every $w \in \mathbb{C}$ we have*

$$\sum_{z \in \mathcal{F}_w} \operatorname{res}_z(f) = 0.$$

Proof In case there is no pole on the boundary of $\mathcal{F}_w$, the assertion follows from the last proposition together with the residue theorem. It follows in general, since the sum does not depend on w, as congruent points have equal residues. Hence we have

$$\sum_{z \in \mathcal{F}_w} \operatorname{res}_z(f) = \sum_{z \in \mathbb{C} \bmod \Lambda} \operatorname{res}_z(f).$$ □

Proposition 1.1.7 *Let $\mathcal{F}$ be a fundamental mesh for the lattice $\Lambda \subset \mathbb{C}$ and let $f \neq 0$ be a Λ-periodic meromorphic function. Then for every $w \in \mathbb{C}$ the number of zeros of f in $\mathcal{F}_w$ equals the number of poles of f in $\mathcal{F}_w$. Here zeros and poles are both counted with multiplicities, so a double pole, for instance, is counted twice.*

Proof A complex number z_0 is a zero or a pole of f of order $k \in \mathbb{Z}$ if the function $\frac{f'}{f}$ has a pole at z_0 of residue k. Hence the assertion follows from the last proposition, since the function $\frac{f'}{f}$ is doubly periodic with respect to the lattice Λ as well.   □

1.2 The ℘-Function of Weierstrass

Except for constant functions, we have not yet seen any doubly periodic function. In this section we are going to construct some by giving Mittag-Leffler sums which have poles at the lattice points.

We first need a criterion for the convergence of the series that we consider. We prove this in a sharper form than is needed now, which will turn out useful later. Let $b \in \mathbb{C} \setminus \{0\}$ be a fixed number. For every $a \in \mathbb{C} \setminus \mathbb{R}b$ the set $\Lambda_a = \mathbb{Z}a \oplus \mathbb{Z}b$ is a lattice.

Lemma 1.2.1 *Let $\Lambda \subset \mathbb{C}$ be a lattice and let $s \in \mathbb{C}$. The series*

$$\sum_{\substack{\lambda \in \Lambda \\ \lambda \neq 0}} \frac{1}{|\lambda|^s}$$

converges absolutely if $\mathrm{Re}(s) > 2$. *Furthermore, fix* $b \in \mathbb{C} \smallsetminus \{0\}$ *and consider the lattice* Λ_a *for* $a \in \mathbb{C} \smallsetminus \mathbb{R}b$. *The sum* $\sum_{\lambda \in \Lambda_a, \lambda \neq 0} \frac{1}{|\lambda|^s}$ *converges uniformly for all* $(a, s) \in C \times \{\mathrm{Re}(s) \geq \alpha\}$, *where* C *is a compact subset of* $\mathbb{C} \smallsetminus \mathbb{R}b$ *and* $\alpha > 2$.

Proof Let α and C be as in the lemma. We can assume $\mathrm{Re}(s) > 0$, because otherwise the series cannot converge as the sequence of its summands does not tend to zero. Further it suffices to consider the case $s \in \mathbb{R}$ since for $s \in \mathbb{C}$ the absolute value of $|\lambda|^{-s}$ equals $|\lambda|^{-\mathrm{Re}(s)}$. So, assuming $s > 0$, the function $x \mapsto x^s$ is monotonically increasing for $x > 0$. Let $\mathcal{F}(a)$ be a fundamental mesh for the lattice Λ_a and let

$$\psi_{a,s}(z) = \sum_{\substack{\lambda \in \Lambda_a \\ \lambda \neq 0}} \frac{1}{|\lambda|^s} \mathbf{1}_{\mathcal{F}(a)+\lambda}(z).$$

We then have

$$|\mathcal{F}(a)| \sum_{\substack{\lambda \in \Lambda_a \\ \lambda \neq 0}} \frac{1}{|\lambda|^s} = \int_{\mathbb{C}} \psi_{a,s}(x + iy)\, dx\, dy,$$

where $|\mathcal{F}(a)|$ is the area of the fundamental mesh $\mathcal{F}(a)$. The continuous map $a \mapsto |\mathcal{F}(a)|$ assumes its minimum and maximum values on the compact set C. One has $\psi_{a,s} \leq \psi_{a,\alpha}$ if $s \geq \alpha$, so it suffices to show the uniform convergence of $\int_{\mathbb{C}} \psi_{a,\alpha}(z)\, dx\, dy$ in a.

Let $r > 0$ be so large that for every $a \in C$ the *diameter* of the fundamental mesh $\mathcal{F}(a)$,

$$\mathrm{diam}(\mathcal{F}(a)) = \sup\{|z - w| : z, w \in \mathcal{F}(a)\}$$

is less than r. For every $z \in \mathbb{C}$ we have $\psi_{a,\alpha}(z) = \frac{1}{|\lambda_{a,z}|^s}$ for some $\lambda_{a,z} \in \Lambda_a$ with $|z - \lambda_{a,z}| < r$. For every $a \in C$ and $z \in \mathbb{C}$ with $|z| \geq r$ one has the inequality

$$|\lambda_{a,z}| = |\lambda_{a,z} - z + z| \leq |\lambda_{a,z} - z| + |z| < r + |z| \leq 2|z|.$$

On the other hand, for $|z| \geq 2r$ we have

$$|\lambda_{a,z}| = |\lambda_{a,z} - z - (-z)| \geq ||\lambda_{a,z} - z| - |z|| \geq \frac{1}{2}|z|.$$

Let $R = 2r$. For $|z| \geq R$ we have $\frac{1}{2^s}|z|^{-s} \leq \psi_{a,\alpha}(z) \leq 2^s|z|^{-s}$ for every $a \in C$. The continuous map $a \mapsto \int_{|z| \leq R} \psi_{a,\alpha}(z)\, dx\, dy$ is bounded on the compact set C. Therefore the series converges uniformly for $a \in C$, if $\int_{|z|>R} \frac{1}{|z|^{\alpha}}\, dx\, dy < \infty$. We now use *polar coordinates* on $\mathbb{C}$. Recall that the map $P : (0, \infty) \times (-\pi, \pi] \to \mathbb{C}$, given by

$$P(r, \theta) = re^{i\theta} = r\cos\theta + ir\sin\theta$$

is a bijection onto the image $\mathbb{C} \smallsetminus \{0\}$. The Jacobian determinant of this map is r, so we get, by the change of variables formula,

$$\int_{\mathbb{C}^{\times}} f(x + iy)\, dx\, dy = \int_{-\pi}^{\pi} \int_0^{\infty} f(re^{i\theta}) r\, dr\, d\theta$$

for every integrable function f. Therefore

$$\int_{|z|>R} \frac{1}{|z|^\alpha}\, dx\, dy = 2\pi \int_R^\infty r^{1-\alpha}\, dr,$$

which gives the claim. $\square$

The following theorem contains the definition of the Weierstrass $\wp$-function.

Theorem 1.2.2 *Let Λ be a lattice in $\mathbb{C}$. The series*

$$\wp(z) \overset{\text{def}}{=} \frac{1}{z^2} + \sum_{\lambda \in \Lambda \smallsetminus \{0\}} \frac{1}{(z-\lambda)^2} - \frac{1}{\lambda^2}$$

converges locally uniformly absolutely in $\mathbb{C} \smallsetminus \Lambda$. It defines a meromorphic Λ-periodic function, called the Weierstrass $\wp$-function.

Proof For $|z| < \frac{1}{2}|\lambda|$ we have $|\lambda - z| \geq \frac{1}{2}|\lambda|$. Further it holds that $|2\lambda - z| \leq \frac{5}{2}|\lambda|$. So that

$$\left| \frac{1}{(z-\lambda)^2} - \frac{1}{\lambda^2} \right| = \left| \frac{\lambda^2 - (z-\lambda)^2}{\lambda^2 (z-\lambda)^2} \right| = \left| \frac{z(2\lambda - z)}{\lambda^2 (z-\lambda)^2} \right| \leq \frac{|z|\frac{5}{2}|\lambda|}{|\lambda|^2 \frac{1}{4}|\lambda|^2} = \frac{10|z|}{|\lambda|^3}.$$

Using Lemma 1.2.1, we get locally uniform convergence.

The way the sum is formed, it is not immediate that the $\wp$-function is actually periodic. For this we first show that it is an even function:

$$\wp(-z) = \frac{1}{z^2} + \sum_{\lambda \in \Lambda \smallsetminus \{0\}} \frac{1}{(z+\lambda)^2} - \frac{1}{\lambda^2} = \frac{1}{z^2} + \sum_{\lambda \in \Lambda \smallsetminus \{0\}} \frac{1}{(z-\lambda)^2} - \frac{1}{\lambda^2} = \wp(z),$$

by replacing λ in the sum with $-\lambda$. Since the series converges locally uniformly, and the summands are holomorphic, we are allowed to differentiate the series term-wise. Its derivative,

$$\wp'(z) = -2 \sum_{\lambda \in \Lambda} \frac{1}{(z-\lambda)^3},$$

is Λ-periodic. Hence for $\lambda \in \Lambda \smallsetminus 2\Lambda$ the function $\wp(z+\lambda) - \wp(z)$ is constant. We compute this constant by setting $z = -\frac{\lambda}{2}$ to get $\wp(\frac{\lambda}{2}) - \wp(-\frac{\lambda}{2}) = 0$, as $\wp$ is even. $\square$

Theorem 1.2.3 (Laurent-expansion of $\wp$) *Let $r = \min\{|\lambda| : \lambda \in \Lambda \smallsetminus \{0\}\}$. For $0 < |z| < r$ one has*

$$\wp(z) = \frac{1}{z^2} + \sum_{n=1}^{\infty} (2n+1) G_{2n+2} z^{2n},$$

where the sum $G_k = G_k(\Lambda) = \sum_{\lambda \in \Lambda \smallsetminus \{0\}} \frac{1}{\lambda^k}$ converges absolutely for $k \geq 4$.

Proof For $0 < |z| < r$ and $\lambda \in \Lambda \setminus \{0\}$ we have $|z/\lambda| < 1$, so

$$\frac{1}{(z-\lambda)^2} = \frac{1}{\lambda^2(1-\frac{z}{\lambda})^2} = \frac{1}{\lambda^2}\left(1 + \sum_{k=1}^{\infty}(k+1)\left(\frac{z}{\lambda}\right)^k\right),$$

and so

$$\frac{1}{(z-\lambda)^2} - \frac{1}{\lambda^2} = \sum_{k=1}^{\infty}\frac{k+1}{\lambda^{k+2}}z^k.$$

We sum over all λ and find

$$\wp(z) = \frac{1}{z^2} + \sum_{k=1}^{\infty}(k+1)\sum_{\lambda \neq 0}\frac{1}{\lambda^{k+2}}z^k = \frac{1}{z^2} + \sum_{k=1}^{\infty}(k+1)G_{k+2}z^k,$$

where we have changed the order of summation, as we may by absolute convergence. This absolute convergence follows from

$$\sum_{k=1}^{\infty}\frac{k+1}{|\lambda|^{k+2}}|z|^k \leq \frac{1}{|\lambda|^3}\sum_{k=1}^{\infty}\frac{k+1}{|\lambda|^{k-1}}|z|^k$$

and Lemma 1.2.1. Since $\wp$ is even, the G_{k+2} vanish for odd k. $\square$

1.3 The Differential Equation of the $\wp$-Function

The differential equation of the $\wp$-function connects doubly periodic functions to elliptic curves, as explained in the notes at the end of this chapter.

Theorem 1.3.1 *The $\wp$-function satisfies the differential equation*

$$\left(\wp'(z)\right)^2 = 4\wp^3(z) - 60G_4\wp(z) - 140G_6.$$

Proof We show that the difference of the two sides has no pole, i.e. is a holomorphic Λ-periodic function, hence constant.

If $z \neq 0$ is small we have

$$\wp'(z) = -\frac{2}{z^3} + 6G_4z + 20G_6z^3 + \cdots,$$

so

$$\left(\wp'(z)\right)^2 = \frac{4}{z^6} - \frac{24G_4}{z^2} - 80G_6 + \cdots.$$

On the other hand,

$$4\wp^3(z) = \frac{4}{z^6} + \frac{36G_4}{z^2} + 60G_6 + \cdots,$$

so that

$$\left(\wp'(z)\right)^2 - 4\wp^3(z) = -\frac{60G_4}{z^2} - 140G_6 + \cdots.$$

We finally get

$$\left(\wp'(z)\right)^2 - 4\wp^3(z) + 60G_4\wp(z) = -140G_6 + \cdots$$

where the left-hand side is a holomorphic Λ-periodic function, hence constant. If on this right-hand side we put $z = 0$, we see that this constant is $-140G_6$. ☐

1.4 Eisenstein Series

For a ring R we denote by $M_2(R)$ the set of all 2×2 matrices with entries from R. In a linear algebra course you prove that a matrix $\begin{pmatrix} a & b \\ c & d \end{pmatrix} \in M_2(R)$ is invertible if and only if its determinant is invertible in R, i.e. if $ad - bd \in R^\times$. You may have done this for R being a field only, but for a ring it is just the same proof. Let $GL_2(R)$ be the group of all invertible matrices in $M_2(R)$. It contains the subgroup $SL_2(R)$ of all matrices of determinant 1. Consider the example $R = \mathbb{Z}$. We have $\mathbb{Z}^\times = \{1, -1\}$. So $GL_2(\mathbb{Z})$ is the group of all integer matrices with determinant ± 1. The subgroup $SL_2(\mathbb{Z})$ is therefore a subgroup of index 2.

For $k \in \mathbb{N}$, $k \geq 4$ the series $G_k(\Lambda) = \sum_{\lambda \in \Lambda \setminus \{0\}} \lambda^{-k}$ converges. The set $w\Lambda$ is again a lattice if $w \in \mathbb{C}^\times$ and we have

$$G_k(w\Lambda) = w^{-k}G_k(\Lambda).$$

Recall for $\alpha, \beta \in \mathbb{C}$, linearly independent over $\mathbb{R}$ we have the lattice

$$\Lambda(\alpha, \beta) = \mathbb{Z}\alpha \oplus \mathbb{Z}\beta.$$

If z is a complex number with $\mathrm{Im}(z) > 0$, then z and 1 are linearly independent over $\mathbb{R}$. We define the *Eisenstein series* as a function on the *upper half plane*

$$\mathbb{H} = \{z \in \mathbb{C} : \mathrm{Im}(z) > 0\}$$

by

$$G_k(z) = G_k(\Lambda(z, 1)) = \sum_{(m,n) \neq (0,0)} \frac{1}{(mz + n)^k},$$

where the sum runs over all $m, n \in \mathbb{Z}$ which are not both zero. Using matrix multiplication, we can write $mz + n = (z \; 1)\binom{m}{n}$. The group $\Gamma_0 = \mathrm{SL}_2(\mathbb{Z})$ acts on the set of pairs $\binom{m}{n}$ by multiplication from the left. For $\gamma = \left(\begin{smallmatrix} a & b \\ c & d \end{smallmatrix}\right) \in \Gamma_0$ we have

$$
\begin{aligned}
G_k(z) &= \sum_{m,n} \left((z \; 1)\binom{m}{n} \right)^{-k} &&= \sum_{m,n} \left((z \; 1)\gamma^t\binom{m}{n} \right)^{-k} \\
&= \sum_{m,n} \left((z \; 1)\binom{a & c}{b & d}\binom{m}{n} \right)^{-k} &&= \sum_{m,n} \left((az+b, cz+d)\binom{m}{n} \right)^{-k} \\
&= (cz+d)^{-k} \sum_{m,n} \left(\left(\frac{az+b}{cz+d}, 1\right)\binom{m}{n} \right)^{-k} &&= (cz+d)^{-k} G_k\left(\frac{az+b}{cz+d}\right),
\end{aligned}
$$

or

$$
G_k\left(\frac{az+b}{cz+d}\right) = (cz+d)^k G_k(z).
$$

Proposition 1.4.1 *If $k \geq 4$ is even, then*

$$
\lim_{y \to \infty} G_k(iy) = 2\zeta(k),
$$

where

$$
\zeta(s) = \sum_{n=1}^{\infty} \frac{1}{n^s}, \quad \mathrm{Re}(s) > 1,
$$

is the Riemann zeta function. (See Exercise 1.4.)

Proof One has

$$
G_k(iy) = 2\zeta(k) + \sum_{\substack{(m,n) \\ m \neq 0}} \frac{1}{(miy+n)^k}.
$$

We show that the second summand tends to zero for $y \to \infty$. Consider the estimate

$$
\left| \sum_{\substack{(m,n) \\ m \neq 0}} \frac{1}{(miy+n)^k} \right| \leq \sum_{\substack{(m,n) \\ m \neq 0}} \frac{1}{n^k + m^k y^k}.
$$

Here, every summand on the right-hand side is monotonically decreasing in as $y \to \infty$ and tends to zero. Further, the right-hand side converges for every $y > 0$, so we conclude by dominated convergence, that the entire sum tends to zero as $y \to \infty$. $\square$

1.5 Bernoulli Numbers and Values of the Zeta Function

We have seen that Eisenstein series assume zeta-values 'at infinity'. We shall need the following exact expressions for these zeta-values later. We now define the Bernoulli numbers B_k.

Lemma 1.5.1 *For $k = 1, 2, 3, \ldots$ there are uniquely determined rational numbers B_k such that for $|z| < 2\pi$ one has*

$$\frac{z}{e^z - 1} + \frac{z}{2} = \frac{z}{2}\frac{e^z + 1}{e^z - 1} = 1 - \sum_{k=1}^{\infty}(-1)^k B_k \frac{z^{2k}}{(2k)!}.$$

The first of these numbers are $B_1 = \frac{1}{6}$, $B_2 = \frac{1}{30}$, $B_3 = \frac{1}{42}$, $B_4 = \frac{1}{30}$, $B_5 = \frac{5}{66}$.

Proof Let $f(z) = \frac{z}{e^z - 1} + \frac{z}{2} = \frac{z}{2}\frac{e^z + 1}{e^z - 1}$. Then f is holomorphic in $\{|z| < 2\pi\}$, so its power series expansion converges in this circle. We show that f is an even function:

$$f(-z) = -\frac{z}{2}\frac{e^{-z} + 1}{e^{-z} - 1} = -\frac{z}{2}\frac{1 + e^z}{1 - e^z} = f(z).$$

Therefore there is such an expression with $B_k \in \mathbb{C}$.

Let $g(z) = \frac{z}{e^z - 1} = \sum_{k=0}^{\infty} c_k z^k$. We show that the c_k are all rational numbers. The equation $z = g(z)(e^z - 1)$ gives

$$z = \sum_{n=0}^{\infty} z^n \left(\sum_{j=0}^{n-1} \frac{c_j}{(n-j)!} \right).$$

So $c_0 = 1$ and for every $n \geq 2$ the number c_{n-1} is a rational linear combination of the c_j with $j < n - 1$. Inductively we conclude $c_j \in \mathbb{Q}$. □

Proposition 1.5.2 *For every natural number k one has*

$$\zeta(2k) = \frac{2^{2k-1}}{(2k)!} B_k \pi^{2k}.$$

The first values are $\zeta(2) = \frac{\pi^2}{6}$, $\zeta(4) = \frac{\pi^4}{90}$, $\zeta(6) = \frac{\pi^6}{945}$.

Proof By definition, the cotangent function satisfies

$$z \cot z = zi \frac{e^{iz} + e^{-iz}}{e^{iz} - e^{-iz}}.$$

Replacing z by $z/2i$, this becomes

$$\frac{z}{2i} \cot\left(\frac{z}{2i}\right) = \frac{z}{2}\frac{e^z + 1}{e^z - 1} = f(z),$$

so that

$$z \cot z = 1 - \sum_{k=1}^{\infty} B_k \frac{2^{2k} z^{2k}}{(2k)!}.$$

The partial fraction expansion of the cotangent function (Exercise 1.8) is

$$\pi \cot(\pi z) = \frac{1}{z} + \sum_{m=1}^{\infty} \left(\frac{1}{z+m} + \frac{1}{z-m} \right).$$

Therefore

$$z \cot z = 1 + 2 \sum_{n=1}^{\infty} \frac{z^2}{z^2 - n^2 \pi^2} = 1 - 2 \sum_{n=1}^{\infty} \sum_{k=1}^{\infty} \frac{z^{2k}}{n^{2k} \pi^{2k}},$$

from which we get

$$\sum_{k=1}^{\infty} B_k \frac{2^{2k} z^{2k}}{(2k)!} = 2 \sum_{n=1}^{\infty} \sum_{k=1}^{\infty} \frac{z^{2k}}{n^{2k} \pi^{2k}}.$$

Comparing coefficients gives the claim. $\square$

1.6 Exercises and Remarks

Exercise 1.1 The *one-point compactification* $\widehat{\mathbb{C}} = \mathbb{C} \cup \{\infty\}$ of $\mathbb{C}$ has the following topology. Open sets are

- open subsets of $\mathbb{C}$, or
- sets V which contain ∞ and have the property that $\mathbb{C} \smallsetminus V$ is a compact subset of $\mathbb{C}$.

Show that $\widehat{\mathbb{C}}$ is compact. Consider the three-dimensional space $\mathbb{C} \times \mathbb{R}$ and let

$$S = \left\{ (z, t) \in \mathbb{C} \times \mathbb{R} : |z|^2 + t^2 = 1 \right\}.$$

Then S is the two-dimensional sphere. Consider the point $N = (0, 1) \in S$, called the *north pole*. Show that for every $z \in \mathbb{C}$ the line through z and N meets the sphere S in exactly one other point $\phi(z)$. Show that the resulting map $\phi : \mathbb{C} \to S \smallsetminus \{N\}$ is a homeomorphism which extends to a homeomorphism $\widehat{\mathbb{C}} \to S$ by sending ∞ to N. This homeomorphism is the reason why $\widehat{\mathbb{C}}$ is also called the *Riemann sphere*.

Exercise 1.2 Let $a \in \mathbb{C} \smallsetminus \{0\}$. A function f on $\mathbb{C}$ is called *simply periodic* of period a, or a-periodic, if $f(z + a) = f(z)$ for every $z \in \mathbb{C}$. Show that if $a, b \in \mathbb{C}$ are linearly independent over $\mathbb{R}$, then a function f is $\Lambda(a, b)$-periodic if and only if it is a-periodic and b-periodic simultaneously. This explains the notion *doubly periodic*.

Exercise 1.3 A subgroup $\Lambda \subset \mathbb{C}$ of the additive group $(\mathbb{C}, +)$ is called a *discrete subgroup* if Λ is discrete in the subset topology, i.e. if for every $\lambda \in \Lambda$ there exists an open set $U_\lambda \subset \mathbb{C}$ such that $\Lambda \cap U_\lambda = \{\lambda\}$. Show

1. A subgroup $\Lambda \subset \mathbb{C}$ is discrete if and only if there is an open set $U_0 \subset \mathbb{C}$ with $U_0 \cap \Lambda = \{0\}$.
2. If $\Lambda \subset \mathbb{C}$ is a discrete subgroup, then there are three possibilities: either $\Lambda = \{0\}$, or there is a $\lambda_0 \in \Lambda$ with $\Lambda = \mathbb{Z}\lambda_0$, or Λ is a lattice.
3. A discrete subgroup $\Lambda \subset \mathbb{C}$ is a lattice if and only if the quotient group $\mathbb{C}/\Lambda$ is compact in the quotient topology.
4. If $\Lambda \subset \mathbb{C}$ is a lattice, then a subgroup $\Sigma \subset \Lambda$ is a lattice if and only if it has finite index, i.e. if the group Λ/Σ is finite.

Exercise 1.4 Show that the sum defining the Riemann zeta function, $\zeta(s) = \sum_{n=1}^{\infty} n^{-s}$, converges absolutely for $\mathrm{Re}(s) > 1$. One can adapt the proof of Lemma 1.2.1.

Exercise 1.5 Show that the Riemann zeta function $\zeta(s) = \sum_{n=1}^{\infty} n^{-s}$ has the Euler product

$$\zeta(s) = \prod_{p} \frac{1}{1 - p^{-s}}, \quad \mathrm{Re}(s) > 1,$$

where the product runs over all prime numbers p. (Hint: consider the sequence $s_N(s) = \prod_{p \leq N} \frac{1}{1-p^{-s}}$. Using the geometric series, write $\frac{1}{1-p^{-s}} = \sum_{k=0}^{\infty} p^{-ks}$ and use absolute convergence of the Dirichlet series defining $\zeta(s)$.)

Exercise 1.6 Let $\alpha, \beta, \alpha' \beta' \in \mathbb{C}$ with $\mathbb{C} = \mathbb{R}\alpha + \mathbb{R}\beta = \mathbb{R}\alpha' + \mathbb{R}\beta'$. Show that the lattices $\Lambda(\alpha, \beta)$ and $\Lambda(\alpha', \beta')$ coincide if and only if there is $\left(\begin{smallmatrix} a & b \\ c & d \end{smallmatrix} \right) \in \mathrm{GL}_2(\mathbb{Z})$ with

$$\begin{pmatrix} \alpha' \\ \beta' \end{pmatrix} = \begin{pmatrix} a & b \\ c & d \end{pmatrix} \begin{pmatrix} \alpha \\ \beta \end{pmatrix}.$$

Exercise 1.7 Let f be meromorphic on $\mathbb{C}$ and Λ-periodic for a lattice Λ. Let $\mathcal{F}_w = \mathcal{F} + w$ a translated fundamental mesh for Λ, such that there are no poles or zeros of f on the boundary $\partial \mathcal{F}_w$. Let $S(0)$ be the sum of all zeros of f in $\mathcal{F}$, counted with multiplicities. Let $S(\infty)$ be the sum of all poles of f in $\mathcal{F}$, also with multiplicities. Show:

$$S(0) - S(\infty) \in \Lambda.$$

(Integrate the function $zf'(z)/f(z)$.)

Exercise 1.8 Prove the partial fraction expansion of the cotangent:

$$\pi \cot(\pi z) = \frac{1}{z} + \sum_{m=1}^{\infty} \left(\frac{1}{z+m} + \frac{1}{z-m} \right).$$

(The difference of the two sides is periodic and entire. Show that it is bounded and odd.)

Exercise 1.9 Let $z, w \in \mathbb{C}$ and let $\wp$ be the Weierstrass $\wp$-function for a lattice Λ. Show that $\wp(z) = \wp(w)$ if and only if $z + w$ or $z - w$ is in the lattice Λ.

Exercise 1.10 Let $\wp$ be the Weierstrass $\wp$-function for a lattice Λ.

(a) Let $a_1, \ldots, a_n$ and $b_1, \ldots, b_m$ be complex numbers. Show that the function

$$f(z) = \frac{\prod_{i=1}^{n} \wp(z) - \wp(a_i)}{\prod_{j=1}^{m} \wp(z) - \wp(b_j)}$$

is even and Λ-periodic.

(b) Show that every even Λ-periodic function is a rational function of $\wp$.

(c) Show that every Λ-periodic meromorphic function is of the form $R(\wp(z)) + \wp'(z)Q(\wp(z))$, where R and Q are rational functions.

Exercise 1.11 (Residue theorem for circle segments) Let $r_0 > 0$ and let f be a holomorphic function on the set $\{z \in \mathbb{C} : 0 < |z| < r_0\}$, which has a simple pole at $z = 0$. Let $a, b : [0, r_0) \to (-\pi, \pi)$ be continuous functions with $a(r) \leq b(r)$ for every $0 \leq r < r_0$ and for $0 < r < r_0$ let $\gamma_r : (a(r), b(r)) \to \mathbb{C}$ the circle segment $\gamma_r(t) = re^{it}$. Show:

$$\lim_{r \to 0} \frac{1}{2\pi i} \int_{\gamma_r} f(z)\, dz = \frac{b(0) - a(0)}{2\pi}\, \mathrm{res}_{z=0}\, f(z).$$

Exercise 1.12 Let (a_n) be a sequence in $\mathbb{C}$. Show that there exists an $r \in \mathbb{R} \cup \{\pm\infty\}$, such that for every $s \in \mathbb{C}$ with $\mathrm{Re}(s) > r$ and for no s with $\mathrm{Re}(s) < r$ the Dirichlet series $\sum_{n=1}^{\infty} a_n n^{-s}$ converges absolutely.

Remarks Putting $g_4 = 15G_4$ and $g_6 = 35G_6$ one sees that $(x, y) = (\wp, \wp'/2)$ satisfies the polynomial equation

$$y^2 = x^3 - g_4 x - g_6.$$

This means that the map $z \mapsto (\wp(z), \wp'(z)/2)$ maps the complex manifold $\mathbb{C}/\Lambda$ bijectively onto the *elliptic curve* given by this equation. Indeed, every elliptic curve is obtained in this way, so elliptic curves are parametrized by lattices. The book [Sil09] gives a good introduction to elliptic curves.

The Riemann zeta function featured in this section has a meromorphic extension to all of $\mathbb{C}$ and satisfies a functional equation, as shown in Theorem 6.1.3. The famous *Riemann hypothesis* says that every zero of the function $\zeta(s)$ in the strip $0 < \mathrm{Re}(s) < 1$ has real part $\frac{1}{2}$. This hypothesis is considered the hardest problem of all mathematics.

Chapter 2
Modular Forms for $SL_2(\mathbb{Z})$

In this chapter we introduce the notion of a modular form and its L-function. We determine the space of modular forms by giving an explicit basis. We define Hecke operators and we show that the L-function of a Hecke eigenform admits an Euler product.

2.1 The Modular Group

Recall the notion of an action of a group G on a set X. This is a map $G \times X \to X$, written $(g, x) \mapsto gx$, such that $1x = x$ and $g(hx) = (gh)x$, where $x \in X$ and $g, h \in G$ are arbitrary elements and 1 is the neutral element of the group G.

Two points $x, y \in X$ are called *conjugate modulo G*, if there exists a $g \in G$ with $y = gx$. The *orbit* of a point $x \in X$ is the set Gx of all gx, where $g \in G$, so the orbit is the set of all points conjugate to x. We write $G \backslash X$ or X/G for the set of all G-orbits.

Example 2.1.1 Let G be the group of all complex numbers of absolute value one, also known as the *circle group*

$$G = \mathbb{T} = \{ z \in \mathbb{C} : |z| = 1 \}.$$

The group G acts on the set $\mathbb{C}$ by multiplication. The map

$$[0, \infty) \to G \backslash \mathbb{C},$$
$$x \mapsto Gx$$

is a bijection.

An action of a group is said to be *transitive* if there is only one orbit, i.e. if any two elements are conjugate.

This is the usual notion of a group action from the left, or left action. Later, in Lemma 2.2.2, we shall also define a group action from the right.

A. Deitmar, *Automorphic Forms*, Universitext,
DOI 10.1007/978-1-4471-4435-9_2, © Springer-Verlag London 2013

For given $g \in G$ the map $x \mapsto gx$ is invertible, as its inverse is $x \mapsto g^{-1}x$.

The group $GL_2(\mathbb{C})$ acts on the set $\mathbb{C}^2 \smallsetminus \{0\}$ by matrix multiplication. Since this action is by linear maps, the group also acts on the projective space $\mathbb{P}^1(\mathbb{C})$, which we define as the set of all one-dimensional subspaces of the vector space $\mathbb{C}^2$. Every non-zero vector in $\mathbb{C}^2$ spans such a vector space and two vectors give the same space if and only if one is a multiple of the other, which means that they are in the same $\mathbb{C}^\times$-orbit. So we have a canonical bijection

$$\mathbb{P}^1(\mathbb{C}) \cong \left(\mathbb{C}^2 \smallsetminus \{0\}\right)/\mathbb{C}^\times.$$

We write the elements of $\mathbb{P}^1(\mathbb{C})$ in the form $[z, w]$, where $(z, w) \in \mathbb{C}^2 \smallsetminus \{0\}$ and

$$[z, w] = [z', w'] \quad \Leftrightarrow \quad \exists \lambda \in \mathbb{C}^\times : (z', w') = (\lambda z, \lambda w).$$

For $w \neq 0$ there exists exactly one representative of the form $[z, 1]$, and the map $z \mapsto [z, 1]$ is an injection $\mathbb{C} \hookrightarrow \mathbb{P}^1(\mathbb{C})$, so that we can view $\mathbb{C}$ as a subset of $\mathbb{P}^1(\mathbb{C})$. The complement of $\mathbb{C}$ in $\mathbb{P}^1(\mathbb{C})$ is a single point $\infty = [1, 0]$, so that $\mathbb{P}^1(\mathbb{C})$ is the one-point compactification $\widehat{\mathbb{C}}$ of $\mathbb{C}$, the *Riemann sphere*. We consider the action of $GL_2(\mathbb{C})$ given by $g.(z, w) = (z, w)g^t$; then with $g = \left(\begin{smallmatrix} a & b \\ c & d \end{smallmatrix}\right)$ we have

$$g.[z, 1] = [az + b, cz + d] = \left[\frac{az + b}{cz + d}, 1\right],$$

if $cz + d \neq 0$. The rational function $\frac{az+b}{cz+d}$ has exactly one pole in the set $\widehat{\mathbb{C}}$, so we define an action of $GL_2(\mathbb{C})$ on the Riemann sphere by

$$g.z = \begin{cases} \frac{az+b}{cz+d} & \text{if } cz + d \neq 0, \\ \infty & \text{if } cz + d = 0, \end{cases}$$

if $z \in \mathbb{C}$. Note that $cz + d$ and $az + b$ cannot both be zero (Exercise 2.1). We finalize the definition of this action with

$$g.\infty = \lim_{\mathrm{Im}(z) \to \infty} g.z = \begin{cases} \frac{a}{c} & \text{if } c \neq 0, \\ \infty & \text{otherwise.} \end{cases}$$

Any matrix of the form $\left(\begin{smallmatrix} \lambda & \\ & \lambda \end{smallmatrix}\right)$ with $\lambda \neq 0$ acts trivially, so it suffices to consider the action on the subgroup $SL_2(\mathbb{C}) = \{g \in GL_2(\mathbb{C}) : \det(g) = 1\}$.

Lemma 2.1.2 *The group* $SL_2(\mathbb{C})$ *acts transitively on the Riemann sphere* $\widehat{\mathbb{C}}$. *The element* $\left(\begin{smallmatrix} -1 & \\ & -1 \end{smallmatrix}\right)$ *acts trivially. If we restrict the action to the subgroup* $G = SL_2(\mathbb{R})$, *the set* $\widehat{\mathbb{C}}$ *decomposes into three orbits:* $\mathbb{H}$ *and* $-\mathbb{H}$, *as well as the set* $\widehat{\mathbb{R}} = \mathbb{R} \cup \{\infty\}$.

Proof For given $z \in \mathbb{C}$ one has $z = \left(\begin{smallmatrix} z & z-1 \\ 1 & 1 \end{smallmatrix}\right).\infty$, so the action is transitive. In particular it follows that $\widehat{\mathbb{R}}$ lies in the G-orbit of the point ∞.

For $g = \left(\begin{smallmatrix} a & b \\ c & d \end{smallmatrix}\right) \in G$ and $z \in \mathbb{C}$ one computes

$$\mathrm{Im}(g.z) = \frac{\mathrm{Im}(z)}{|cz + d|^2}.$$

This implies that G leaves the three sets mentioned invariant. We have $\left(\begin{smallmatrix} 1 & x \\ & 1 \end{smallmatrix}\right).0 = x \in \mathbb{R}$ and $\left(\begin{smallmatrix} & -1 \\ 1 & \end{smallmatrix}\right).0 = \infty$, therefore $\widehat{\mathbb{R}}$ is one G-orbit. We show that G acts transitively on $\mathbb{H}$. For a given $z = x + iy \in \mathbb{H}$ one has

$$z = \begin{pmatrix} \sqrt{y} & \frac{x}{\sqrt{y}} \\ 0 & \frac{1}{\sqrt{y}} \end{pmatrix} i.$$

$\square$

Definition 2.1.3 We denote by LATT the set of all lattices in $\mathbb{C}$. Let BAS be the set of all $\mathbb{R}$-bases of $\mathbb{C}$, i.e. the set of all pairs $(z, w) \in \mathbb{C}^2$, which are linearly independent over $\mathbb{R}$. Let BAS^+ be the subset of all bases that are clockwise-oriented, i.e. the set of all $(z, w) \in \text{BAS}$ with $\text{Im}(z/w) > 0$. There is a natural map

$$\Psi : \text{BAS}^+ \to \text{LATT},$$

defined by

$$\Psi(z, w) = \mathbb{Z}z \oplus \mathbb{Z}w.$$

This map is surjective but not injective, since for example $\Psi(z + w, w) = \Psi(z, w)$. The group $\Gamma_0 = \text{SL}_2(\mathbb{Z})$ acts on BAS^+ by $\gamma.(z, w) = (z, w)\gamma^t = (az + bw, cz + dw)$ if $\gamma = \left(\begin{smallmatrix} a & b \\ c & d \end{smallmatrix}\right)$. Here we remind the reader that an invertible real matrix preserves the orientation of a basis if and only if the determinant of the matrix is positive.

The group $\Gamma_0 = \text{SL}_2(\mathbb{Z})$ is called the *modular group*.

Lemma 2.1.4 *Two bases are mapped to the same lattice under Ψ if and only if they lie in the same Γ_0-orbit. So Ψ induces a bijection*

$$\Psi : \Gamma_0 \backslash \text{BAS}^+ \xrightarrow{\cong} \text{LATT}.$$

Proof Let (z, w) and (z', w') be two clockwise-oriented bases such that $\Psi(z, w) = \Lambda = \Psi(z', w')$. Since z', w' are elements of the lattice generated by z and w, there are $a, b, c, d \in \mathbb{Z}$ with $(z', w') = (az + cw, bz + dw) = (z, w)\left(\begin{smallmatrix} a & b \\ c & d \end{smallmatrix}\right)$. Since, on the other hand, z and w lie in the lattice generated by z' and w', there are $\alpha, \beta, \gamma, \delta \in \mathbb{Z}$ with $(z, w) = (z', w')\left(\begin{smallmatrix} \alpha & \beta \\ \gamma & \delta \end{smallmatrix}\right)$, so $(z, w)\left(\begin{smallmatrix} a & b \\ c & d \end{smallmatrix}\right)\left(\begin{smallmatrix} \alpha & \beta \\ \gamma & \delta \end{smallmatrix}\right) = (z, w)$. As z and w are linearly independent over $\mathbb{R}$, it follows that $\left(\begin{smallmatrix} a & b \\ c & d \end{smallmatrix}\right)\left(\begin{smallmatrix} \alpha & \beta \\ \gamma & \delta \end{smallmatrix}\right) = \left(\begin{smallmatrix} 1 & \\ & 1 \end{smallmatrix}\right)$ and so $g = \left(\begin{smallmatrix} a & b \\ c & d \end{smallmatrix}\right)$ is an element of $\text{GL}_2(\mathbb{Z})$. In particular one gets $\det(g) = \pm 1$. Since g maps the clockwise-oriented basis (z, w) to the clockwise-oriented basis (z', w'), one concludes $\det(g) > 0$, i.e. $\det(g) = 1$ and so $g \in \Gamma_0$, which means that the two bases are in the same Γ_0-orbit. The converse direction is trivial. $\square$

The set BAS^+ is a bit unwieldy, so one divides out the action of the group $\mathbb{C}^\times$. This action of $\mathbb{C}^\times$ on the set BAS^+ is defined by $\xi(a, b) = (\xi a, \xi b)$. One has $(a, b) = b(a/b, 1)$, so every $\mathbb{C}^\times$-orbit contains exactly one element of the form $(z, 1)$ with $z \in \mathbb{H}$. The action of $\mathbb{C}^\times$ commutes with the action of Γ_0, so $\mathbb{C}^\times$ acts on $\Gamma_0 \backslash \text{BAS}^+$. On the other hand, $\mathbb{C}^\times$ acts on LATT by multiplication and the map Ψ translates one action into the other, which means $\Psi(\lambda(z, w)) = \lambda \Psi(z, w)$. As

Ψ is bijective, the two $\mathbb{C}^\times$-actions are isomorphic and Ψ maps orbits bijectively to orbits, so giving a bijection

$$\Psi : \Gamma_0 \backslash \mathrm{BAS}^+ / \mathbb{C}^\times \xrightarrow{\cong} \mathrm{LATT}/\mathbb{C}^\times.$$

Now let $z \in \mathbb{H}$. Then $(z, 1) \in \mathrm{BAS}^+$. For $\gamma = \left(\begin{smallmatrix} a & b \\ c & d \end{smallmatrix}\right) \in \Gamma_0$ one has, modulo the $\mathbb{C}^\times$-action:

$$(z, 1)\gamma^t \mathbb{C}^\times = (az + b, cz + d)\mathbb{C}^\times = \left(\frac{az + b}{cz + d}, 1\right)\mathbb{C}^\times.$$

Letting Γ_0 act on $\mathbb{H}$ by linear fractionals, the map $z \mapsto (z, 1)\mathbb{C}^\times$ is thus equivariant with respect to the actions of Γ_0.

Theorem 2.1.5 *The map $z \mapsto \mathbb{Z}z + \mathbb{Z}$ induces a bijection*

$$\Gamma_0 \backslash \mathbb{H} \xrightarrow{\cong} \mathrm{LATT}/\mathbb{C}^\times.$$

Proof The map is a composition of the maps

$$\Gamma_0 \backslash \mathbb{H} \xrightarrow{\varphi} \Gamma_0 \backslash \mathrm{BAS}^+ / \mathbb{C}^\times \xrightarrow{\cong} \mathrm{LATT}/\mathbb{C}^\times,$$

so it is well defined. We have to show that φ is bijective.

To show surjectivity, let $(v, w) \in \mathrm{BAS}^+$. Then $(v, w)\mathbb{C}^\times = (v/w, 1)\mathbb{C}^\times$ and $v/w \in \mathbb{H}$, so φ is surjective. For injectivity, assume $\varphi(\Gamma_0 z) = \varphi(\Gamma_0 w)$. This means $\Gamma_0(z, 1)\mathbb{C}^\times = \Gamma_0(w, 1)\mathbb{C}^\times$, so there are $\gamma = \left(\begin{smallmatrix} a & b \\ c & d \end{smallmatrix}\right) \in \Gamma_0$ and $\lambda \in \mathbb{C}^\times$ with $(w, 1) = \gamma(z, 1)\lambda$. The right-hand side is

$$\gamma(z, 1)\lambda = \lambda(az + b, cz + d) = (w, 1).$$

Comparing the second coordinates, we get $\lambda = (cz + d)^{-1}$ and so $w = \frac{az+b}{cz+d} = \gamma.z$, as claimed. $\square$

The element $-1 = \left(\begin{smallmatrix} -1 & \\ & -1 \end{smallmatrix}\right)$ acts trivially on the upper half plane $\mathbb{H}$. This motivates the following definition.

Definition 2.1.6 Let $\overline{\Gamma}_0 = \Gamma_0/\pm 1$. For a subgroup Γ of Γ_0 let $\overline{\Gamma}$ be the image of Γ in $\overline{\Gamma}_0$. Then we have

$$[\overline{\Gamma}_0 : \overline{\Gamma}] = \begin{cases} [\Gamma_0 : \Gamma] & \text{if } -1 \in \Gamma, \\ \frac{1}{2}[\Gamma_0 : \Gamma] & \text{otherwise.} \end{cases}$$

Let

$$S \stackrel{\text{def}}{=} \begin{pmatrix} 0 & -1 \\ 1 & 0 \end{pmatrix}, \qquad T \stackrel{\text{def}}{=} \begin{pmatrix} 1 & 1 \\ 0 & 1 \end{pmatrix}.$$

One has

$$Sz = \frac{-1}{z}, \qquad Tz = z + 1,$$

as well as $S^2 = -1 = (ST)^3$. Denote by D the set of all $z \in \mathbb{H}$ with $|\operatorname{Re}(z)| < \frac{1}{2}$ and $|z| > 1$, as depicted in the next figure. Let $\overline{D}$ be the closure of D in $\mathbb{H}$. The set D is a so-called *fundamental domain* for the group $\mathrm{SL}_2(\mathbb{Z})$; see Definition 2.5.17.

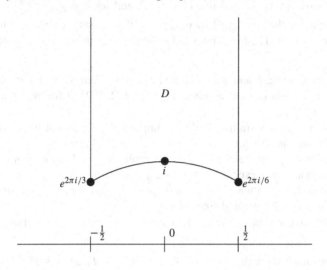

Theorem 2.1.7

(a) *For every $z \in \mathbb{H}$ there exists a $\gamma \in \Gamma_0$ with $\gamma z \in \overline{D}$.*
(b) *If $z, w \in \overline{D}$, with $z \neq w$, lie in the same Γ_0-orbit, then we have $\operatorname{Re}(z) = \pm\frac{1}{2}$ and $z = w \pm 1$, or $|z| = 1$ and $w = -1/z$. In any case the two points lie on the boundary of D.*
(c) *For $z \in \mathbb{H}$ let $\Gamma_{0,z}$ be the stabilizer of z in Γ_0. For $z \in \overline{D}$ we have $\Gamma_{0,z} = \{\pm 1\}$ except when*

  - *$z = i$, then $\Gamma_{0,z}$ is a group of order four, generated by S,*
  - *$z = \rho = e^{2\pi i/3}$, then $\Gamma_{0,z}$ is of order six, generated by ST,*
  - *$z = -\overline{\rho} = e^{\pi i/3}$, then $\Gamma_{0,z}$ is of order six, generated by TS.*

(d) *The group Γ_0 is generated by S and T.*

Proof Let Γ' be the subgroup of Γ_0 generated by S and T. We show that for every $z \in \mathbb{H}$ there is a $\gamma' \in \Gamma'$ with $\gamma' z \in \overline{D}$. So let $g = \left(\begin{smallmatrix} a & b \\ c & d \end{smallmatrix}\right)$ in Γ'. For $z \in \mathbb{H}$ one has

$$\operatorname{Im}(gz) = \frac{\operatorname{Im}(z)}{|cz + d|^2}.$$

Since c and d are integers, for every $M > 0$ the set of all pairs (c, d) with $|cz + d| < M$ is finite. Therefore there exists $\gamma \in \Gamma'$ such that $\operatorname{Im}(\gamma z)$ is maximal. Choose an integer n such that $T^n \gamma z$ has real part in $[-1/2, 1/2]$. We claim that the element $w = T^n \gamma z$ lies in $\overline{D}$. It suffices to show that $|w| \geq 1$. Assuming

$|w| < 1$, we conclude that the element $-1/w = Sw$ has imaginary part strictly bigger than $\text{Im}(w)$, which contradicts our choices. So indeed we get $w = T^n \gamma z$ in $\overline{D}$ and part (a) is proven.

We now show parts (b) and (c). Let $z \in \overline{D}$ and let $1 \neq \gamma = \left(\begin{smallmatrix} a & b \\ c & d \end{smallmatrix}\right) \in \Gamma_0$ with $\gamma z \in \overline{D}$. Replacing the pair (z, γ) by $(\gamma z, \gamma^{-1})$, if necessary, we assume $\text{Im}(\gamma z) \geq \text{Im}(z)$, so $|cz + d| \leq 1$. This cannot hold for $|c| \geq 2$, so we have the cases $c = 0, 1, -1$.

- If $c = 0$, then $d = \pm 1$ and we can assume $d = 1$. Then $\gamma z = z + b$ and $b \neq 0$. Since the real parts of both numbers lie in $[-1/2, 1/2]$, it follows that $b = \pm 1$ and $\text{Re}(z) = \pm 1/2$.
- If $c = 1$, then the assertion $|z + d| \leq 1$ implies $d = 0$, except if $z = \rho, -\overline{\rho}$, in which case we can also have $d = 1, -1$.
  - If $d = 0$, then $|z| = 1$ and $ad - bc = 1$ implies $b = -1$, so $gz = a - 1/z$ and we conclude $a = 0$, except if $\text{Re}(z) = \pm \frac{1}{2}$, so $z = \rho, -\overline{\rho}$.
  - If $z = \rho$ and $d = 1$, then $a - b = 1$ and $g\rho = a - 1/(1 + \rho) = a + \rho$, so $a = 0, 1$. The case $z = -\overline{\rho}$ is treated similarly.
- If $c = -1$, one can replace the whole matrix with its negative and thus can apply the case $c = 1$.

Finally, we must show that $\Gamma_0 = \Gamma'$. For this let $\gamma \in \Gamma_0$ and $z \in D$. Then there is $\gamma' \in \Gamma'$ with $\gamma' \gamma z = z$, so $\gamma = \gamma'^{-1} \in \Gamma'$. $\qquad\Box$

2.2 Modular Forms

In this section we introduce the protagonists of this chapter. Before that, we start with weakly modular functions.

Definition 2.2.1 Let $k \in \mathbb{Z}$. A meromorphic function f on the upper half plane $\mathbb{H}$ is called *weakly modular of weight k* if

$$f\left(\frac{az + b}{cz + d}\right) = (cz + d)^k f(z)$$

holds for every $z \in \mathbb{H}$, in which f is defined and every $\left(\begin{smallmatrix} a & b \\ c & d \end{smallmatrix}\right) \in SL_2(\mathbb{Z})$.

Note: for such a function $f \neq 0$ to exist, k must be even, since the matrix $\left(\begin{smallmatrix} -1 & \\ & -1 \end{smallmatrix}\right)$ lies in $SL_2(\mathbb{Z})$.

For $\sigma = \left(\begin{smallmatrix} a & b \\ c & d \end{smallmatrix}\right) \in G$ we denote the induced map $z \mapsto \sigma z = \frac{az+b}{cz+d}$ again by σ. Then

$$\frac{d(\sigma z)}{dz} = \frac{1}{(cz + d)^2}.$$

We deduce from this that a holomorphic function f is weakly modular of weight 2 if and only if the differential form $\omega = f(z)\,dz$ on $\mathbb{H}$ is invariant under Γ_0, i.e. if $\gamma^* \omega = \omega$ holds for every $\gamma \in \Gamma_0$, where $\gamma^* \omega$ is the pullback of the form ω under the map $\gamma : \mathbb{H} \to \mathbb{H}$.

More generally, we define for $k \in \mathbb{Z}$ and $f : \mathbb{H} \to \mathbb{C}$:

$$f|_k\sigma(z) \overset{\text{def}}{=} (cz+d)^{-k} f\left(\frac{az+b}{cz+d}\right),$$

where $\sigma = \left(\begin{smallmatrix} a & b \\ c & d \end{smallmatrix}\right) \in G$. If k is fixed, we occasionally leave the index out, i.e. we write $f|\sigma = f|_k\sigma$.

Lemma 2.2.2 *The maps $f \mapsto f|\sigma$ define a linear (right-)action of the group G on the space of functions $f : \mathbb{H} \to \mathbb{C}$, i.e.*

- *for every $\sigma \in G$ the map $f \mapsto f|\sigma$ is linear,*
- *one has $f|1 = f$ and $f|(\sigma\sigma') = (f|\sigma)|\sigma'$ for all $\sigma, \sigma' \in G$.*

Every right-action can be made into a left-action by inversion, i.e. one defines $\sigma f = f|\sigma^{-1}$ and one then gets $(\sigma\sigma')f = \sigma(\sigma'f)$.

Proof The only non-trivial assertion is $f|(\sigma\sigma') = (f|\sigma)|\sigma'$. For $k=0$ this is simply:

$$f|(\sigma\sigma')(z) = f(\sigma\sigma'z) = f|\sigma(\sigma'z) = (f|\sigma)|\sigma'(z).$$

Let $j(\sigma, z) = (cz+d)$. One verifies that this 'factor of automorphy' satisfies a so-called cocycle relation:

$$j(\sigma\sigma', z) = j(\sigma, \sigma'z) j(\sigma', z).$$

As $f|_k\sigma(z) = j(\sigma, z)^{-k} f|_0\sigma(z)$, we conclude

$$f|_k(\sigma\sigma')(z) = j(\sigma\sigma', z)^{-k} f|_0(\sigma\sigma')(z)$$
$$= j(\sigma, \sigma'z)^{-k} j(\sigma', z)^{-k} (f|_0\sigma)|_0\sigma'(z) = (f|_k\sigma)|_k\sigma'(z). \qquad \square$$

Lemma 2.2.3 *Let $k \in 2\mathbb{Z}$. A meromorphic function f on $\mathbb{H}$ is weakly modular of weight k if and only if for every $z \in \mathbb{H}$ one has*

$$f(z+1) = f(z) \quad \text{and} \quad f(-1/z) = z^k f(z).$$

Proof By definition, f is weakly modular if and only if $f|_k\gamma = f$ for every $\gamma \in \Gamma_0$, which means that f is invariant under the group action of Γ_0. It suffices to check invariance on the two generators S and T of the group. $\qquad \square$

We now give the definition of a modular function. Let f be a weakly modular function. The map $q : z \mapsto e^{2\pi i z}$ maps the upper half plane surjectively onto the pointed unit disk $\mathbb{D}^* = \{z \in \mathbb{C} : 0 < |z| < 1\}$. Two points z, w in $\mathbb{H}$ have the same image under q if and only if there is $m \in \mathbb{Z}$ such that $w = z + m$. So q induces a bijection $q : \mathbb{Z}\backslash\mathbb{H} \to \mathbb{D}^*$. In particular, for every weakly modular function f on $\mathbb{H}$ there is a function $\tilde{f}$ on $\mathbb{D}^* \smallsetminus q(\{\text{poles}\})$ with

$$f(z) = \tilde{f}(q(z)).$$

This means that for $w \in \mathbb{D}^*$ we have

$$\tilde{f}(w) = f\left(\frac{\log w}{2\pi i}\right),$$

where $\log w$ is an arbitrary branch of the holomorphic logarithm, being defined in a neighborhood of w. Then $\tilde{f}$ is a meromorphic function on the pointed unit disk.

Definition 2.2.4 A weakly modular function f of weight k is called a *modular function* of weight k if the induced function $\tilde{f}$ is meromorphic on the entire unit disk $\mathbb{D} = \{z \in \mathbb{C} : |z| < 1\}$.

Suggestively, in this case one also says that f is 'meromorphic at infinity'. This means that $\tilde{f}(q)$ has at most a pole at $q = 0$. It follows that poles of $\tilde{f}$ in $\mathbb{D}^*$ cannot accumulate at $q = 0$, because that would imply an essential singularity at $q = 0$. For the function f it means that there exists a bound $T = T_f > 0$ such that f has no poles in the region $\{z \in \mathbb{H} : \text{Im}(z) > T\}$.

The Fourier expansion of the function f is of particular importance. Next we show that the Fourier series converges uniformly. In the next lemma we write $C^\infty(\mathbb{R}/\mathbb{Z})$ for the set of all infinitely often differentiable functions $g : \mathbb{R} \to \mathbb{C}$, which are periodic of period 1, which means that one has $g(x + 1) = g(x)$ for every $x \in \mathbb{R}$.

Definition 2.2.5 Let $D \subset \mathbb{R}$ be an unbounded subset. A function $f : D \to \mathbb{C}$ is said to be *rapidly decreasing* if for every $N \in \mathbb{N}$ the function $x^N f(x)$ is bounded on D.

For $D = \mathbb{N}$ one gets the special case of a rapidly decreasing sequence.

Examples 2.2.6

- For $D = \mathbb{N}$ the sequence $a_k = \frac{1}{k!}$ is rapidly decreasing.
- For $D = [0, \infty)$ the function $f(x) = e^{-x}$ is rapidly decreasing.
- For $D = \mathbb{R}$ the function $f(x) = e^{-x^2}$ is rapidly decreasing.

Proposition 2.2.7 (Fourier series) *If g is in $C^\infty(\mathbb{R}/\mathbb{Z})$, then for every $x \in \mathbb{R}$ one has*

$$g(x) = \sum_{k \in \mathbb{Z}} c_k(g) e^{2\pi i k x},$$

where $c_k(g) = \int_0^1 g(t) e^{-2\pi i k t}\, dt$ and the sum converges uniformly. The Fourier coefficients $c_k = c_k(g)$ are rapidly decreasing as functions in $k \in \mathbb{Z}$.

The Fourier coefficients $c_k(g)$ are uniquely determined in the following sense: Let $(a_k)_{k \in \mathbb{Z}}$ be a family of complex numbers such that for every $x \in \mathbb{R}$ the identity

$$g(x) = \sum_{k=-\infty}^{\infty} a_k e^{2\pi i k x}$$

holds with locally uniform convergence of the series. Then it follows that $a_k = c_k(g)$ for every $k \in \mathbb{Z}$.

Proof Using integration by parts repeatedly, we get for $k \neq 0$,

$$|c_k(g)| = \left| \int_0^1 g(t)e^{-2\pi itk}\, dt \right| = \left| \frac{1}{-2\pi ik} \int_0^1 g'(t)e^{-2\pi ikt}\, dt \right|$$

$$= \left| \frac{1}{-4\pi^2 k^2} \int_0^1 g''(t)e^{-2\pi ikt}\, dt \right| \leq \cdots$$

$$\leq \frac{1}{(4\pi^2 k^2)^n} \left| \int_0^1 g^{(2n)}(t)e^{-2\pi ikt}\, dt \right|.$$

So the sequence $(c_k(g))$ is rapidly decreasing. Consequently, the sum $\sum_{k \in \mathbb{Z}} |c_k(g)|$ converges, so the series $\sum_{k \in \mathbb{Z}} c_k(g)e^{2\pi ikx}$ converges uniformly. We only have to show that it converges to g. It suffices to do that at the point $x = 0$, since, assuming we have this convergence at $x = 0$, we can set $g_x(t) = g(x + t)$ and we see

$$g(x) = g_x(0) = \sum_k c_k(g_x).$$

By $c_k(g_x) = \int_0^1 g(t + x)e^{-2\pi ikt}\, dt = e^{2\pi ikx}c_k(g)$ we get the claim. So we only have to show $g(0) = \sum_k c_k(g)$. Replacing $g(x)$ with $g(x) - g(0)$, we can assume $g(0) = 0$, in which case we have to show that $\sum_k c_k(g) = 0$. Let

$$h(x) = \frac{g(x)}{e^{2\pi ix} - 1}.$$

As $g(0) = 0$, it follows that $h \in C^\infty(\mathbb{R}/\mathbb{Z})$ and we have

$$c_k(g) = \int_0^1 h(x)\big(e^{2\pi ix} - 1\big)e^{-2\pi ikx}\, dx = c_{k-1}(h) - c_k(h).$$

Since $h \in C^\infty(\mathbb{R}/\mathbb{Z})$, the series $\sum_k c_k(h)$ converges absolutely as well and $\sum_k c_k(g) = \sum_k (c_{k-1}(h) - c_k(h)) = 0$.

Now for the uniqueness of the Fourier coefficients. Let $(a_k)_{k \in \mathbb{Z}}$ be as in the proposition. By locally uniform convergence the following interchange of integration and summation is justified. For $l \in \mathbb{Z}$ we have

$$c_l(g) = \int_0^1 g(t)e^{-2\pi ilt}\, dt = \int_0^1 \sum_{k=-\infty}^{\infty} a_k e^{2\pi kt}e^{-2\pi ilt}\, dt$$

$$= \sum_{k=-\infty}^{\infty} a_k \int_0^1 e^{2\pi kt}e^{-2\pi ilt}\, dt.$$

One has

$$\int_0^1 e^{2\pi kt}e^{-2\pi ilt}\, dt = \int_0^1 e^{2\pi(k-l)t}\, dt = \begin{cases} 1 & \text{if } k = l, \\ 0 & \text{otherwise.} \end{cases}$$

This implies $c_l(g) = a_l$.   □

This nice proof of the convergence of Fourier series is, to the author's knowledge, due to H. Jacquet.

Let f be a weakly modular function of weight k. As $f(z) = f(z+1)$ and f is infinitely differentiable (except at the poles), one can write it as a Fourier series:

$$f(x+iy) = \sum_{n=-\infty}^{+\infty} c_n(y)e^{2\pi i n x},$$

if there is no pole of f on the line $\mathrm{Im}(w) = y$, which holds true for all but countably many values of $y > 0$. For such y the sequence $(c_n(y))_{n\in\mathbb{Z}}$ is rapidly decreasing.

Lemma 2.2.8 *Let f be a modular function on the upper half plane $\mathbb{H}$ and let $T > 0$ such that f has no poles in the set $\{\mathrm{Im}(z) > T\}$. For every $n \in \mathbb{Z}$ and $y > T$ one has $c_n(y) = a_n e^{-2\pi n y}$ for a constant a_n. Then*

$$f(z) = \sum_{n=-N}^{+\infty} a_n e^{2\pi i n z},$$

where $-N$ is the pole-order of the induced meromorphic function $\tilde{f}$ at $q = 0$. For every $y > 0$, the sequence $a_n e^{-yn}$ is rapidly decreasing.

Proof The induced function $\tilde{f}$ with $f(z) = \tilde{f}(q(z))$ or $\tilde{f}(q) = f(\frac{\log q}{2\pi i})$ is meromorphic around $q = 0$. In a pointed neighborhood of zero, the function $\tilde{f}$ therefore has a Laurent expansion

$$\tilde{f}(w) = \sum_{n=-\infty}^{\infty} a_n w^n.$$

Replacing w by $q(z)$, one gets

$$f(z) = \sum_{n=-\infty}^{+\infty} a_n e^{2\pi i n z}.$$

The claim follows from the uniqueness of the Fourier coefficients.   □

Note, in particular, that the Fourier expansion of a modular function f equals the Laurent expansion of the induced function $\tilde{f}$.

Definition 2.2.9 A modular function f is called a *modular form* if it is holomorphic in the upper half plane $\mathbb{H}$ and *holomorphic at ∞*, i.e. $a_n = 0$ holds for every $n < 0$.

A modular form f is called *cusp form* if additionally $a_0 = 0$. In that case one says that f *vanishes at ∞*.

As an example, consider Eisenstein series G_k for $k \geq 4$. Write $q = e^{2\pi i z}$.

Proposition 2.2.10 *For even $k \geq 4$ we have*

$$G_k(z) = 2\zeta(k) + 2\frac{(2\pi i)^k}{(k-1)!} \sum_{n=1}^{\infty} \sigma_{k-1}(n)q^n,$$

where $\sigma_k(n) = \sum_{d|n} d^k$ is the kth divisor sum.

Proof On the one hand we have the partial fraction expansion of the cotangent function

$$\pi \cot(\pi z) = \frac{1}{z} + \sum_{m=1}^{\infty} \left(\frac{1}{z+m} + \frac{1}{z-m} \right),$$

and on the other

$$\pi \cot(\pi z) = \pi \frac{\cos(\pi z)}{\sin(\pi z)} = i\pi \frac{q+1}{q-1} = \pi i - \frac{2\pi i}{1-q} = \pi i - 2\pi i \sum_{n=0}^{\infty} q^n.$$

So

$$\frac{1}{z} + \sum_{m=1}^{\infty} \left(\frac{1}{z+m} + \frac{1}{z-m} \right) = \pi i - 2\pi i \sum_{n=0}^{\infty} q^n.$$

We repeatedly differentiate both sides to get for $k \geq 4$,

$$\sum_{m\in\mathbb{Z}} \frac{1}{(z+m)^k} = \frac{1}{(k-1)!}(-2\pi i)^k \sum_{n=1}^{\infty} n^{k-1} q^n.$$

The Eisenstein series is

$$G_k(z) = \sum_{(n,m)\neq(0,0)} \frac{1}{(nz+m)^k} = 2\zeta(k) + 2\sum_{n=1}^{\infty}\sum_{m\in\mathbb{Z}} \frac{1}{(nz+m)^k}$$

$$= 2\zeta(k) + \frac{2(-2\pi i)^k}{(k-1)!} \sum_{d=1}^{\infty}\sum_{a=1}^{\infty} d^{k-1} q^{ad}$$

$$= 2\zeta(k) + \frac{2(2\pi i)^k}{(k-1)!} \sum_{n=1}^{\infty} \sigma_{k-1}(n) q^n.$$

The proposition is proven. $\qquad\qquad\qquad\qquad\qquad\qquad\qquad\qquad\qquad\square$

Let f be a modular function of weight k. For $\gamma \in \Gamma_0$ the formula $f(\gamma z) = (cz+d)^k f(z)$ shows that the orders of vanishing of f at the points z and γz agree. So the order $\mathrm{ord}_z f$ depends only on the image of z in $\Gamma_0\backslash\mathbb{H}$.

We further define $\mathrm{ord}_\infty(f)$ as the order of vanishing of $\tilde{f}(q)$ at $q = 0$, where $\tilde{f}(e^{2\pi i z}) = f(z)$. Finally let $z \in \mathbb{H}$ be equal to the number $2e_z$, the order of the stabilizer group of z in Γ_0, so $e_z = \frac{|\Gamma_{0,z}|}{2}$. Then

$$e_z = \begin{cases} 2 & \text{if } z \text{ lies in the } \Gamma_0\text{-orbit of } i, \\ 3 & \text{if } z \text{ lies in the } \Gamma_0\text{-orbit of } \rho = e^{2\pi i/3}, \\ 1 & \text{otherwise.} \end{cases}$$

Here we recall that the *orbit* of an element $w \in \mathbb{H}$ is defined as

$$\Gamma_0\text{-orbit}(w) = \Gamma_0 w = \{\gamma.w : \gamma \in \Gamma_0\}.$$

Theorem 2.2.11 *Let $f \neq 0$ be a modular function of weight k. Then*

$$\mathrm{ord}_\infty(f) + \sum_{z \in \Gamma_0 \backslash \mathbb{H}} \frac{1}{e_z} \mathrm{ord}_z(f) = \frac{k}{12}.$$

Proof Note first that the sum is finite, as f has only finitely many zeros and poles modulo Γ_0. Indeed, in $\Gamma_0 \backslash \mathbb{H}$ these cannot accumulate, by the identity theorem. Also at ∞ they cannot accumulate, as f is meromorphic at ∞ as well.

We write the claim as

$$\mathrm{ord}_\infty(f) + \frac{1}{2}\mathrm{ord}_i(f) + \frac{1}{3}\mathrm{ord}_\rho(f) + \sum_{\substack{z \in \Gamma_0 \backslash \mathbb{H} \\ z \neq i, \rho}} \mathrm{ord}_z(f) = \frac{k}{12}.$$

Let D be the fundamental domain of Γ_0 as in Sect. 2.1. We integrate the function $\frac{1}{2\pi i} \frac{f'}{f}$ along the positively oriented boundary of D, as in the following figure.

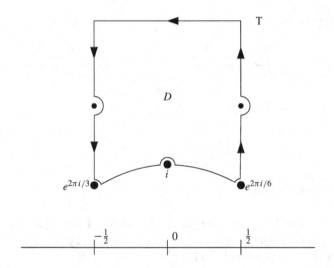

Assume first that f has neither a zero or a pole on the boundary of D, with the possible exception of i or $\rho, -\overline{\rho}$. Let C be the positively oriented boundary of D, except for $i, \rho, -\overline{\rho}$, which we circumvent by circular segments as in the figure. Further, we cut off the domain D at $\mathrm{Im}(z) = T$ for some $T > 0$ which is bigger than the imaginary part of any zero or pole of f. By the residue theorem we get

$$\frac{1}{2\pi i} \int_C \frac{f'}{f} = \sum_{\substack{z \in \Gamma_0 \backslash \mathbb{H} \\ z \neq i, \rho}} \mathrm{ord}_z(f).$$

On the other hand:

(a) Substituting $q = e^{2\pi i z}$ we transform the line $\frac{1}{2} + iT$, $-\frac{1}{2} + iT$ into a circle ω around $q = 0$ of negative orientation. So

$$\frac{1}{2\pi i} \int_{\frac{1}{2}+iT}^{-\frac{1}{2}+iT} \frac{f'}{f} = \frac{1}{2\pi i} \int_\omega \frac{\tilde{f}'}{\tilde{f}} = -\operatorname{ord}_\infty(f).$$

(b) The circular segment $k(\rho)$ around ρ has angle $\frac{2\pi}{6}$. By Exercise 1.11 we conclude:

$$\frac{1}{2\pi i} \int_{k(\rho)} \frac{f'}{f} \to -\frac{1}{6} \operatorname{ord}_\rho(f),$$

as the radius of the circular segment tends to zero. Analogously, one treats the circular segments $k(i)$ and $k(-\bar{\rho})$,

$$\frac{1}{2\pi i} \int_{k(i)} \frac{f'}{f} \to -\frac{1}{2} \operatorname{ord}_i(f), \qquad \frac{1}{2\pi i} \int_{k(-\bar{\rho})} \frac{f'}{f} \to -\frac{1}{6} \operatorname{ord}_\rho(f).$$

(c) The vertical path integrals add up to zero.
(d) The two segments s_1, s_2 of the unit circle map to each other under the transform $z \mapsto Sz = -z^{-1}$. One has

$$\frac{f'}{f}(Sz)S'(z) = \frac{k}{z} + \frac{f'}{f}(z).$$

So

$$\frac{1}{2\pi i} \int_{s_1} \frac{f'}{f} + \frac{1}{2\pi i} \int_{s_2} \frac{f'}{f} = \frac{1}{2\pi i} \int_{s_1} \left(\frac{f'}{f}(z) - \frac{f'}{f}(Sz)S'(z) \right) dz$$

$$= -\frac{1}{2\pi i} \int_{s_1} \frac{k}{z} dz \to \frac{k}{12}.$$

Comparing these two expressions for the integral, letting the radii of the small circular segments shrink to zero, one obtains the result.

If f has more poles or zeros on the boundary, the path of integration may be modified so as to circumvent these, as shown in the figure. □

Let $\mathcal{M}_k = \mathcal{M}_k(\Gamma_0)$ be the complex vector space of all modular forms of weight k and let S_k be the space of cusp forms of weight k. Then $S_k \subset \mathcal{M}_k$ is the kernel of the linear map $f \mapsto f(i\infty)$. By definition, it follows that

$$\mathcal{M}_k \mathcal{M}_l \subset \mathcal{M}_{k+l},$$

which means that if $f \in \mathcal{M}_k$ and $g \in \mathcal{M}_l$, then $fg \in \mathcal{M}_{k+l}$.

Note that a holomorphic function f on $\mathbb{H}$ with $f|_k \gamma = f$ for every $\gamma \in \Gamma_0$ lies in $\mathcal{M}_k$ if and only if the limit

$$\lim_{\operatorname{Im}(z)\to\infty} f(z)$$

exists.

The differential equation of the Weierstrass function $\wp$ features the coefficients

$$g_4 = 60G_4, \qquad g_6 = 140G_6.$$

It follows that $g_4(i\infty) = 120\zeta(4)$ and $g_6(i\infty) = 280\zeta(6)$. By Proposition 1.5.2 we have

$$\zeta(4) = \frac{\pi^4}{90}, \quad \text{and} \quad \zeta(6) = \frac{\pi^6}{945}.$$

So with

$$\Delta = g_4^3 - 27g_6^2,$$

it follows that $\Delta(i\infty) = 0$, i.e. Δ is a cusp form of weight 12.

Theorem 2.2.12 *Let k be an even integer.*

(a) *If $k < 0$ or $k = 2$, then $\mathcal{M}_k = 0$.*

(b) *If $k = 0, 4, 6, 8, 10$, then $\mathcal{M}_k$ is a one-dimensional vector space spanned by $1, G_4, G_6, G_8, G_{10}$, respectively. In these cases the space S_k is zero.*

(c) *Multiplication by Δ defines an isomorphism*

$$\mathcal{M}_{k-12} \xrightarrow{\cong} S_k.$$

Proof Take a non-zero element $f \in \mathcal{M}_k$. All terms on the left of the equation

$$\operatorname{ord}_\infty(f) + \frac{1}{2}\operatorname{ord}_i(f) + \frac{1}{3}\operatorname{ord}_\rho(f) + \sum_{\substack{z \in \Gamma_0 \backslash \mathbb{H} \\ z \neq i, \rho}} \operatorname{ord}_z(f) = \frac{k}{12}$$

are ≥ 0. Therefore $k \geq 0$ and also $k \neq 2$, as $1/6$ cannot be written in the form $a + b/2 + c/3$ with $a, b, c \in \mathbb{N}_0$. This proves (a).

If $0 \leq k < 12$, then $\operatorname{ord}_\infty(f) = 0$, and therefore $S_k = 0$ and $\dim \mathcal{M}_k \leq 1$. This implies (b).

The function Δ has weight 12, so $k = 12$. It is a cusp form, so $\operatorname{ord}_\infty(\Delta) > 0$. The formula implies $\operatorname{ord}_\infty(\Delta) = 1$ and that Δ has no further zeros. The multiplication with Δ gives an injective map $\mathcal{M}_{k-12} \to S_k$ and for $0 \neq f \in S_k$ we have $f/\Delta \in \mathcal{M}_{k-12}$, so the multiplication with Δ is surjective, too. $\qquad\square$

Corollary 2.2.13

(a) *One has*

$$\dim \mathcal{M}_k = \begin{cases} [k/12] & \text{if } k \equiv 2 \bmod 12, k \geq 0, \\ [k/12] + 1 & \text{if } k \not\equiv 2 \bmod 12, k \geq 0. \end{cases}$$

(b) *The space $\mathcal{M}_k$ has a basis consisting of all monomials $G_4^m G_6^n$ with $m, n \in \mathbb{N}_0$ and $4m + 6n = k$.*

Proof (a) follows from Theorem 2.2.12. For (b) we show that these monomials span the space $\mathcal{M}_k$. For $k \leq 6$, this is contained in Theorem 2.2.12. For $k \geq 8$ we use induction. Choose $m, n \in \mathbb{N}_0$ such that $4m + 6n = k$. The modular form $g = G_4^m G_6^n$ satisfies $g(\infty) \neq 0$. Therefore, for given $f \in \mathcal{M}_k$ there is $\lambda \in \mathbb{C}$ such that $f - \lambda g$ is a cusp form, i.e. equal to Δh for some $h \in \mathcal{M}_{k-12}$. By the induction hypothesis the function h lies in the span of the monomials indicated, and so does f.

It remains to show the linear independence of the monomials. Assume the contrary. Then a linear equation among these monomials of a fixed weight would lead to a polynomial equation satisfied by the function G_4^3/G_6^2, which would mean that this function is constant. This, however, is impossible, as the formula of Theorem 2.2.11 shows that G_4 vanishes at ρ, but G_6 does not. $\square$

Let $M = \bigoplus_{k=0}^\infty \mathcal{M}_k$ be the graded algebra of all modular forms. One can formulate the corollary by saying that the map

$$\mathbb{C}[X, Y] \to M, \quad X \mapsto G_4, \ Y \mapsto G_6$$

is an isomorphism of $\mathbb{C}$-algebras.

We have seen that

$$G_k(z) = 2\zeta(k) + 2\frac{(2\pi i)^k}{(k-1)!}\sum_{n=1}^\infty \sigma_{k-1}(n)q^n,$$

where $\sigma_k(n) = \sum_{d\mid n} d^k$. Denote the normalized Eisenstein series by $E_k(z) = G_k(z)/(2\zeta(k))$. With $\gamma_k = (-1)^{k/2}\frac{2k}{B_{k/2}}$ we then have

$$E_k(z) = 1 + \gamma_k \sum_{n=1}^\infty \sigma_{k-1}(n)q^n.$$

Examples

$$E_4 = 1 + 240\sum_{n=1}^\infty \sigma_3(n)q^n, \qquad E_6 = 1 - 504\sum_{n=1}^\infty \sigma_5(n)q^n,$$

$$E_8 = 1 + 480\sum_{n=1}^\infty \sigma_7(n)q^n, \qquad E_{10} = 1 - 264\sum_{n=1}^\infty \sigma_9(n)q^n,$$

$$E_{12} = 1 + \frac{65520}{691}\sum_{n=1}^\infty \sigma_{11}(n)q^n.$$

Remark As the spaces of modular forms of weights 8 and 10 are one-dimensional, we immediately get

$$E_4^2 = E_8, \qquad E_4 E_6 = E_{10}.$$

These formulae are equivalent to

$$\sigma_7(n) = \sigma_3(n) + 120\sum_{m=1}^{n-1} \sigma_3(m)\sigma_3(n-m).$$

and

$$11\sigma_9(n) = 21\sigma_5(n) - 10\sigma_3(n) + 5040 \sum_{m=1}^{n-1} \sigma_3(m)\sigma_5(n-m).$$

It is quite a non-trivial task to find proofs of these number-theoretical statements without using analysis!

2.3 Estimating Fourier Coefficients

Our goal is to attach so-called L-functions to modular forms by feeding their Fourier coefficients into Dirichlet series. In order to show convergence of these Dirichlet series, we must give growth estimates for the Fourier coefficients. Let

$$f(z) = \sum_{n=0}^{\infty} a_n q^n, \quad q = e^{2\pi i z}$$

be a modular form of weight $k \geq 4$.

Proposition 2.3.1 *If $f = G_k$, then the Fourier coefficients a_n grow like n^{k-1}. More precisely: there are constants $A, B > 0$ with*

$$An^{k-1} \leq |a_n| \leq Bn^{k-1}.$$

Proof There is a positive number $A > 0$ such that for $n \geq 1$ we have $|a_n| = A\sigma_{k-1}(n) \geq An^{k-1}$. On the other hand,

$$\frac{|a_n|}{n^{k-1}} = A \sum_{d|n} \frac{1}{d^{k-1}} \leq A \sum_{d=1}^{\infty} \frac{1}{d^{k-1}} = A\zeta(k-1) < \infty. \qquad \square$$

Theorem 2.3.2 (Hecke) *The Fourier coefficients a_n of a cusp form f of weight $k \geq 4$ satisfy*

$$a_n = O\left(n^{k/2}\right).$$

The *O-notation* means that there is a constant $C > 0$ such that

$$|a_n| \leq Cn^{k/2}.$$

Proof Since f is a cusp form, it satisfies the estimate $f(z) = O(q) = O(e^{-2\pi y})$ for $q \to 0$ or $y \to \infty$. Let $\phi(z) = y^{k/2}|f(z)|$. The function ϕ is invariant under the group Γ_0. Furthermore, it is continuous and $\phi(z)$ tends to 0 for $y \to \infty$. So ϕ is bounded on the fundamental domain D of Sect. 2.1, so it is bounded on all of $\mathbb{H}$. This means that there exists a constant $C > 0$ with $|f(z)| \leq Cy^{-k/2}$ for every

$z \in \mathbb{H}$. By definition, $a_n = \int_0^1 f(x+iy)q^{-n}\,dx$, so that $|a_n| \leq Cy^{-k/2}e^{2\pi ny}$, and this estimate holds for every $y > 0$. For $y = 1/n$ one gets $|a_n| \leq e^{2\pi}Cn^{k/2}$. $\qquad\square$

Remark It is possible to improve the exponent. Deligne has shown that the Fourier coefficients of a cusp form satisfy

$$a_n = O\left(n^{\frac{k}{2}-\frac{1}{2}+\varepsilon}\right)$$

for every $\varepsilon > 0$.

Corollary 2.3.3 *For every $f \in \mathcal{M}_k(\Gamma_0)$ with Fourier expansion*

$$f(z) = \sum_{n=0}^{\infty} a_n e^{2\pi i n z}$$

we have the estimate

$$a_n = O\left(n^{k-1}\right).$$

Proof This follows from $\mathcal{M}_k = \mathcal{S}_k + \mathbb{C}G_k$, as well as Proposition 2.3.1 and Theorem 2.3.2. $\qquad\square$

2.4 L-Functions

In this section we encounter the question of why modular forms are so important for number theory. To each modular form f we attach an L-function $L(f, s)$. These L-functions are conjectured to be universal in the sense that L-functions defined in entirely different settings are equal to modular L-functions. In the example of L-functions of (certain) elliptic curves this has been shown by Andrew Wiles, who used it to prove Fermat's Last Theorem [Wil95].

Definition 2.4.1 For a cusp form f of weight k with Fourier expansion

$$f(z) = \sum_{n=1}^{\infty} a_n e^{2\pi i n z},$$

we define its *L-series* or *L-function* by

$$L(f, s) = \sum_{n=1}^{\infty} \frac{a_n}{n^s}, \quad s \in \mathbb{C}.$$

Lemma 2.4.2 *The series $L(f, s)$ converges locally uniformly in the region* $\operatorname{Re}(s) > \frac{k}{2} + 1$.

Proof From $a_n = O(n^{k/2})$, as in Theorem 2.3.2, it follows that

$$a_n n^{-s} = O\left(n^{\frac{k}{2}-\operatorname{Re}(s)}\right),$$

which implies the claim. $\qquad\square$

For the functional equation of the L-function we need the Gamma function, the definition of which we now recall.

Definition 2.4.3 The *Gamma function* is defined for $\mathrm{Re}(z) > 0$ by the integral

$$\Gamma(z) = \int_0^\infty e^{-t} t^{z-1} \, dt.$$

Lemma 2.4.4 *The Gamma integral converges locally uniformly absolutely in the right half plane* $\mathrm{Re}(z) > 0$ *and defines a holomorphic function there. It satisfies the functional equation*

$$\Gamma(z+1) = z\Gamma(z).$$

The Gamma function can be extended to a meromorphic function on $\mathbb{C}$*, with simple poles at* $z = -n, n \in \mathbb{N}_0$ *and holomorphic otherwise. The residue at* $z = -n$ *is* $\frac{(-1)^n}{n!}$*.*

Proof The function e^{-t} decreases faster at $+\infty$ than any power of t. Therefore the integral $\int_1^\infty e^{-t} t^{z-1} \, dt$ converges absolutely for every $z \in \mathbb{C}$ and the convergence is locally uniform in z. For $0 < t < 1$ the integrand is $\leq t^{\mathrm{Re}(z)-1}$, so the integral $\int_0^1 e^{-t} t^{z-1} \, dt$ converges locally uniformly for $\mathrm{Re}(z) > 0$. As $z t^{z-1}$ is the derivative of t^z, we can use integration by parts to compute

$$z\Gamma(z) = \int_0^\infty e^{-t} \left(t^z\right)' dt = \underbrace{-e^{-t} t^z \big|_0^\infty}_{=0} + \underbrace{\int_0^\infty e^{-t} t^z \, dt}_{=\Gamma(z+1)}.$$

The function $\Gamma(z)$ is holomorphic in $\mathrm{Re}(z) > 0$. Using the formula

$$\Gamma(z) = \frac{1}{z} \Gamma(z+1),$$

we can extend the Gamma function to the region $\mathrm{Re}(z) > -1$ with a simple pole at $z = 0$ of residue equal to $\Gamma(1) = \int_0^\infty e^{-t} \, dt = 1$. This argument can be iterated to get the meromorphic continuation to all of $\mathbb{C}$. $\qquad\square$

Theorem 2.4.5 *Let* f *be a cusp form of weight* k*. Then the L-function* $L(f, s)$*, initially holomorphic for* $\mathrm{Re}(s) > \frac{k}{2} + 1$*, has an analytic continuation to an entire function. The extended function*

$$\Lambda(f, s) \stackrel{\mathrm{def}}{=} (2\pi)^{-s} \Gamma(s) L(f, s)$$

is entire as well and satisfies the functional equation

$$\Lambda(f, s) = (-1)^{k/2} \Lambda(f, k - s).$$

The function $\Lambda(f, s)$ *is bounded on every vertical strip, i.e. for every* $T > 0$ *there exists* $C_T > 0$ *such that* $|\Lambda(f, s)| \leq C_T$ *for every* $s \in \mathbb{C}$ *with* $|\mathrm{Re}(s)| \leq T$*.*

Proof Let $f(z) = \sum_{n=1}^{\infty} a_n q^n$ with $q = e^{2\pi i z}$ be the Fourier expansion. According to Theorem 2.3.2 there is a constant $C > 0$ such that $|a_n| \leq C n^{k/2}$ holds for every $n \in \mathbb{N}$. So for given $\varepsilon > 0$ we have for all $y \geq \varepsilon$,

$$\left| f(iy) \right| = \left| \sum_{n=1}^{\infty} a_n e^{-2\pi n y} \right| \leq C \sum_{n=1}^{\infty} n^{k/2} e^{-2\pi n y} \leq D e^{-\pi y},$$

with $D = C \sum_{n=1}^{\infty} n^{k/2} e^{-\varepsilon \pi n} < \infty$. So the function $f(iy)$ is rapidly decreasing as $y \to \infty$. The same estimate holds for the function $y \mapsto \sum_{n=1}^{\infty} |a_n| e^{-2\pi y n}$. Consequently, for every $s \in \mathbb{C}$ we have

$$\int_{\varepsilon}^{\infty} \sum_{n=1}^{\infty} |a_n| e^{-2\pi y n} \left| y^{s-1} \right| dy < \infty.$$

Hence we are allowed to interchange sums and integrals in the following computation due to absolute convergence:

$$\int_{\varepsilon}^{\infty} f(iy) y^{s-1} \, dy = \int_{\varepsilon}^{\infty} \sum_{n=1}^{\infty} a_n e^{-2\pi n y} y^{s-1} \, dy$$

$$= \sum_{n=1}^{\infty} a_n \int_{\varepsilon}^{\infty} e^{-2\pi n y} y^{s-1} \, dy$$

$$= \sum_{n=1}^{\infty} a_n (2\pi n)^{-s} \int_{\varepsilon}^{\infty} e^{-y} y^{s-1} \, dy.$$

For $\mathrm{Re}(s) > \frac{k}{2} + 1$ the right-hand side converges to

$$(2\pi)^{-s} \Gamma(s) L(f, s) = \Lambda(f, s),$$

as ε tends to zero. On the other hand, $f(i\frac{1}{y}) = f(-\frac{1}{iy}) = (yi)^k f(iy)$, so that $f(i/y)$ is also rapidly decreasing, and the left-hand side converges to $\int_0^{\infty} f(iy) y^{s-1} \, dy$, as $\varepsilon \to 0$. Together, for $\mathrm{Re}(s) > \frac{k}{2} + 1$ we get

$$\int_0^{\infty} f(iy) y^{s-1} \, dy = \Lambda(f, s).$$

We write this integral as the sum $\int_0^1 + \int_1^{\infty}$. As $f(iy)$ is rapidly decreasing, the integral $\Lambda_1(f, s) = \int_1^{\infty} f(iy) y^{s-1} \, dy$ converges for every $s \in \mathbb{C}$ and defines an entire function.

Because of

$$|\Lambda_1(f, s)| \leq \int_1^{\infty} |f(iy)| y^{\mathrm{Re}(s)-1} \, dy,$$

the function $\Lambda_1(f, s)$ is bounded on every vertical strip.

For the second integral we have

$$\Lambda_2(f,s) = \int_0^1 f(iy)y^s \frac{dy}{y} = \int_1^\infty f\left(i\frac{1}{y}\right)y^{-s}\frac{dy}{y} = (-1)^{k/2} \int_1^\infty f(iy)y^{k-s}\frac{dy}{y},$$

which means $\Lambda_2(f,s) = (-1)^{k/2}\Lambda_1(f,k-s)$, so the claim follows.   □

Generally, a series of the form

$$L(s) = \sum_{n=1}^\infty \frac{a_n}{n^s}$$

for $s \in \mathbb{C}$, convergent or not, is called a *Dirichlet series*. The following typical convergence behavior of a Dirichlet series will be needed in the sequel.

Lemma 2.4.6 *Let (a_n) be a sequence of complex numbers. If for a given $s_0 \in \mathbb{C}$ the sequence $\frac{a_n}{n^{s_0}}$ is bounded, then the Dirichlet series $L(s) = \sum_{n=1}^\infty \frac{a_n}{n^s}$ converges absolutely uniformly on every set of the form*

$$\{s \in \mathbb{C} : \operatorname{Re}(s) \geq \operatorname{Re}(s_0) + 1 + \varepsilon\},$$

where $\varepsilon > 0$.

This lemma reminds us of the convergence behavior of a power series. This is by no means an accident, as the power series with coefficients (a_n) and the corresponding Dirichlet series are linked via the Mellin transform, as we shall see below.

Proof Suppose that $|a_n n^{-s_0}| \leq M$ for some $M > 0$ and every $n \in \mathbb{N}$. Let $\varepsilon > 0$ be given and let $s \in \mathbb{C}$ with $\operatorname{Re}(s) \geq \operatorname{Re}(s_0) + 1 + \varepsilon$. Then $s = s_0 + \alpha$ with $\operatorname{Re}(\alpha) \geq 1 + \varepsilon$, and so

$$\left|\frac{a_n}{n^s}\right| = \left|\frac{a_n}{n^{s_0}}\right|\frac{1}{n^{\operatorname{Re}(\alpha)}} \leq M\frac{1}{n^{1+\varepsilon}}.$$

As the series over $1/n^{1+\varepsilon}$ converges, the lemma follows.   □

Theorem 2.4.7 (Hecke's converse theorem) *Let a_n be a sequence in $\mathbb{C}$, such that the Dirichlet series $L(s) = \sum_{n=1}^\infty a_n n^{-s}$ converges in the region $\{\operatorname{Re}(s) > C\}$ for some $C \in \mathbb{R}$. If the function $\Lambda(s) = (2\pi)^{-s}\Gamma(s)L(s)$ extends to an entire function, which satisfies the functional equation*

$$\Lambda(s) = (-1)^{k/2}\Lambda(k-s),$$

then there exists a cusp form $f \in S_k$ with $L(s) = L(f,s)$.

Proof We use the inversion formula of the Fourier transform: For $f \in L^1(\mathbb{R})$ let

$$\hat{f}(y) = \int_{\mathbb{R}} f(x)e^{-2\pi ixy}\,dx.$$

Suppose that f is two times continuously differentiable and that the functions f, f', f'' are all in $L^1(\mathbb{R})$. Then $\hat{f}(y) = O((1 + |y|)^{-2})$, so that $\hat{f} \in L^1(\mathbb{R})$. Under these conditions, we have the *Fourier inversion formula*:

$$\hat{\hat{f}}(x) = f(-x).$$

A proof of this fact can be found in any of the books [Dei05, Rud87, SW71]. We use this formula here for the proof of the Mellin inversion formula.

Theorem 2.4.8 (Mellin inversion formula) *Suppose that the function g is two times continuously differentiable on the interval $(0, \infty)$ and for some $c \in \mathbb{R}$ the functions*

$$x^c g(x), \quad x^{c+1} g'(x), \quad x^{c+2} g''(x)$$

are all in $\in L^1(\mathbb{R}_+, \frac{dx}{x})$. Then the Mellin transform

$$\mathcal{M}g(s) \overset{\text{def}}{=} \int_0^\infty x^s g(x) \frac{dx}{x}$$

exists for $\operatorname{Re}(s) = c$, and satisfies the growth estimate $\mathcal{M}g(c + it) = O((1 + |t|)^{-2})$. Finally, for every $x \in (0, \infty)$ one has the inversion formula:

$$g(x) = \frac{1}{2\pi i} \int_{c-i\infty}^{c+i\infty} x^{-s}\, \mathcal{M}g(s)\, ds.$$

Proof A given $s \in \mathbb{C}$ with $\operatorname{Re}(s) = c$ can be written as $s = c - 2\pi i y$ for a unique $y \in \mathbb{R}$. The substitution $x = e^t$ gives

$$\mathcal{M}g(s) = \int_\mathbb{R} e^{st} g(e^t)\, dt = \int_\mathbb{R} e^{ct} g(e^t) e^{-2\pi i y t}\, dt = \hat{F}(y),$$

with $F(t) = e^{ct} g(e^t)$. The conditions imply that F is two times continuously differentiable and that F, F', F'' are all in $L^1(\mathbb{R})$. Further, one has $\hat{F}(y) = \mathcal{M}g(c - 2\pi i y)$. By the Fourier inversion formula we deduce

$$e^{ct} g(e^t) = F(t) = \hat{\hat{F}}(-t) = \int_\mathbb{R} \hat{F}(y) e^{2\pi i y t}\, dy$$

$$= \int_\mathbb{R} \mathcal{M}g(c - 2\pi i y) e^{2\pi i y t}\, dy = \frac{e^{ct}}{2\pi i} \int_{c-i\infty}^{c+i\infty} \mathcal{M}g(s) e^{-st}\, ds.$$

The theorem is proven. □

We now show Hecke's converse theorem. Let a_n be a sequence in $\mathbb{C}$, such that the Dirichlet series $L(s) = \sum_{n=1}^\infty a_n n^{-s}$ converges in the region $\{\operatorname{Re}(s) > C\}$ for a given $C \in \mathbb{R}$. We define

$$f(z) = \sum_{n=1}^\infty a_n e^{2\pi i n z}.$$

According to Lemma 2.4.6 there is a natural number $N \in \mathbb{N}$ such that the Dirichlet series $L(s)$ converges absolutely for $\mathrm{Re}(s) \geq N$. Therefore one has $a_n = O(n^N)$, so the series $f(z)$ converges locally uniformly on the upper half plane $\mathbb{H}$ and defines a holomorphic function there. We intend to show that it is a cusp form of weight k. Since the group Γ is generated by the elements S and T, it suffices to show that $f(-1/z) = z^k f(z)$. As f is holomorphic, it suffices to show that $f(i/y) = (iy)^k f(iy)$ for $y > 0$.

We first show that the Mellin transform of the function $g(y) = f(iy)$ exists and that the Mellin inversion formula holds for g. We have

$$|f(iy)| = \left|\sum_{n=1}^{\infty} a_n e^{-2\pi n y}\right| \leq \text{const.} \sum_{n=1}^{\infty} n^N e^{-2\pi n y}.$$

Denote $g_N(y) = \sum_{n=1}^{\infty} n^N e^{-2\pi n y}$. Let

$$g_0(y) = \sum_{n=0}^{\infty} e^{-2\pi n y} = \frac{1}{1 - e^{-2\pi y}} = \frac{1}{2\pi y} + h(y)$$

for some function h which is holomorphic in $y = 0$. Then

$$g_N(y) = \frac{1}{(-2\pi)^N} g_0^{(N)}(y) = \frac{c_1}{y^{N+1}} + h^{(N)}(y),$$

so $|g_N(y)| \leq \frac{C}{y^{N+1}}$ for $y \to 0$. The same estimate holds for $f(iy)$. For $y > 1$ the function $|f(iy)|$ is less then a constant times

$$g_N(y) = \sum_{n=1}^{\infty} n^N e^{-2\pi n y} \leq e^{-2\pi(y-1)} \sum_{n=1}^{\infty} n^N e^{-2\pi n} = e^{-2\pi y} e^{2\pi} g_N(1).$$

So the function $f(iy)$ is rapidly decreasing for $y \to \infty$. The same estimates hold for every derivative of f, increasing N if necessary. So the Mellin integral $\mathcal{M}g(s)$ converges for $\mathrm{Re}(s) > N + 1$ and since $f(iy)$ is rapidly decreasing for $y \to \infty$, the conditions for the Mellin inversion formula are satisfied. Hence by Theorem 2.4.8 we have for every $c > N + 1$,

$$f(iy) = \frac{1}{2\pi i} \int_{c-i\infty}^{c+i\infty} \Lambda(s) y^{-s} \, ds.$$

We next use a classical result of complex analysis, which itself follows from the maximum principle.

Lemma 2.4.9 (Phragmén–Lindelöf principle) *Let $\phi(s)$ be holomorphic in the strip $a \leq \mathrm{Re}(s) \leq b$ for some real numbers $a < b$. Assume there is $\alpha > 0$, such that for every $a \leq \sigma \leq b$ we have $\phi(\sigma + it) = O(e^{|t|^\alpha})$. Suppose there is $M \in \mathbb{R}$ with $\phi(\sigma + it) = O((1 + |t|)^M)$ for $\sigma = a$ and $\sigma = b$. Then we have $\phi(\sigma + it) = O((1 + |t|)^M)$ uniformly for all $\sigma \in [a, b]$.*

Proof See for instance [Con78], Chap. VI, or [Haz01, SS03].   □

We apply this principle to the case $\phi = \Lambda$ and $a = k - c$ as well as $b = c$. We move the path of integration to $\operatorname{Re}(s) = c' = k - c$, where the integral also converges, according to the functional equation. This move of the integration path is possible by the Phragmén–Lindelöf principle. We infer that

$$f(iy) = \frac{1}{2\pi i} \int_{k-c-i\infty}^{k-c+i\infty} \Lambda(s) y^{-s} \, ds = \frac{(-1)^{k/2}}{2\pi i} \int_{k-c-i\infty}^{k-c+i\infty} \Lambda(k-s) y^{-s} \, ds$$

$$= \frac{(-1)^{k/2}}{2\pi i} \int_{c-i\infty}^{c+i\infty} \Lambda(s) y^{s-k} \, ds = (iy)^{-k} f(i/y).$$

2.5 Hecke Operators

We introduce Hecke operators, which are given by summation over cosets of matrices of fixed determinant. In later chapters, we shall encounter a reinterpretation of these operators in the adelic setting.

For given $n \in \mathbb{N}$ let M_n denote the set of all matrices in $M_2(\mathbb{Z})$ of determinant n. The group $\Gamma_0 = \mathrm{SL}_2(\mathbb{Z})$ acts on M_n by multiplication from the left.

Lemma 2.5.1 *The set M_n decomposes into finitely many Γ_0-orbits under multiplication from the left. More precisely, the set*

$$R_n = \left\{ \begin{pmatrix} a & b \\ & d \end{pmatrix} : a, d \in \mathbb{N}, \ ad = n, \ 0 \le b < d \right\}$$

is a set of representatives of $\Gamma_0 \backslash M_n$.

Notation Here and for the rest of the book we use the convention that a zero entry of a matrix may be left out, so $\begin{pmatrix} a & b \\ & d \end{pmatrix}$ stands for the matrix $\begin{pmatrix} a & b \\ 0 & d \end{pmatrix}$.

Proof We have to show that every Γ_0-orbit meets the set R_n in exactly one element. For this let $\begin{pmatrix} a & b \\ c & d \end{pmatrix} \in M_n$. For $x \in \mathbb{Z}$ we have

$$\begin{pmatrix} 1 & \\ x & 1 \end{pmatrix} \begin{pmatrix} a & b \\ c & d \end{pmatrix} = \begin{pmatrix} a & b \\ c+ax & d+bx \end{pmatrix}.$$

This implies that, modulo Γ_0, we can assume $0 \le c < |a|$. By the identity

$$\begin{pmatrix} & -1 \\ 1 & \end{pmatrix} \begin{pmatrix} a & b \\ c & d \end{pmatrix} = \begin{pmatrix} -c & -d \\ a & b \end{pmatrix}$$

one can interchange a and c, then reduce again by the first step and iterate this process until one gets $c = 0$, which implies that every Γ_0-orbit contains an element of the form $\begin{pmatrix} a & b \\ & d \end{pmatrix}$. Then $ad = \det = n$, and since $-1 \in \Gamma_0$ one can assume $a, d \in \mathbb{N}$. By

$$\begin{pmatrix} 1 & x \\ & 1 \end{pmatrix} \begin{pmatrix} a & b \\ & d \end{pmatrix} = \begin{pmatrix} a & b+dx \\ & d \end{pmatrix}$$

one can finally reduce to $0 \le b < d$, so every Γ_0-orbit meets the set R_n.

In order to show that R_n is a proper set of representatives, it remains to show that two elements in R_n, which lie in the same Γ_0-orbit, are equal. For this let $\begin{pmatrix} a & b \\ & d \end{pmatrix}, \begin{pmatrix} a' & b' \\ & d' \end{pmatrix} \in R_n$ be in the same Γ_0-orbit. This means that there is $\begin{pmatrix} x & y \\ z & w \end{pmatrix} \in \Gamma_0$ with

$$\begin{pmatrix} a' & b' \\ & d' \end{pmatrix} = \begin{pmatrix} x & y \\ z & w \end{pmatrix} \begin{pmatrix} a & b \\ & d \end{pmatrix}.$$

The right-hand side is of the form $\begin{pmatrix} * & * \\ az & * \end{pmatrix}$. Since $a \neq 0$, we infer that $z = 0$. Then $xw = 1$, so $x = w = \pm 1$. Because of

$$\begin{pmatrix} x & y \\ & w \end{pmatrix} \begin{pmatrix} a & b \\ & d \end{pmatrix} = \begin{pmatrix} ax & * \\ * & * \end{pmatrix}$$

one has $a' = ax > 0$, so $x > 0$ and therefore $x = 1 = w$, so $a' = a$ and $d' = d$. It follows that

$$\begin{pmatrix} a & b' \\ & d \end{pmatrix} = \begin{pmatrix} 1 & y \\ & 1 \end{pmatrix} \begin{pmatrix} a & b \\ & d \end{pmatrix} = \begin{pmatrix} a & b+dy \\ & d \end{pmatrix},$$

so that the condition $0 \leq b, b' < d$ finally forces $b = b'$.   □

Let $GL_2(\mathbb{R})^+$ be the set of all $g \in GL_2(\mathbb{R})$ of positive determinants. The group $GL_2(\mathbb{R})^+$ acts on the upper half plane $\mathbb{H}$ by

$$\begin{pmatrix} a & b \\ c & d \end{pmatrix} z = \frac{az+b}{cz+d}.$$

The center $\mathbb{R}^\times \begin{pmatrix} 1 & \\ & 1 \end{pmatrix}$ acts trivially.

For $k \in 2\mathbb{Z}$, a function f on $\mathbb{H}$ and $\gamma = \begin{pmatrix} a & b \\ c & d \end{pmatrix} \in GL_2(\mathbb{R})^+$ we write

$$f|_k \gamma(z) = \det(\gamma)^{k/2}(cz+d)^{-k} f\left(\frac{az+b}{cz+d}\right).$$

If k is fixed, we also use the simpler notation $f|\gamma(z)$. Note that the power $k/2$ of the determinant factor has been chosen so that the center of $GL_n(\mathbb{R})^+$ acts trivially.

We write $\Gamma_0 = SL_2(\mathbb{Z})$. For $n \in \mathbb{N}$ define the Hecke operator T_n as follows.

Definition 2.5.2 Denote by V the vector space of all functions $f : \mathbb{H} \to \mathbb{C}$ with $f|\gamma = f$ for every $\gamma \in \Gamma_0$. Define $T_n : V \to V$ by

$$T_n f = n^{\frac{k}{2}-1} \sum_{y : \Gamma_0 \backslash M_n} f|y,$$

where the colon means that the sum runs over an arbitrary set of representatives of $\Gamma_0 \backslash M_n$ in M_n. The factor $n^{\frac{k}{2}-1}$ is for normalization only. The sum is well defined and finite, as $f|\gamma = f$ for every $\gamma \in \Gamma_0$ and $\Gamma_0 \backslash M_n$ is finite. In order to show that $T_n f$ indeed lies in the space V, we compute for $\gamma \in \Gamma_0$,

$$T_n f|\gamma = n^{\frac{k}{2}-1} \sum_{y : \Gamma_0 \backslash M_n} (f|y)|\gamma = n^{\frac{k}{2}-1} \sum_{y : \Gamma_0 \backslash M_n} f|y\gamma = n^{\frac{k}{2}-1} \sum_{y : \Gamma_0 \backslash M_n} f|y = T_n f.$$

Using Lemma 2.5.1 we can write

$$T_n f(z) = n^{k-1} \sum_{\substack{ad=n \\ 0 \leq b < d}} d^{-k} f\left(\frac{az+b}{d}\right).$$

Lemma 2.5.3 *The Hecke operator T_n preserves the spaces $\mathcal{M}_k(\Gamma_0)$ and $S_k(\Gamma_0)$.*

Proof We have just shown that for a given $f \in \mathcal{M}_k(\Gamma_0)$ the function $T_n f$ is invariant under the action of Γ_0. Being a finite sum of holomorphic functions, the function $T_n f$ is holomorphic on $\mathbb{H}$. To show that $T_n f$ is a modular form, we write

$$T_n f(z) = n^{k-1} \sum_{\substack{ad=n \\ 0 \leq b < d}} d^{-k} f\left(\frac{az+b}{d}\right).$$

This formula shows that $T_n f(z)$ converges as $\text{Im}(z) \to \infty$, since $f(z)$ does. This means that $T_n f \in \mathcal{M}_k(\Gamma_0)$. If f is a cusp form, the limit is zero and the same holds for $T_n f$. □

Proposition 2.5.4 *The Hecke operators satisfy the equations*

- $T_1 = \text{Id}$,
- $T_{mn} = T_m T_n$, *if* $\gcd(m, n) = 1$,
- *for every prime number p and every $n \in \mathbb{N}$ one has* $T_p T_{p^n} = T_{p^{n+1}} + p^{k-1} T_{p^{n-1}}$.

Together these equations imply that $T_n T_m = T_m T_n$ always, i.e. all Hecke operators commute with each other.

Proof The first assertion is trivial. For the second note

$$|R_n| = \sum_{d|n} d = \sigma_1(n).$$

If $m, n \in \mathbb{N}$ are coprime, then it follows that $|R_{mn}| = |R_m||R_n|$. To ease the presentation we will, in the following calculations, in an integer matrix $\begin{pmatrix} a & b \\ & d \end{pmatrix}$, consider the number b only modulo d. Under this proviso, we show that the map

$$R_n \times R_m \to R_{mn}, \quad (A, B) \mapsto AB$$

is a bijection, where we still assume that m and n are coprime. As both sets have the same cardinality, it suffices to show injectivity. So let

$$\begin{pmatrix} aa' & ab'+bd' \\ & dd' \end{pmatrix} = \begin{pmatrix} a & b \\ & d \end{pmatrix}\begin{pmatrix} a' & b' \\ & d' \end{pmatrix} = \begin{pmatrix} \alpha & \beta \\ & \delta \end{pmatrix}\begin{pmatrix} \alpha' & \beta' \\ & \delta' \end{pmatrix}$$
$$= \begin{pmatrix} \alpha\alpha' & \alpha\beta'+\beta\delta' \\ & \delta\delta' \end{pmatrix}.$$

Then $aa' = \alpha\alpha'$ and since $(m, n) = 1$, it follows that $a = \alpha$ and $a' = \alpha'$. Analogously for d and δ. So we have

$$ab' + bd' \equiv \alpha\beta' + \beta d' \bmod(dd').$$

Reduction modulo d' gives

$$ab' \equiv a\beta' \bmod (d').$$

Being a divisor of n, the number a is coprime to d', so $b' \equiv \beta' \bmod(d')$. In the same way we get $b \equiv \beta \bmod d$. Hence $R_m R_n = R_{mn}$ and so

$$T_m T_n f = m^{\frac{k}{2}-1} \sum_{y \in R_m} T_n f \mid y = (mn)^{\frac{k}{2}-1} \sum_{y \in R_m} \sum_{z \in R_n} f \mid (yz)$$

$$= (mn)^{\frac{k}{2}-1} \sum_{w \in R_{mn}} f \mid w = T_{mn} f.$$

For the last point note

$$R_p = \left\{ \begin{pmatrix} p & \\ & 1 \end{pmatrix} \right\} \cup \left\{ \begin{pmatrix} 1 & b \\ & p \end{pmatrix} : b \bmod p \right\},$$

as well as

$$R_{p^n} = \left\{ \begin{pmatrix} p^a & x \\ & p^b \end{pmatrix} : \begin{matrix} a,b \geq 0,\; a+b=n \\ x \bmod(p^b) \end{matrix} \right\}.$$

It follows that

$$R_p R_{p^n} = \left\{ \begin{pmatrix} p^{a+1} & px \\ & p^b \end{pmatrix} : \begin{matrix} a,b\geq 0,\; a+b=n \\ x \bmod(p^b) \end{matrix} \right\} \cup \left\{ \begin{pmatrix} p^a & x+yp^b \\ & p^{b+1} \end{pmatrix} : \begin{matrix} a,b\geq 0, a+b=n \\ x \bmod(p^b) \\ y \bmod p \end{matrix} \right\}.$$

The second set, together with $\left\{ \begin{pmatrix} p^{n+1} & \\ & 1 \end{pmatrix} \right\}$, is a set of representatives $R_{p^{n+1}}$. The sum over this gives the term $T_{p^{n+1}}$. The first set minus $\left\{ \begin{pmatrix} p^{n+1} & \\ & 1 \end{pmatrix} \right\}$ is

$$\left\{ \begin{pmatrix} p^{a+1} & px \\ & p^b \end{pmatrix} : \begin{matrix} a,b\geq 0,\; a+b=n \\ x \bmod(p^b) \\ b \geq 1 \end{matrix} \right\}$$

$$= \left\{ p \begin{pmatrix} p^a & x \\ & p^{b-1} \end{pmatrix} : \begin{matrix} a,b\geq 0,\; a+b=n \\ x \bmod(p^b) \\ b \geq 1 \end{matrix} \right\}.$$

Denote this last set by S. Since the central p acts trivially, one gets

$$(p^{n+1})^{\frac{k}{2}-1} \sum_{y \in S} f \mid y = (p^{n+1})^{\frac{k}{2}-1} p \sum_{y \in R_{p^{n-1}}} f \mid y = p^{k-1} T_{p^{n-1}} f. \qquad \square$$

We now want to see how the application of a Hecke operator changes the Fourier expansion of a modular form.

Proposition 2.5.5 *For a given form* $f(z) = \sum_{m\geq 0} c(m) q^m \in \mathcal{M}_k$ *and* $n \in \mathbb{N}$ *the Fourier expansion of* $T_n f$ *is*

$$T_n f(z) = \sum_{m \geq 0} \gamma(m) q^m$$

with

$$\gamma(m) = \sum_{\substack{a|(m,n) \\ a \geq 1}} a^{k-1} c\left(\frac{mn}{a^2}\right).$$

Proof By definition we have

$$T_n f(z) = n^{k-1} \sum_{\substack{ad=n, \ a \geq 1 \\ 0 \leq b < d}} d^{-k} \sum_{m \geq 0} c(m) e^{2\pi i m(az+b)/d}.$$

The sum $\sum_{0 \leq b < d} e^{2\pi i bm/d}$ equals d if $d|m$ and 0 otherwise. Setting $m' = m/d$ one gets

$$T_n f(z) = n^{k-1} \sum_{\substack{ad=n \\ a \geq 1, \ m' \geq 0}} d^{-k+1} c(m'd) q^{am'}.$$

Sorting this by powers of q results in

$$T_n f(z) = \sum_{\mu \geq 0} q^{\mu} \sum_{\substack{a|(n,\mu) \\ a \geq 1}} a^{k-1} c\left(\frac{\mu n}{a^2}\right).$$

The proposition is proven.   □

The following two corollaries are simple consequences of the proposition.

Corollary 2.5.6 *One has $\gamma(0) = \sigma_{k-1}(n)c(0)$ and $\gamma(1) = c(n)$.*

Corollary 2.5.7 *If p is a prime number, then*

$$\gamma(m) = c(pm) \qquad\qquad\qquad \text{if } m \not\equiv 0 \bmod(p),$$

$$\gamma(m) = c(pm) + p^{k-1} c(m/p), \quad \text{if } m \equiv 0 \bmod(p).$$

In Proposition 2.5.4 we have shown that Hecke operators commute with each other. We next show that they can be diagonalized simultaneously.

Lemma 2.5.8 *A set of commuting self-adjoint operators on a finite-dimensional unitary space can be simultaneously diagonalized.*

We elaborate the formulation of this lemma as follows: let V be a finite-dimensional complex vector space equipped with an inner product $\langle .,. \rangle$ and let $E \subset \text{End}(V)$ be a set of self-adjoint operators on V. Suppose that any two elements $S, T \in E$ commute, i.e. $ST = TS$. Then there exists a basis of V such that all elements of E are represented by diagonal matrices with respect to that basis. More precisely, this basis, say $v_1, \ldots, v_n$, consists of simultaneous eigenvectors, so for each $1 \leq j \leq n$ there exists a map $\chi_j : E \to \mathbb{C}$ such that

$$T v_j = \chi_j(T) v_j$$

holds for every $T \in E$.

Proof We prove the lemma by induction on the dimension of V. If $\dim(V) = 1$, then there is nothing to show. So suppose $\dim(V) > 1$ and that the claim is proven for all spaces of smaller dimension. If all $T \in E$ are multiples of the identity, i.e. $T = \lambda\,\mathrm{Id}$ for some $\lambda = \lambda(T) \in \mathbb{C}$, then the claim follows. So assume there exists a $T \in E$ not a multiple of the identity. Since T is self-adjoint, it is diagonalizable, so V is the direct sum of the eigenspaces of T, and each eigenspace is of dimension strictly smaller than $\dim(V)$. Let $S \in E$ and let V_λ be the T-eigenspace for the eigenvalue λ. We claim that $S(V_\lambda) \subset V_\lambda$. For a given $v \in V_\lambda$ we have

$$T\big(S(v)\big) = S\big(T(v)\big) = S(\lambda v) = \lambda S(v),$$

i.e. $S(v) \in V_\lambda$ and the space V_λ is stable under all $S \in E$ and by the induction hypothesis, V_λ has a basis of simultaneous eigenvectors. As this holds for all eigenvalues of T, the entire space V has such a basis.   □

Definition 2.5.9 Let E be as in the lemma. Then V has a basis $v_1, \ldots, v_n$ such that for every $S \in E$,

$$Sv_j = \chi_j(S)v_j$$

for a scalar $\chi_j(S) \in \mathbb{C}$. We say, the v_j are *simultaneous eigenvectors* of E.

Recall the notion of a complex *algebra*. This is a $\mathbb{C}$-vector space A with a bilinear map $A \times A \to A$ written $(a, b) \mapsto ab$, which is *associative*, i.e. one has

$$(ab)c = a(bc)$$

for all $a, b, c \in A$.

Examples 2.5.10

- The set $M_n(\mathbb{C})$ of complex $n \times n$ matrices is a complex algebra which is isomorphic to the algebra $\mathrm{End}(V)$ of linear endomorphisms of a complex vector space of dimension n. Giving an isomorphism $\mathrm{End}(V) \cong M_n(\mathbb{C})$ is equivalent to choosing a basis of V.
- The set $\mathcal{B}(V)$ of bounded linear operators on a Banach space V is a complex algebra.
- Let $\emptyset \neq E \subset \mathrm{End}(V)$ for a vector space V. The *algebra generated by E* is the set of all linear combinations of operators of the form $S_1 \cdots S_n$, where $S_1, \ldots, S_n \in E$. It is the smallest algebra which contains E.

Denote by $\mathcal{A}$ the *algebra* generated by E. Then the v_j are simultaneous eigenvectors for the whole of $\mathcal{A}$, and the maps χ_j can be extended to maps $\chi_j : \mathcal{A} \to \mathbb{C}$, such that for every operator $T \in \mathcal{A}$ the eigen-equation $Tv_j = \chi_j(T)v$ holds. Note that for $S, T \in \mathcal{A}$ one has

$$\chi_j(S+T)v_j = (S+T)v_j = Sv_j + Tv_j = \chi(S)v_j + \chi(T)v_j,$$

so $\chi_j(S+T) = \chi_j(S) + \chi_j(T)$. Further $\chi_j(\lambda T) = \lambda \chi_j(T)$ for every $\lambda \in \mathbb{C}$; this means that each χ_j is a linear map. More than that, one has

$$\chi_j(ST)v_j = ST v_j = S(T(v_j)) = S(\chi_j(T)v_j) = \chi_j(T)S(v_j) = \chi_j(T)\chi_j(S)v_j,$$

so it even follows $\chi_j(ST) = \chi_j(S)\chi_j(T)$, i.e. the map χ_j is multiplicative. Together this means: every χ_j is an *algebra homomorphism* of the algebra $\mathcal{A}$ to $\mathbb{C}$.

In the sequel, we shall need to following theorem, known as the *Elementary Divisor Theorem*.

Theorem 2.5.11 (Elementary Divisor Theorem) *For a given integer matrix $A \in M_n(\mathbb{Z})$ with $\det(A) \neq 0$ there exist invertible matrices $S, T \in GL_n(\mathbb{Z})$ and natural numbers $d_1, d_2, \dots, d_n$ with $d_j | d_{j+1}$ such that*

$$A = S \begin{pmatrix} d_1 & & \\ & \ddots & \\ & & d_n \end{pmatrix} T.$$

The numbers $d_1, \dots, d_n$ are uniquely determined by A and are called the elementary divisors of A.

Proof For example in [HH80]. □

Definition 2.5.12 Denote by $GL_2(\mathbb{Q})^+$ the set of all matrices $g \in GL_2(\mathbb{Q})$ with $\det(g) > 0$. This is a subgroup of the group $GL_2(\mathbb{Q})$ of index 2.

Proposition 2.5.13 *We continue to write $\Gamma_0 = SL_2(\mathbb{Z})$. A complete set of representatives of the double quotient*

$$\Gamma_0 \backslash GL_2(\mathbb{Q})^+ / \Gamma_0$$

is given by the set of all diagonal matrices $\begin{pmatrix} a & \\ & an \end{pmatrix}$, where $a \in \mathbb{Q}$ and $n \in \mathbb{N}$.

Proof For a given $\alpha \in GL_2(\mathbb{Q})^+$ there exists $N \in \mathbb{N}$, such that $N\alpha$ is an integer matrix. By the Elementary Divisor Theorem there are $S, T \in GL_2(\mathbb{Z})$ such that $N\alpha = SDT$, where $D = \begin{pmatrix} d_1 & \\ & nd_1 \end{pmatrix}$ with $d_1, n \in \mathbb{N}$. If necessary, one can multiply S and T with the matrix $\begin{pmatrix} -1 & \\ & 1 \end{pmatrix}$, so that $S, T \in SL_2(\mathbb{Z})$ can be assumed. Therefore we find $\Gamma_0 \alpha \Gamma_0 = \Gamma_0 \begin{pmatrix} d_1/N & \\ & nd_1/N \end{pmatrix} \Gamma_0$. The uniqueness of the representative follows from the Elementary Divisor Theorem, if one chooses N as the unique smallest $N \in \mathbb{N}$ making $N\alpha$ an integer matrix. □

Corollary 2.5.14 *For given $g \in GL_2(\mathbb{Q})^+$ and $\Gamma_0 = SL_2(\mathbb{Z})$ one has*

$$\Gamma_0 g^{-1} \Gamma_0 = \frac{1}{\det(g)} \Gamma_0 g \Gamma_0.$$

Proof By the proposition we can assume that g is a diagonal matrix $\begin{pmatrix} a & \\ & an \end{pmatrix}$. Then $g^{-1} = \begin{pmatrix} 1/a & \\ & 1/an \end{pmatrix} = \frac{1}{\det(g)} \begin{pmatrix} an & \\ & a \end{pmatrix}$ and this last matrix lies in the same double Γ_0-coset as g, since

$$\begin{pmatrix} & -1 \\ 1 & \end{pmatrix} \begin{pmatrix} an & \\ & a \end{pmatrix} \begin{pmatrix} & 1 \\ -1 & \end{pmatrix} = \begin{pmatrix} a & \\ & an \end{pmatrix},$$

so the corollary is proven.   □

We have seen that the group $G = SL_2(\mathbb{R})$ acts on the upper half plane $\mathbb{H}$ via

$$\begin{pmatrix} a & b \\ c & d \end{pmatrix} z = \frac{az + b}{cz + d}.$$

Lemma 2.5.15 *The measure* $d\mu = \frac{dx\,dy}{y^2}$ *on* $\mathbb{H}$ *is invariant under the action of* G, *i.e. we have*

$$\int_{\mathbb{H}} f(z)\,d\mu(z) = \int_{\mathbb{H}} f(gz)\,d\mu(z)$$

for every integrable function f *and every* $g \in G$.

Proof Every $g \in G$ defines a holomorphic map $z \mapsto gz$ on $\mathbb{H}$. We compute its differential as

$$g'z = \frac{d(gz)}{dz} = \frac{a(cz + d) - c(az + b)}{(cz + d)^2} = \frac{1}{(cz + d)^2}.$$

This is equivalent to the identity of differential forms

$$d(gz) = \frac{1}{(cz + d)^2} dz,$$

where $dz = dx + idy$ and $d(gz)$ is the pullback of dz under g. Applying complex conjugation yields $\overline{dz} = dx - idy$, so $dz \wedge \overline{dz} = -2i(dx \wedge dy)$. Further, by the above,

$$d(gz) \wedge \overline{d(gz)} = \frac{1}{|cz + d|^4} dz \wedge \overline{dz} = \frac{\text{Im}(gz)^2}{\text{Im}(z)^2} dz \wedge \overline{dz},$$

or

$$\frac{d(gz) \wedge \overline{d(gz)}}{\text{Im}(gz)^2} = \frac{dz \wedge \overline{dz}}{\text{Im}(z)^2},$$

which is to say that the differential form $\frac{dz \wedge \overline{dz}}{\text{Im}(z)^2}$ is invariant under G. This implies the claim.   □

This lemma can also be proved without the use of differential forms; see Exercise 2.8.

Theorem 2.5.16 *The spaces $\mathcal{M}_k$ and S_k have bases consisting of simultaneous eigenvectors of all Hecke operators.*

Proof We want to apply the lemma with $E = \{T_n : n \in \mathbb{N}\}$. For this we have to define an inner product on $\mathcal{M}_k$. For given $f, g \in \mathcal{M}_k$ the function $f(z)\overline{g(z)}y^k$ is invariant under the group Γ_0. It is a continuous, hence measurable, function on the quotient $\Gamma_0\backslash\mathbb{H}$. The measure $\frac{dx\,dy}{y^2}$ is Γ_0-invariant as well, and hence defines a measure μ on $\Gamma_0\backslash\mathbb{H}$. This is an important point, so we will explain it a bit further. One way to view this measure on the quotient $\Gamma_0\backslash\mathbb{H}$ is to identify $\Gamma_0\backslash\mathbb{H}$ with a measurable set of representatives R with $D \subset R \subset \overline{D}$, where D is the standard fundamental domain of Definition 2.1.6. Then any measurable subset $A \subset \Gamma_0\backslash\mathbb{H}$ can be viewed as a subset of $R \subset \mathbb{H}$ and the measure $\frac{dx\,dy}{y^2}$ can be applied. Interestingly, the measure μ on $\Gamma_0\backslash\mathbb{H}$ is a *finite measure*, i.e.

$$\mu(\Gamma_0\backslash\mathbb{H}) < \infty,$$

as the $\frac{dx\,dy}{y^2}$-measure of $\overline{D}$ is finite by Exercise 2.9. According to Exercise 2.15 the integral

$$\langle f, g \rangle_{\text{Pet}} = \int_{\Gamma_0\backslash\mathbb{H}} f(z)\overline{g(z)}y^k \frac{dx\,dy}{y^2}$$

exists if one of the two functions f, g is a cusp form. This integral defines an inner product on the space S_k, which is called the *Petersson inner product*. We show that $\langle T_n f, g \rangle_{\text{Pet}} = \langle f, T_n g \rangle_{\text{Pet}}$, so the T_n are self-adjoint on the space S_k. This implies the claim on S_k. The space $S_k^{\perp} = \{f \in \mathcal{M}_k : \langle f, g \rangle_{\text{Pet}} = 0 \ \forall g \in S_k\}$ is one-dimensional if $\mathcal{M}_k \neq 0$. By the self-adjointness of the Hecke operators, this space is T_n-invariant as well, so, being one-dimensional, it is a simultaneous eigenspace. It only remains to show the claimed self-adjointness.

We do this by extending the Petersson inner product to functions which are not necessarily invariant under Γ_0, but only under a subgroup of finite index in Γ_0. We first consider the case $k = 0$. Take two continuous and bounded functions f, g on $\mathbb{H}$, which are invariant under Γ_0, so they satisfy $f(\gamma z) = f(z)$ for every $z \in \mathbb{H}$ and every $\gamma \in \Gamma_0$, and the same for the function g. Then we define

$$\langle f, g \rangle = \int_{\Gamma_0\backslash\mathbb{H}} f(z)\overline{g(z)}\,d\mu(z),$$

where μ is the measure $\frac{dx\,dy}{y^2}$. The integral exists, since f and g are bounded and $\Gamma_0\backslash\mathbb{H}$ has finite measure, as we have seen above. We now make a crucial observation: If $\Gamma \subset \Gamma_0$ is a subgroup of finite index, then

$$\langle f, g \rangle = \frac{1}{[\overline{\Gamma}_0 : \overline{\Gamma}]} \int_{\overline{\Gamma}\backslash\mathbb{H}} f(z)\overline{g(z)}\,d\mu(z),$$

where, as in Definition 2.1.6, the group $\overline{\Gamma}_0$ is $\Gamma_0 / \pm 1$ and $\overline{\Gamma}$ is the image of Γ in $\overline{\Gamma}_0$. If the functions f and g are continuous and bounded, but only invariant under

Γ and no longer invariant under Γ_0, then the last expression still does make sense. This means that we can define $\langle f, g \rangle$ in this more general situation by the expression

$$\langle f, g \rangle \overset{\text{def}}{=} \frac{1}{[\overline{\Gamma}_0 : \overline{\Gamma}]} \int_{\Gamma \backslash \mathbb{H}} f(z)\overline{g(z)}\, d\mu(z).$$

In this way we extend the definition of the Petersson inner product in the case $k = 0$. In the case $k > 0$ we consider two continuous functions f, g with $f|_k \sigma = f$ for every $\sigma \in \Gamma$, and the same for g. We assume that the Γ-invariant function $|f(z)y^{k/2}|$ is bounded on the upper half plane $\mathbb{H}$ and the same for g. We then define

$$\langle f, g \rangle_k \overset{\text{def}}{=} \frac{1}{[\overline{\Gamma}_0 : \overline{\Gamma}]} \int_{\Gamma \backslash \mathbb{H}} f(z)\overline{g(z)}y^k\, d\mu(z).$$

We claim that for a given $\alpha \in GL_2(\mathbb{Q})^+$ the group $\Gamma = \alpha^{-1}\Gamma_0\alpha \cap \Gamma_0$ is a subgroup of Γ_0 of finite index.

Proof of This Claim By Proposition 2.5.13 we can assume $\alpha = \left(\begin{smallmatrix} r & \\ & rn \end{smallmatrix} \right)$ with $r \in \mathbb{Q}$ and $n \in \mathbb{N}$. Then

$$\alpha^{-1} \begin{pmatrix} a & b \\ c & d \end{pmatrix} \alpha = \begin{pmatrix} a & nb \\ \frac{c}{n} & d \end{pmatrix}.$$

So a given $\left(\begin{smallmatrix} a & b \\ c & d \end{smallmatrix} \right) \in \Gamma_0$ lies in Γ if and only if $c/n \in \mathbb{Z}$, i.e. if n divides c. Therefore the group Γ contains the group $\Gamma(n)$ of all matrices $\gamma \in SL_2(\mathbb{Z})$ with $\gamma \equiv \left(\begin{smallmatrix} 1 & \\ & 1 \end{smallmatrix} \right) \bmod n$. This group is by definition the kernel of the group homomorphism $SL_2(\mathbb{Z}) \to SL_2(\mathbb{Z}/n\mathbb{Z})$, which comes from the reduction homomorphism $\mathbb{Z} \to \mathbb{Z}/n\mathbb{Z}$. As the group $SL_2(\mathbb{Z}/n\mathbb{Z})$ is finite, the group Γ has finite index in Γ_0.

Definition 2.5.17 Let $\Gamma \subset SL_2(\mathbb{Z})$ be a subgroup. A *fundamental domain* for Γ is an open subset $F \subset \mathbb{H}$, such that there is a set $R \subset \mathbb{H}$ of representatives for $\Gamma \backslash \mathbb{H}$ with

$$F \subset R \subset \overline{F} \quad \text{and} \quad \mu(\overline{F} \smallsetminus F) = 0,$$

where μ is the measure $\frac{dx\,dy}{y^2}$.

In particular, if F is a fundamental domain for Γ, then $\bigcup_{\sigma \in \Gamma} \sigma\overline{F} = \mathbb{H}$, so every point in $\mathbb{H}$ lies in a Γ-translate of $\overline{F}$.

Lemma 2.5.18 *Let $F \subset \mathbb{H}$ be a fundamental domain for the group $\Gamma \subset SL_2(\mathbb{Z})$. For every measurable, Γ-invariant function f on $\mathbb{H}$ one has*

$$\int_F f(z)\,d\mu(z) = \int_{\Gamma \backslash \mathbb{H}} f(z)\,d\mu(z),$$

where $\mu = \frac{dx\,dy}{y^2}$ is the invariant measure. So in particular, the first integral exists if and only if the second does.

Proof The projection $p : \mathbb{H} \to \Gamma \backslash \mathbb{H}$ maps F injectively onto a subset, whose complement is of measure zero. Therefore $\int_{\Gamma \backslash \mathbb{H}} f(z) \, d\mu(z) = \int_{p(F)} f(z) \, d\mu(z)$. Since the measure on the quotient is defined by the measure on $\mathbb{H}$, the bijection $p : F \to p(F)$ preserves measures. This implies the claim. □

Lemma 2.5.19

(a) D is a fundamental domain for $\Gamma_0 = \mathrm{SL}_2(\mathbb{Z})$.

(b) If Γ is a subgroup of $\Gamma_0 = \mathrm{SL}_2(\mathbb{Z})$ of finite index and S is a set of representatives of $\overline{\Gamma} \backslash \overline{\Gamma}_0$, then

$$SD = \bigcup_{\gamma \in S} \gamma D$$

is a fundamental domain for the group Γ. The set $S \subset \overline{\Gamma}_0$ is uniquely determined by the fundamental domain SD.

Proof Part (a) follows from Theorem 2.1.7.

(b) The set S is finite, as Γ has finite index in Γ_0. Hence it follows that $\overline{SD} = \bigcup_{\gamma \in S} \gamma \overline{D}$. Now let R_{Γ_0} be a set of representatives of $\Gamma_0 \backslash \mathbb{H}$ with $D \subset R_{\Gamma_0} \subset \overline{D}$. Then $R_{\Gamma} = \bigcup_{\gamma \in S} \gamma R_{\Gamma_0}$ is a set of representatives of $\Gamma \backslash \mathbb{H}$ with $SD \subset R_{\Gamma} \subset \overline{SD}$. Further one has

$$\mu(\overline{SD} \smallsetminus SD) = \mu\left(\bigcup_{\gamma \in S} \gamma \overline{D} \smallsetminus \bigcup_{\gamma \in S} \gamma D \right) \le \mu\left(\bigcup_{\gamma \in S} \gamma \overline{D} \smallsetminus \gamma D \right)$$

$$= \mu\left(\bigcup_{\gamma \in S} \gamma (\overline{D} \smallsetminus D) \right) \le \sum_{\gamma \in S} \mu(\overline{D} \smallsetminus D) = 0.$$

The last assertion follows from the fact that for $\gamma \ne \tau$ in $\overline{\Gamma}_0$ the translates γD and τD are disjoint. □

The points $\gamma \infty \in \widehat{\mathbb{R}}$ for $\gamma \in S$ are called the *cusps* of the fundamental domain SD. These lie in $\widehat{\mathbb{Q}} = \mathbb{Q} \cup \{\infty\}$. The wording becomes clearer, when one considers the unit disk instead of the upper half plane. So let $\mathbb{E} = \{z \in \mathbb{C} : |z| < 1\}$ be the open unit disk. The *Cayley map*:

$$\tau(z) \stackrel{\text{def}}{=} \frac{z - i}{z + i}$$

is a bijection from $\mathbb{H}$ to $\mathbb{E}$ such that τ as well as its inverse τ^{-1} are both holomorphic. Transporting the fundamental domain SD into E by means of the map τ, the cusps are the points where the fundamental domain touches the boundary of the disk, i.e. the unit circle. Each cusp is the endpoint of two circles which lie inside E and are orthogonal to the unit circle, so they are tangential at the cusp, i.e. the cusp is 'infinitesimally sharp', which explains the name 'cusp'. The next figure shows a fundamental domain F with one cusp.

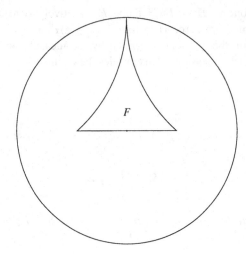

As the specific choice of a set of representatives S is not important, we frequently write D_Γ for the fundamental domain SD.

Lemma 2.5.20 *The Petersson inner product is invariant under $GL_2(\mathbb{Q})^+$, which means the following: For given $f, g \in M_k$, one of them in S_k and for each $\alpha \in GL_2(\mathbb{Q})^+$, the inner product $\langle f|\alpha, g|\alpha \rangle_k$ is defined in the above sense with $\Gamma = \alpha \Gamma_0 \alpha^{-1} \cap \Gamma_0$, and it holds that*

$$\langle f|\alpha, g|\alpha \rangle_k = \langle f, g \rangle_k.$$

Proof Let $\Gamma_0 = SL_2(\mathbb{Z})$ and $\Gamma = \alpha \Gamma_0 \alpha^{-1} \cap \Gamma_0$, as well as $\Gamma' = \alpha^{-1} \Gamma \alpha = \alpha^{-1} \Gamma_0 \alpha \cap \Gamma_0$. For given $f \in M_k$ the function $h = f|\alpha$ has the property that $h|\sigma = h$ for every $\sigma \in \Gamma'$, since $\sigma = \alpha^{-1} \gamma \alpha$ for some $\gamma \in \Gamma_0$, so

$$h|\sigma = f|\alpha\sigma = f|\gamma\alpha = f|\alpha = h.$$

The same holds for g, so the inner product $\langle f|\alpha, g|\alpha \rangle$ is well defined. Note that for $\alpha = \begin{pmatrix} * & * \\ c & d \end{pmatrix} \in GL_2(\mathbb{Q})^+$ we have

$$\text{Im}(\alpha z) = \det \alpha \frac{\text{Im}(z)}{|cz + d|^2}.$$

In the following calculation we use the $GL_2(\mathbb{Q})^+$-invariance of the measure μ together with the fact that we may replace integration over $\Gamma \backslash \mathbb{H}$ with integration over a fundamental domain according to Lemma 2.5.18. We further use that $\alpha^{-1} D_\Gamma$ is a fundamental domain for Γ' to get

$$\langle f, g \rangle_k = \frac{1}{[\overline{\Gamma}_0 : \overline{\Gamma}]} \int_{\Gamma \backslash \mathbb{H}} f(z)\overline{g(z)} \, \text{Im}(z)^k \, d\mu(z)$$

$$= \frac{1}{[\overline{\Gamma}_0 : \overline{\Gamma}]} \int_{D_\Gamma} f(z)\overline{g(z)} \, \text{Im}(z)^k \, d\mu(z)$$

$$= \frac{1}{[\overline{\Gamma}_0 : \overline{\Gamma}]} \int_{\alpha^{-1} D_\Gamma} f(\alpha z)\overline{g(\alpha z)} \, \mathrm{Im}(\alpha z)^k \, d\mu(z)$$

$$= \frac{1}{[\overline{\Gamma}_0 : \overline{\Gamma}]} \int_{\Gamma' \backslash \mathbb{H}} f|\alpha(z)\overline{g}|\alpha(z) \, \mathrm{Im}(z)^k \, d\mu(z) = \langle f|\alpha, g|\alpha \rangle_k.$$

Finally we have $[\overline{\Gamma}_0 : \overline{\Gamma}] = [\overline{\Gamma}_0 : \overline{\Gamma}']$, since $[\overline{\Gamma}_0 : \overline{\Gamma}] = \mu(D_\Gamma)/\mu(D) = \mu(\alpha^{-1} D_\Gamma)/\mu(D) = [\overline{\Gamma}_0 : \overline{\Gamma}']$. □

The lemma implies for $y \in GL_2(\mathbb{Q})^+$,

$$\langle f|y, g \rangle = \langle f|y|y^{-1}, g|y^{-1} \rangle = \langle f, g|y^{-1} \rangle,$$

hence,

$$\langle T_n f, g \rangle = n^{k-1} \sum_{y : \Gamma_0 \backslash M_n} \langle f|y, g \rangle = n^{k-1} \sum_{y : \Gamma_0 \backslash M_n} \langle f, g|y^{-1} \rangle.$$

As f and g are both invariant under Γ_0, the expression $\langle f, g|y^{-1} \rangle$ depends only on the double coset $\Gamma_0 y^{-1} \Gamma_0$. By Corollary 2.5.14, this double coset equals $\Gamma_0 \frac{1}{\det(y)} y \Gamma_0$. The center acting trivially on $\mathcal{M}_k$, this matrix acts like y. Therefore,

$$\langle T_n f, g \rangle = n^{k-1} \sum_{y : \Gamma_0 \backslash M_n} \langle f, g|y \rangle = \langle f, T_n g \rangle.$$

It follows that there are bases of $\mathcal{M}_k$ and $\mathcal{S}_k$ consisting of simultaneous eigenvectors of all Hecke operators. Theorem 2.5.16 follows. □

Theorem 2.5.21 *Let $f(z) = \sum_{n=0}^{\infty} c(n)q^n$ be a non-constant simultaneous eigenfunction of all Hecke operators, i.e. for every $n \in \mathbb{N}$ there is a number $\lambda(n) \in \mathbb{C}$ such that $T_n f = \lambda(n)f$.*

(a) *The coefficient $c(1)$ is not zero.*
(b) *If $c(1) = 1$, which can be reached by scaling f, then $c(n) = \lambda(n)$ for every $n \in \mathbb{N}$.*

Proof By Corollary 2.5.6 the coefficient of q in $T_n f$ equals $c(n)$. On the other hand, this coefficient equals $\lambda(n)c(1)$. Therefore, $c(1) = 0$ would lead to $c(n) = 0$ for all n, hence $f = 0$. Both claims follow. □

A Hecke eigenform $f \in \mathcal{M}_k$ is called *normalized* if the coefficient $c(1)$ is equal to 1.

Corollary 2.5.22 *Let $k > 0$. Two normalized Hecke eigenforms, which share the same Hecke eigenvalues, coincide.*

Proof Let $f, g \in \mathcal{M}_k$ with $T_n f = \lambda(n) f$ and $T_n g = \lambda(n) g$ for every $n \in \mathbb{N}$. By the theorem, all coefficients of the q-expansions of f and g coincide, with the possible exception of the zeroth coefficients. This means that $f - g$ is constant. As $k > 0$, there are no constant modular forms of weight k other than zero. We conclude $f = g$. $\qquad\square$

Corollary 2.5.23 *For a normalized Hecke eigenform $f(z) = \sum_{n=0}^{\infty} c(n) q^n$ we have*

- $c(mn) = c(m) c(n)$ *if* $\gcd(m, n) = 1$,
- $c(p) c(p^n) = c(p^{n+1}) + p^{k-1} c(p^{n-1})$, $n \geq 1$.

Proof The assertion follows from the corresponding relations for Hecke operators in Proposition 2.5.4. $\qquad\square$

Definition 2.5.24 We say that a Dirichlet series $L(s) = \sum_{n=1}^{\infty} a_n n^{-s}$, which converges in some half plane $\{\mathrm{Re}(s) > a\}$, has an *Euler product* of degree $k \in \mathbb{N}$, if for every prime p there is a polynomial

$$Q_p(x) = 1 + a_{p,1} x + \cdots + a_{p,k} x^k$$

such that in the domain $\mathrm{Re}(s) > a$ one has

$$L(s) = \prod_p \frac{1}{Q_p(p^{-s})}.$$

Example 2.5.25 The Riemann zeta function $\zeta(s) = \sum_{n=1}^{\infty} n^{-s}$, convergent for $\mathrm{Re}(s) > 1$, has the Euler product

$$\zeta(s) = \prod_p \frac{1}{1 - p^{-s}};$$

see Exercise 1.5.

Corollary 2.5.26 *The L-function $L(f, s) = \sum_{n=1}^{\infty} c(n) n^{-s}$ of a normalized Hecke eigenform $f(z) = \sum_{n=0}^{\infty} c(n) q^n \in \mathcal{M}_k$ has an Euler product:*

$$L(f, s) = \prod_p \frac{1}{1 - c(p) p^{-s} + p^{k-1-2s}},$$

which converges locally uniformly absolutely for $\mathrm{Re}(s) > k$.

Proof By Corollary 2.3.3 the coefficients grow at most like $c(n) = O(n^{k-1})$. So the L-series converges locally uniformly absolutely for $\mathrm{Re}(s) > k$. The partial sum

$$\sum_{n \in p^{\mathbb{N}_0}} c(n) n^{-s} = \sum_{n=0}^{\infty} c(p^n) p^{-sn}$$

also converges absolutely. Denote by $\prod_{p \leq N}$ the finite product over all primes $p \leq N$ for a given $N \in \mathbb{N}$. For coprime $m, n \in \mathbb{N}$ we have $c(mn) = c(m)c(n)$, so that

$$\prod_{p \leq N} \sum_{n=0}^{\infty} c(p^n) p^{-sn} = \sum_{\substack{n \in \mathbb{N} \\ p | n \Rightarrow p \leq N}} c(n) n^{-s},$$

where the sum on the right-hand side runs over all natural numbers whose prime divisors are all $\leq N$. As the L-series converges absolutely, the right-hand side converges to $L(f, s)$ for $N \to \infty$, and we have

$$L(f, s) = \sum_{m=1}^{\infty} c(m) m^{-s} = \prod_{p} \sum_{n=0}^{\infty} c(p^n) p^{-sn}.$$

It remains to show

$$\sum_{n=0}^{\infty} c(p^n) p^{-ns} = \frac{1}{1 - c(p)p^{-s} + p^{k-1-2s}}.$$

We expand

$$\left(\sum_{n=0}^{\infty} c(p^n) p^{-ns} \right) \left(1 - c(p)p^{-s} + p^{k-1-2s} \right)$$

$$= \sum_{n=0}^{\infty} p^{-ns} \left(c(p^n) - \underbrace{c(p)c(p^n)}_{=c(p^{n+1})+p^{k-1}c(p^{n-1}),\ n \geq 1} p^{-s} + p^{k-1} c(p^n) p^{-2s} \right)$$

$$= 1 - c(p)p^{-s} + p^{k-1-2s} + \sum_{n=1}^{\infty} c(p^n) p^{-ns}$$

$$\qquad - c(p^{n+1}) p^{-(n+1)s} - p^{k-1} \left(c(p^{n-1}) \right) p^{(n+1)s} - c(p^n) p^{-(n+2)s} \right)$$

$$= 1 - c(p)p^{-s} + p^{k-1-2s} + c(p)p^{-s} - p^{k-1} p^{-2s} = 1. \qquad \square$$

2.6 Congruence Subgroups

In the theory of automorphic forms one also considers functions which satisfy the modularity condition not for the full modular group $SL_2(\mathbb{Z})$, but only for subgroups of finite index. The most important subgroups are the *congruence subgroups*.

Definition 2.6.1 Fix a natural number N. The reduction map $\mathbb{Z} \to \mathbb{Z}/N\mathbb{Z}$ is a ring homomorphism and it induces a group homomorphism $SL_2(\mathbb{Z}) \to SL_2(\mathbb{Z}/N\mathbb{Z})$. The group $\Gamma(N) = \ker(SL_2(\mathbb{Z}) \to SL_2(\mathbb{Z}/N\mathbb{Z}))$ is called the *principal congruence subgroup* of $\Gamma_0 = SL_2(\mathbb{Z})$ of level N. So we have

$$\Gamma(N) = \left\{ \begin{pmatrix} a & b \\ c & d \end{pmatrix} : a \equiv d \equiv 1 \bmod N,\ b \equiv c \equiv 0 \bmod N \right\}.$$

A subgroup $\Gamma \subset SL_2(\mathbb{Z})$ is called a *congruence subgroup* if it contains a principal congruence subgroup, i.e. if there is a natural number $N \in \mathbb{N}$ with $\Gamma(N) \subset \Gamma$.

Note the special case

$$\Gamma(1) = \Gamma_0 = SL_2(\mathbb{Z}).$$

Note that for $N \geq 3$ the group $\Gamma(N)$ does not contain the element -1. Therefore, for such a group Γ there can exist non-zero modular forms of odd weight.

Lemma 2.6.2

(a) *The intersection of two congruence groups is a congruence group.*
(b) *Let Γ be a congruence subgroup and let $\alpha \in GL_2(\mathbb{Q})$. Then $\Gamma \cap \alpha \Gamma \alpha^{-1}$ is also a congruence subgroup.*

Proof (a) Let $\Gamma, \Gamma' \subset \Gamma_0$ be congruence subgroups. By definition, there are $M, N \in \mathbb{N}$ with $\Gamma(M) \subset \Gamma$, $\Gamma(N) \subset \Gamma'$. Then $\Gamma(MN) \subset (\Gamma(M) \cap \Gamma(N)) \subset (\Gamma \cap \Gamma')$.

(b) Fix $N \geq 2$ such that $\Gamma(N) \subset \Gamma$. There are natural numbers M_1, M_2 such that $M_1\alpha, M_2\alpha^{-1} \in M_2(\mathbb{Z})$. Set $M = M_1 M_2 N$. We claim that $\Gamma(M) \subset \alpha \Gamma_0 \alpha^{-1}$ or equivalently $\alpha^{-1}\Gamma(M)\alpha \subset \Gamma_0$. For $\gamma \in \Gamma(M)$ we write $\gamma = I + Mg$ with $g \in M_2(\mathbb{Z})$. It follows that $\alpha^{-1}\gamma\alpha = I + N(M_2\alpha^{-1})g(M_1\alpha) \in \Gamma(N) \subset \Gamma$. □

Let D_Γ be a fundamental domain for the congruence subgroup Γ as constructed in Lemma 2.5.19. The cusps of the fundamental domain D_Γ lie in the set

$$\Gamma(1)\infty = \mathbb{Q} \cup \{\infty\}.$$

The stabilizer group $\Gamma(1)_\infty$ of the point ∞ in $\Gamma(1)$ is $\pm \left(\begin{smallmatrix} 1 & \mathbb{Z} \\ & 1 \end{smallmatrix} \right)$.

Lemma 2.6.3 *Let Γ be a subgroup of finite index in $\Gamma_0 = SL_2(\mathbb{Z})$. For every $c \in \mathbb{Q} \cup \{\infty\}$ there exists a $\sigma_c \in GL_2(\mathbb{Q})^+$ such that*

- $\sigma_c \infty = c$ *and*

- $\sigma_c^{-1}\Gamma_c\sigma_c = \begin{cases} \left(\begin{smallmatrix} 1 & \mathbb{Z} \\ & 1 \end{smallmatrix} \right) & \text{if } -1 \notin \Gamma, \\ \pm\left(\begin{smallmatrix} 1 & \mathbb{Z} \\ & 1 \end{smallmatrix} \right) & \text{if } -1 \in \Gamma. \end{cases}$

The element σ_c is uniquely determined up to multiplication from the right by a matrix of the form $a\left(\begin{smallmatrix} 1 & x \\ & 1 \end{smallmatrix} \right)$ with an $x \in \mathbb{Q}$ and $a \in \mathbb{Q}^\times$.

Proof A given $c \in \mathbb{Q}$ can be written as $c = \alpha/\gamma$ with coprime integers α and γ. There then exist $\beta, \delta \in \mathbb{Z}$ with $\alpha\delta - \beta\gamma = 1$, so $\sigma = \left(\begin{smallmatrix} \alpha & \beta \\ \gamma & \delta \end{smallmatrix} \right) \in SL_2(\mathbb{Z})$. It follows that $\sigma \infty = c$. Replacing Γ with the group $\sigma^{-1}\Gamma\sigma$ we reduce the claim to the case $c = \infty$.

So we can assume $c = \infty$. Since Γ has finite index in $\Gamma(1)$, there exists $n \in \mathbb{N}$ with $\left(\begin{smallmatrix} 1 & n \\ & 1 \end{smallmatrix} \right)\Gamma = \Gamma$, so $\left(\begin{smallmatrix} 1 & n \\ & 1 \end{smallmatrix} \right) \in \Gamma$. Let $n \in \mathbb{N}$ be the smallest with this property. This means $\Gamma_\infty = \left(\begin{smallmatrix} 1 & n\mathbb{Z} \\ & 1 \end{smallmatrix} \right)$ or $\pm\left(\begin{smallmatrix} 1 & n\mathbb{Z} \\ & 1 \end{smallmatrix} \right)$, so the claim follows with $\sigma_c = \left(\begin{smallmatrix} 1/n & \\ & 1 \end{smallmatrix} \right)$.

For the uniqueness of σ_c, let σ_c' be another element of $GL_2^+(\mathbb{Q})$ with the same properties. Let $g = \sigma_c^{-1}\sigma_c'$, so $\sigma_c' = \sigma_c g$. The first property implies $g\infty = \infty$, so $g = \begin{pmatrix} a & b \\ & c \end{pmatrix}$ is an upper triangular matrix. Consider the case $-1 \notin \Gamma$. The second property implies $g^{-1}\begin{pmatrix} 1 & \mathbb{Z} \\ & 1 \end{pmatrix}g = \begin{pmatrix} 1 & \mathbb{Z} \\ & 1 \end{pmatrix}$. In particular, one gets $g^{-1}\begin{pmatrix} 1 & 1 \\ & 1 \end{pmatrix}g = \begin{pmatrix} 1 & \pm 1 \\ & 1 \end{pmatrix}$, which implies the claim. The case $-1 \in \Gamma$ is similar. $\qquad\square$

Definition 2.6.4 Let Γ be a subgroup of finite index in $SL_2(\mathbb{Z})$. A meromorphic function f on $\mathbb{H}$ is called *weakly modular* of weight k with respect to Γ, if $f|_k\gamma = f$ holds for every $\gamma \in \Gamma$.

A weakly modular function f is called *modular* if for every cusp $c \in \mathbb{Q} \cup \{\infty\}$ there exists $T_c > 0$ and some $N_c \in \mathbb{N}$ such that

$$f|\sigma_c(z) = \sum_{n \geq -N_c} a_{c,n} e^{2\pi inz}$$

holds for every $z \in \mathbb{H}$ with $\mathrm{Im}(z) > T_c$. In other words this means that the Fourier expansion is bounded below at every cusp. One also expresses this by saying that f is meromorphic at every cusp. By Lemma 2.6.3 this condition does not depend on the choice of the element σ_c, whereas the Fourier coefficients do depend on this choice.

The function f is called a *modular form* of weight k for the group Γ, if f is modular and holomorphic everywhere, including the cusps, which means that $a_{c,n} = 0$ for $n < 0$ at every cusp c. A modular form is called a *cusp form* if the zeroth Fourier coefficients $a_{c,0}$ vanish for all cusps c. The vector spaces of modular forms and cusp forms are denoted by $\mathcal{M}_k(\Gamma)$ and $\mathcal{S}_k(\Gamma)$.

As already mentioned in the proof of Theorem 2.5.16, the *Petersson inner product* can be defined for cusp forms of any congruence group Γ as follows: for $f, g \in S_k(\Gamma)$ one sets

$$\langle f, g \rangle_{\mathrm{Pet}} = \frac{1}{[\overline{\Gamma}(1) : \overline{\Gamma}]} \int_{\Gamma \backslash \mathbb{H}} f(z)\overline{g(z)} y^k \frac{dx\,dy}{y^2}.$$

2.7 Non-holomorphic Eisenstein Series

In the theory of automorphic forms one also considers non-holomorphic functions of the upper half plane, besides the holomorphic ones. These so-called *Maaß wave forms* will be introduced properly in the next section. In this section, we start with a special example, the non-holomorphic Eisenstein series. We introduce a fact, known as the *Rankin–Selberg method*, which says that the inner product of a non-holomorphic Eisenstein series and a Γ_0-automorphic function equals the Mellin integral transform of the zeroth Fourier coefficient of the automorphic function. This in particular implies that the Eisenstein series is orthogonal to the space of cusp forms, a fact of central importance in the spectral theory of automorphic forms.

Definition 2.7.1 The non-holomorphic *Eisenstein series* for $\Gamma_0 = SL_2(\mathbb{Z})$ is for $z = x + iy \in \mathbb{H}$ and $s \in \mathbb{C}$ defined by

$$E(z, s) = \pi^{-s} \Gamma(s) \frac{1}{2} \sum_{\substack{m,n \in \mathbb{Z} \\ (m,n) \neq (0,0)}} \frac{y^s}{|mz + n|^{2s}}.$$

By Lemma 1.2.1 the series $E(z, s)$ converges locally uniformly in $\mathbb{H} \times \{\operatorname{Re}(s) > 1\}$. Therefore the Eisenstein series is a continuous function, holomorphic in s, by the convergence theorem of Weierstrass.

Definition 2.7.2 By a *smooth function* we mean an infinitely often differentiable function.

Lemma 2.7.3 *For fixed s with $\operatorname{Re}(s) > 2$ the Eisenstein series $E(z, s)$ is a smooth function in $z \in \mathbb{H}$.*

Proof We divide the sum that defines $E(z, s)$ into two parts. One part with $m = 0$ and the other with $m \neq 0$. For $m = 0$ the sum does not depend on z, so the claim follows trivially. Consider the case $m \neq 0$ and let log be the principal branch of the logarithm, i.e. it is defined on $\mathbb{C} \smallsetminus (-\infty, 0]$ by $\log(re^{i\theta}) = \log(r) + i\theta$, if $r > 0$ and $-\pi < \theta < \pi$. For $z \in \mathbb{H}$ and $w \in \overline{\mathbb{H}}$, the lower half plane, we have

$$\log(zw) = \log(z) + \log(w).$$

For $m \neq 0$, $n \in \mathbb{Z}$, and $z \in \mathbb{H}$, one of the two complex numbers $mz + n$, $m\bar{z} + n$ is in $\mathbb{H}$, the other in $\overline{\mathbb{H}}$. Hence

$$|mz + n|^{-2s} = e^{-s \log((mz+n)(m\bar{z}+n))} = e^{-s \log(mz+n)} e^{-s \log(m\bar{z}+n)}.$$

Write $\log(mz + n) = \log(|mz + n|) + i\theta$ for some $|\theta| < \pi$. Then

$$\operatorname{Re}\big(-s \log(mz + n)\big) = -\operatorname{Re}(s) \log\big(|mz + n|\big) + \operatorname{Im}(s)\theta$$
$$\leq -\operatorname{Re}(s) \log\big(|mz + n|\big) + \big|\operatorname{Im}(s)\big|\pi,$$

so that

$$\big|e^{-s \log(mz+n)}\big| = e^{\operatorname{Re}(-s \log(mz+n))} \leq e^{|\operatorname{Im}(s)|\pi} |mz + n|^{-\operatorname{Re}(s)}.$$

For $z \in \mathbb{H}$ and $w \in \overline{\mathbb{H}}$ define

$$F(z, w, s) = \pi^{-s} \Gamma(s) y^s \frac{1}{2} \sum_{\substack{m,n \in \mathbb{Z} \\ (m,n) \neq (0,0)}} e^{-s \log(mz+n)} e^{-s \log(mw+n)}.$$

Keep w fixed and estimate the summand of the series $F(z, w, s)$ as follows

$$\big|e^{-s \log(mz+n)} e^{-s \log(mw+n)}\big| \leq C e^{2|\operatorname{Im}(s)|\pi} |mz + n|^{-\operatorname{Re}(s)},$$

with a constant $C > 0$, which depends on w. According to Lemma 1.2.1, the series $F(z, w, s)$ converges locally uniformly in z, for fixed w and s with $\mathrm{Re}(s) > 2$. As the summands are holomorphic, the function $F(z, w, s)$ is holomorphic in z. The same argument shows that F is holomorphic in w for fixed z. By Exercise 2.20 the function $F(z, w, s)$ can locally be written as a power series in z and w simultaneously, which means that $F(z, w, s)$ is a smooth function in (z, w) for fixed s with $\mathrm{Re}(s) > 2$. Therefore $F(z, \overline{z}, s) = E(z, s)$ is a smooth function, too. $\qquad\square$

Lemma 2.7.4 *Let* $\Gamma_0 = \mathrm{SL}_2(\mathbb{Z})$ *and let* $\Gamma_{0,\infty}$ *be the stabilizer group of* ∞, *so* $\Gamma_{0,\infty} = \pm\left(\begin{smallmatrix} 1 & \mathbb{Z} \\ & 1 \end{smallmatrix}\right)$. *Then the map*

$$\Gamma_{0,\infty} \backslash \Gamma_0 \to \{\pm(x, y) \in \mathbb{Z}^2 / \pm 1 : x, y \text{ coprime}\}$$

$$\Gamma_{0,\infty} \begin{pmatrix} a & b \\ c & d \end{pmatrix} \mapsto \pm(c, d)$$

is a bijection.

Proof If $c, d \in \mathbb{Z}$ are coprime, then there exist $a, b \in \mathbb{Z}$ such that $ad - bc = 1$. If (a, b) is one such pair, then every other is of the form $(a + cx, b + dx)$ for some $x \in \mathbb{Z}$. (Idea of proof: Assume $1 < c \le d$. After division with remainder there is $0 \le r < c$ with $d = r + cq$. Then divide c by r with remainder and so on. This algorithm will stop. Plugging in the solutions backwards gives a pair (a, b).)

For $\left(\begin{smallmatrix} 1 & x \\ & 1 \end{smallmatrix}\right) \in \Gamma_\infty$ and $\left(\begin{smallmatrix} a & b \\ c & d \end{smallmatrix}\right) \in \Gamma$ one has

$$\begin{pmatrix} 1 & x \\ & 1 \end{pmatrix} \begin{pmatrix} a & b \\ c & d \end{pmatrix} = \begin{pmatrix} a + cx & b + dx \\ c & d \end{pmatrix}.$$

This implies the lemma. $\qquad\square$

Definition 2.7.5 An *automorphic function* on $\mathbb{H}$ with respect to the congruence subgroup $\Gamma \subset \mathrm{SL}_2(\mathbb{Z})$ is a function $\phi : \mathbb{H} \to \mathbb{C}$, which is invariant under the operation of Γ, so that $\phi(\gamma z) = \phi(z)$ holds for every $\gamma \in \Gamma$.

Proposition 2.7.6

(a) *The series* $\tilde{E}(z, s) = \sum_{\gamma : \Gamma_\infty \backslash \Gamma} \mathrm{Im}(\gamma z)^s$ *converges for* $\mathrm{Re}(s) > 1$ *and we have*

$$E(z, s) = \pi^{-s} \Gamma(s) \zeta(2s) \tilde{E}(z, s),$$

where $\zeta(s)$ *is the Riemann zeta function.*

(b) *The functions* $E(z, s)$ *and* $\tilde{E}(z, s)$ *are automorphic under* $\Gamma = \mathrm{SL}_2(\mathbb{Z})$, *i.e. we have*

$$E(\gamma z, s) = E(z, s)$$

for every $\gamma \in \Gamma$. *The same holds for* $\tilde{E}$.

Proof (a) With $\gamma = \begin{pmatrix} * & * \\ c & d \end{pmatrix}$ we have $\operatorname{Im}(\gamma z) = \operatorname{Im}(z)/|cz + d|^2$. According to Lemma 2.7.4 it holds that

$$\tilde{E}(z, s) = \sum_{\substack{(c,d)=1 \\ \bmod \pm 1}} \frac{y^s}{|cz + d|^{2s}}.$$

Hence we get convergence with $E(z, s)$ as a majorant. We conclude

$$\zeta(2s)\tilde{E}(z, s) = \sum_{n=1}^{\infty} \sum_{\substack{(c,d)=1 \\ \bmod \pm 1}} \frac{y^s}{|ncz + nd|^{2s}} = \frac{1}{2} \sum_{\substack{m,n\in\mathbb{Z} \\ (m,n)\neq(0,0)}} \frac{y^s}{|mz + n|^{2s}}.$$

(b) It suffices to show the claim for $\tilde{E}$. We compute

$$\tilde{E}(\gamma z, s) = \sum_{\tau:\Gamma_\infty\backslash\Gamma} \operatorname{Im}(\tau\gamma z)^s = \sum_{\tau:\Gamma_\infty\backslash\Gamma} \operatorname{Im}(\tau z)^s,$$

since if τ runs through a set of representatives for $\Gamma_\infty\backslash\Gamma$, then so does $\tau\gamma$. □

In particular it follows that

$$E(z + 1, s) = E(z, s).$$

It follows that for $\operatorname{Re}(s) > 2$ the smooth function $E(z, s)$ has a Fourier expansion in z. We will examine this Fourier expansion more closely.

The integral

$$K_s(y) = \frac{1}{2} \int_0^\infty e^{-y(t+t^{-1})/2} t^s \frac{dt}{t}$$

converges locally uniformly absolutely for $y > 0$ and $s \in \mathbb{C}$. The function K_s so defined is called the *K-Bessel function*. It satisfies the estimate

$$|K_s(y)| \leq e^{-y/2} K_{\operatorname{Re}(s)}(2), \quad \text{if } y > 4.$$

Proof For two real numbers a, b we have

$$a > b > 2 \quad \Rightarrow \quad \begin{Bmatrix} ab > 2a \\ 2a > a + b \end{Bmatrix} \quad \Rightarrow \quad ab > a + b.$$

The last assertion is symmetric in a and b, so it holds for all $a, b > 2$. Therefore one has $e^{-ab} < e^{-a}e^{-b}$. Applying this to $a = y/2 > 2$ and $b = t + t^{-1}$ and integrating along t gives

$$|K_s(y)| \leq \frac{1}{2} \int_0^\infty e^{-y/2} e^{-(t+t^{-1})} t^{\operatorname{Re}(s)} \frac{dt}{t} = e^{-y/2} K_{\operatorname{Re}(s)}(2).$$

We also note that the integrand in the Bessel integral is invariant under $t \mapsto t^{-1}$, $s \mapsto -s$, so that

$$K_{-s}(y) = K_s(y).$$

Theorem 2.7.7 *The Eisenstein series $E(z, s)$ has a Fourier expansion*

$$E(z, s) = \sum_{r=-\infty}^{\infty} a_r(y, s)e^{2\pi i r x},$$

where

$$a_0(y, s) = \pi^{-s}\Gamma(s)\zeta(2s)y^s + \pi^{s-1}\Gamma(1-s)\zeta(2-2s)y^{1-s}$$

and for $r \neq 0$,

$$a_r(y, s) = 2|r|^{2-1/2}\sigma_{1-2s}(|r|)\sqrt{y}K_{s-\frac{1}{2}}(2\pi|r|y).$$

One reads off that the Eisenstein series $E(z, s)$, as a function in s, has a mero-morphic expansion to all of $\mathbb{C}$. It is holomorphic except for simple poles at $s = 0, 1$. It is a smooth function in z for all $s \neq 0, 1$. Every derivative in z is holomorphic in $s \neq 0, 1$. The residue at $s = 1$ is constant in z and takes the value $1/2$. The Eisenstein series satisfies the functional equation

$$E(z, s) = E(z, 1-s).$$

Locally uniformly in $x \in \mathbb{R}$ one has

$$E(x + iy) = O(y^\sigma), \quad \text{for } y \to \infty,$$

where $\sigma = \max(\mathrm{Re}(s), 1 - \mathrm{Re}(s))$.

Proof The claims all follow from the explicit Fourier expansion, which remains to be shown.

Definition 2.7.8 A function $f : \mathbb{R} \to \mathbb{C}$ is called a *Schwartz function* if f is in-finitely differentiable and every derivative $f^{(k)}$, $k \geq 0$ is rapidly decreasing. Let $\mathcal{S}(\mathbb{R})$ denote the vector space of all Schwartz functions on $\mathbb{R}$. If f is in $\mathcal{S}(\mathbb{R})$, then its Fourier transform $\hat{f}$ also lies in $\mathcal{S}(\mathbb{R})$; see [Dei05, Rud87, SW71].

Lemma 2.7.9 *If $\mathrm{Re}(s) > \frac{1}{2}$ and $r \in \mathbb{R}$, then*

$$\left(\frac{y}{\pi}\right)^s \Gamma(s)\int_{\mathbb{R}}(x^2 + y^2)^{-s}e^{2\pi i r x}\,dx = \begin{cases} \pi^{-s+1/2}\Gamma(s - \frac{1}{2})y^{1-s}, & \text{if } r = 0, \\ 2|r|^{s-1/2}\sqrt{y}K_{s-1/2}(2\pi|r|y), & \text{if } r \neq 0. \end{cases}$$

Proof We plug in the Γ-integral on the left-hand side to get

$$\int_{\mathbb{R}}\int_0^\infty e^{-t}\left(\frac{ty}{\pi(x^2 + y^2)}\right)^s e^{2\pi i r x}\frac{dt}{t}\,dx = \int_0^\infty\int_{\mathbb{R}} e^{-\pi t(x^2 + y^2)/y}t^s e^{2\pi i r x}\,dx\,\frac{dt}{t},$$

where we have substituted $t \mapsto \pi t (x^2 + y^2)/y$. The function $f(x) = e^{-\pi x^2}$ is its own Fourier transform: $\hat{f} = f$. To see this, note that f is, up to scaling, uniquely determined as the solution of the differential equation

$$f'(x) = -2\pi x f(x).$$

By induction one shows that for every natural n there is a polynomial $p_n(x)$, such that $f^{(n)}(x) = p_n(x) e^{-\pi x^2}$. So f lies in the Schwartz space $S(\mathbb{R})$ and so does its Fourier transform $\hat{f}$, and one computes

$$(\hat{f})'(y) = \int_{\mathbb{R}} (-2\pi i x) e^{-\pi x^2} e^{-2\pi xy} dx = i \int_{\mathbb{R}} (e^{-\pi x^2})' e^{-2\pi ixy} dx = -2\pi y \hat{f}(y).$$

Therefore $\hat{f} = cf$ and $\hat{\hat{f}} = c\hat{f} = c^2 f$. Since, on the other hand, $\hat{\hat{f}}(x) = f(-x) = f(x)$, we infer that $c^2 = 1$, so $c = \pm 1$. By $\hat{f}(0) = \int_{\mathbb{R}} e^{-\pi x^2} dx > 0$ it follows that $c = 1$.

By a simple substitution one gets from this

$$\int_{\mathbb{R}} e^{-t\pi x^2/y} e^{2\pi irx} dx = \sqrt{\frac{y}{t}} e^{-y\pi r^2/t}.$$

We see that the left-hand side of the lemma equals

$$\int_0^{\infty} e^{-\pi ty} \sqrt{\frac{y}{t}} e^{-y\pi r^2/t} t^s \frac{dt}{t},$$

which gives the claim.   □

We now compute the Fourier expansion of the Eisenstein series $E(z, s)$. The coefficients are given by

$$a_r(y, s) = \int_0^1 E(x + iy, s) e^{-2\pi irx} dx$$

$$= \pi^{-s} \Gamma(s) \frac{1}{2} \int_0^1 \sum_{\substack{m,n \in \mathbb{Z} \\ (m,n) \neq (0,0)}} \frac{y^s}{|mx + imy + n|^{2s}} e^{-2\pi irx} dx.$$

The summands with $m = 0$ only give a contribution in the case $r = 0$. This contribution is

$$\pi^{-s} \Gamma(s) y^s \sum_{n=1}^{\infty} n^{-2s} = \pi^{-s} \Gamma(s) \zeta(2s) y^s.$$

For $m \neq 0$ note that the contribution for (m, n) and $(-m, -n)$ are equal. Therefore it suffices to sum over $m > 0$. The contribution to a_r is

$$\pi^{-s} \Gamma(s) y^s \sum_{m=1}^{\infty} \sum_{n=-\infty}^{\infty} \int_0^1 [(mx + n)^2 + m^2 y^2]^{-s} e^{-2\pi irx} dx$$

$$= \pi^{-s} \Gamma(s) y^s \sum_{m=1}^{\infty} \sum_{n \bmod m} \int_{-\infty}^{\infty} [(mx + n)^2 + m^2 y^2]^{-s} e^{-2\pi irx} dx.$$

The substitution $x \mapsto x - n/m$ yields

$$\pi^{-s}\Gamma(s)y^s \sum_{m=1}^{\infty} m^{-2s} \sum_{n \bmod m} e^{2\pi irn/m} \int_{-\infty}^{\infty} (x^2 + y^2)^{-s} e^{-2\pi irx} \, dx.$$

Because of

$$\sum_{n \bmod m} e^{2\pi irn/m} = \begin{cases} m & \text{if } m|r, \\ 0 & \text{otherwise} \end{cases}$$

the contribution is

$$\pi^{-s}\Gamma(s)y^s \sum_{m|r} m^{1-2s} \int_{-\infty}^{\infty} (x^2 + y^2)^{-s} e^{2\pi irx} \, dx.$$

There are two cases. Firstly, if $r = 0$, the condition $m|r$ is vacuous and we get

$$\pi^{-s}\Gamma(s)y^s \zeta(2s-1) \int_{-\infty}^{\infty} (x^2 + y^2)^{-s} \, dx = \pi^{-s+1/2}\Gamma\left(s - \frac{1}{2}\right)\zeta(2s-1)y^{1-s},$$

where we have used Lemma 2.7.9. The Riemann zeta function satisfies the functional equation

$$\hat{\zeta}(s) = \hat{\zeta}(1-s),$$

with $\hat{\zeta}(s) = \pi^{-s/2}\Gamma(s/2)\zeta(s)$, as is shown in Theorem 6.1.3. Therefore the zeroth term a_0 is as claimed. Secondly, in the case $r \neq 0$ we get the claim again by Lemma 2.7.9. □

We now explain the Rankin–Selberg method. Let $\Gamma = SL_2(\mathbb{Z})$ and let $\phi : \mathbb{H} \to \mathbb{C}$ be a smooth, Γ-automorphic function. We assume that ϕ is rapidly decreasing at the cusp ∞, i.e. that

$$\phi(x + iy) = O(y^{-N}), \quad y \geq 1$$

holds for every $N \in \mathbb{N}$. Because of $\phi(z + 1) = \phi(z)$ the function ϕ has a Fourier expansion

$$\phi(z) = \sum_{n=-\infty}^{\infty} \phi_n(y) e^{2\pi inx}$$

with $\phi_n(y) = \int_0^1 \phi(x + iy) e^{-2\pi inx} \, dx$. The term ϕ_0 is called the *constant term* of the Fourier expansion. Let

$$M\phi_0(s) = \int_0^{\infty} \phi_0(y) y^s \frac{dy}{y}$$

be the Mellin transform of the zeroth term. We shall show that this integral converges for $\mathrm{Re}(s) > 0$. Put

$$\Lambda(s) = \pi^{-s}\Gamma(s)\zeta(2s)M\phi_0(s-1).$$

Proposition 2.7.10 (Rankin–Selberg method) *The integral $\mathcal{M}\phi_0(s)$ converges locally uniformly absolutely in the domain $\mathrm{Re}(s) > 0$. One has*

$$\Lambda(s) = \int_{\Gamma(1)\backslash\mathbb{H}} E(z,s)\phi(z)\frac{dx\,dy}{y^2}.$$

The function $\Lambda(s)$, defined for $\mathrm{Re}(s) > 0$, extends to a meromorphic function on $\mathbb{C}$ with at most simple poles at $s = 0$ and $s = 1$. It satisfies the functional equation

$$\Lambda(s) = \Lambda(1 - s).$$

The residue at $s = 1$ equals

$$\mathrm{res}_{s=1}\,\Lambda(s) = \frac{1}{2}\int_{\Gamma(1)\backslash\mathbb{H}}\phi(z)\frac{dx\,dy}{y^2}.$$

Proof The proof relies on an unfolding trick as follows

$$\int_{\Gamma\backslash\mathbb{H}} \tilde{E}(z,s)\phi(z)\,d\mu(z) = \int_D \sum_{\gamma:\Gamma_\infty\backslash\Gamma} \mathrm{Im}(\gamma z)^s \phi(z)\,d\mu(z)$$

$$= \sum_{\gamma:\Gamma_\infty\backslash\Gamma} \int_D \mathrm{Im}(\gamma z)^s \phi(z)\,d\mu(z)$$

$$= \sum_{\gamma:\Gamma_\infty\backslash\Gamma} \int_{\gamma D} \mathrm{Im}(z)^s \phi(z)\,d\mu(z)$$

$$= \int_{\bigcup_{\gamma:\Gamma_\infty\backslash\Gamma}\gamma D} \mathrm{Im}(z)^s \phi(z)\,d\mu(z)$$

$$= \int_{\Gamma_\infty\backslash\mathbb{H}} \mathrm{Im}(z)^s \phi(z)\,d\mu(z)$$

$$= \int_0^\infty \int_0^1 y^{s-1} \phi(x+iy)\,dx\,\frac{dy}{y}$$

$$= \int_0^\infty \phi_0(y)y^{s-1}\frac{dy}{y} = \mathcal{M}\phi_0(s-1).$$

The claims now follow from Theorem 2.7.7. $\square$

We apply the Rankin–Selberg method to show that the Rankin–Selberg convolution of modular L-functions is meromorphic. Let $k \in 2\mathbb{N}_0$ and let $f, g \in M_k$ be normalized Hecke eigenforms. Denote the Fourier coefficients of f and g by a_n and b_n for $n \geq 0$, respectively. We define the *Rankin–Selberg convolution* of $L(f, s)$ and $L(g, s)$ by

$$L(f \times g, s) \stackrel{\mathrm{def}}{=} \zeta(2s - 2k + 2)\sum_{n=1}^{\infty} a_n b_n n^{-s}.$$

By Proposition 2.3.1 and Theorem 2.3.2 one has $a_n, b_n = O(n^{k-1})$. Therefore the series $L(f \times g, s)$ converges absolutely for $\mathrm{Re}(s) > 2k - 1$. We put

$$\Lambda(f \times g, s) = (2\pi)^{-2s} \Gamma(s) \Gamma(s - k + 1) L(f \times g, s).$$

Theorem 2.7.11 *Suppose that one of the functions f, g is a cusp form. Then $\Lambda(f \times g, s)$ extends to a meromorphic function on $\mathbb{C}$. It is holomorphic except for possible simple poles at $s = k$ and $s = k - 1$. It satisfies the functional equation*

$$\Lambda(f \times g, s) = \Lambda(f \times g, 2k - 1 - s).$$

The residue at $s = k$ is $\frac{1}{2}\pi^{1-k}\langle f, g\rangle_k$.

Proof We apply Proposition 2.7.10 to the function $\phi(z) = f(z)\overline{g(z)}y^k$. Then

$$
\begin{aligned}
\phi_0(y) &= \int_0^1 f(x + iy)\overline{g(x + iy)}y^k \, dx \\
&= \sum_{n=0}^{\infty} \sum_{m=0}^{\infty} \int_0^1 a_n e^{2\pi inx} e^{-2\pi ny} \overline{b_m} e^{-2\pi imx} e^{-2\pi my} y^k \, dx.
\end{aligned}
$$

Since $\int_0^1 e^{2\pi i(n-m)x} \, dx = 0$ except for $n = m$, we get $\phi_0(y) = \sum_{n=1}^{\infty} a_n \overline{b_n} e^{-4\pi ny} y^k$. So

$$M\phi_0(s) = \sum_{n=0}^{\infty} a_n \overline{b_n} \int_0^{\infty} e^{-4\pi ny} y^{s+k} \frac{dy}{y} = (4\pi)^{-s-k} \Gamma(s + k) \sum_{n=1}^{\infty} a_n \overline{b_n} n^{-s-k}.$$

The number b_n is the eigenvalue of the Hecke operator T_n. As T_n is self-adjoint, b_n is real. Therefore,

$$M\phi_0(s - 1) = (4\pi)^{-s-k+1} \Gamma(s - 1 + k) \frac{1}{\zeta(2s)} L(f \times g, s - 1 + k).$$

Let $\Lambda(s)$ be as in Proposition 2.7.10. It follows that

$$\Lambda(s) = 4^{-s-k+1} \pi^{-2s-k+1} \Gamma(s) \Gamma(s - 1 + k) L(f \times g, s - 1 + k),$$

or

$$\Lambda(s + 1 - k) = \pi^{k-1} (2\pi)^{-2s} \Gamma(s) \Gamma(s + 1 - k) L(f \times g, s) = \pi^{k-1} \Lambda(f \times g, s).$$

By Proposition 2.7.10 one has $\Lambda(s + 1 - k) = \Lambda(1 - (s + 1 - k))$, which implies the claimed functional equation. Finally one has

$$
\mathrm{res}_{s=k} \Lambda(f \times g, s) = \mathrm{res}_{s=k} \pi^{1-k} \Lambda(s + 1 - k) = \pi^{1-k} \mathrm{res}_{s=1} \Lambda(s)
$$

$$
= \pi^{1-k} \frac{1}{2} \int_{\Gamma(1)\backslash\mathbb{H}} \phi(z) \, d\mu(z) = \pi^{1-k} \frac{1}{2}\langle f, g\rangle_k. \qquad \square
$$

Next we show that the L-function $L(f \times g, s)$ has an Euler product. We factorize the polynomials

$$1 - a_n X + p^{k-1} X^2 = (1 - \alpha_1(p)X)(1 - \alpha_2(p)X),$$
$$1 - b_n X + p^{k-1} X^2 = (1 - \beta_1(p)X)(1 - \beta_2(p)X).$$

Theorem 2.7.12 *Let $k \in 2\mathbb{N}_0$ and let $f, g \in \mathcal{M}_k$ be normalized Hecke eigenforms. The Rankin–Selberg L-function has the Euler product expansion*

$$L(f \times g, s) = \prod_p \prod_{i=1}^{2} \prod_{j=1}^{2} \left(1 - \alpha_i(p)\beta_j(p)p^{-s}\right)^{-1}.$$

Proof This is a consequence of the following lemma.

Lemma 2.7.13 *Let $\alpha_1, \alpha_2, \beta_1, \beta_2$ be complex numbers with $\alpha_1\alpha_2\beta_1\beta_2 \neq 0$ and suppose that the equalities*

$$\sum_{r=0}^{\infty} a_r z^r = (1 - \alpha_1 z)^{-1}(1 - \alpha_2 z)^{-1},$$

$$\sum_{r=0}^{\infty} b_r z^r = (1 - \beta_1 z)^{-1}(1 - \beta_2 z)^{-1}$$

hold for small complex numbers z. Then for small z one has

$$\sum_{r=0}^{\infty} a_r b_r z^r = \left(1 - \alpha_1\alpha_2\beta_1\beta_2 z^2\right) \prod_{i=1}^{2} \prod_{j=1}^{2} (1 - \alpha_i \beta_j z)^{-1}.$$

Proof Let $\phi(z) = \sum_{r=0}^{\infty} a_r z^r$ and $\psi(z) = \sum_{r=0}^{\infty} b_r z^r$. Consider the path integral

$$\frac{1}{2\pi i} \int_{\partial K} \phi(qz)\psi(q^{-1}) \frac{dq}{q},$$

where K is a circle around zero such that the poles of $q \mapsto \phi(zq)$ are outside K, and the poles of $q \mapsto \psi(q^{-1})$ are inside. This is possible for z small enough. The integral is equal to

$$\sum_{r,r'=0}^{\infty} a_r b_{r'} x^r \frac{1}{2\pi i} \int_{\partial K} q^{r-r'-1} \, dq = \sum_{r=0}^{\infty} a_r b_r x^r.$$

On the other hand, the integral equals

$$\frac{1}{2\pi i} \int_{\partial K} \frac{1}{(1 - \alpha_1 x q)(1 - \alpha_2 x q)(1 - \beta_1 q^{-1})(1 - \beta_2 q^{-1})} \frac{dq}{q},$$

which we calculate by the residue theorem as

$$\left(1 - \alpha_1\alpha_2\beta_1\beta_2 x^2\right)\prod_{i=1}^{2}\prod_{j=1}^{2}(1 - \alpha_i\beta_j x)^{-1}. \qquad \square$$

2.8 Maaß Wave Forms

This section is not strictly necessary for the rest of the book, but we include it for completeness. In this section we shall not give full proofs all the time, but rather sketch the arguments.

The group $G = SL_2(\mathbb{R})$ acts on $\mathbb{H}$ by diffeomorphisms, so it acts on $C^\infty(\mathbb{H})$ by $L_g : C^\infty(\mathbb{H}) \to C^\infty(\mathbb{H})$, where for $g \in G$ the operator L_g is defined by

$$L_g\varphi(z) = \varphi\left(g^{-1}z\right).$$

On the upper half plane $\mathbb{H}$ we have the *hyperbolic Laplace operator*, which is a differential operator defined by

$$\Delta = -y^2\left(\frac{\partial^2}{\partial x^2} + \frac{\partial^2}{\partial y^2}\right).$$

Lemma 2.8.1 *The hyperbolic Laplacian is invariant under G, i.e. one has*

$$L_g\Delta L_{g^{-1}} = \Delta$$

for every $g \in G$.

Proof The assertion is equivalent to $L_g\Delta = \Delta L_g$. It suffices to show this assertion for generators of the group $SL_2(\mathbb{R})$. Such generators are given in Exercise 2.6. We leave the explicit calculation for this invariance to the reader, see Exercise 2.7. $\square$

Definition 2.8.2 A *Maaß wave form* or *Maaß form* for the group $\Gamma(1)$ is a smooth function f on $\mathbb{H}$ such that

- $f(\gamma z) = f(z)$ for every $\gamma \in \Gamma(1)$,
- $\Delta f = \lambda f$ for some $\lambda \in \mathbb{C}$,
- there exists $N \in \mathbb{N}$ with $f(x + iy) = O(y^N)$ for $y \geq 1$.

If additionally one has

$$\int_0^1 f(z + t)\,dt = 0$$

for every $z \in \mathbb{H}$, then f is called a *Maaß cusp form*.

Proposition 2.8.3 *The non-holomorphic Eisenstein series*

$$E(z, s) = \pi^{-s}\Gamma(s)\frac{1}{2}\sum_{\substack{m,n\in\mathbb{Z}\\(m,n)\neq(0,0)}}\frac{y^s}{|mz + n|^{2s}}$$

is a Maaß form; more precisely it holds that

$$\Delta E(z, s) = s(1 - s)E(z, s), \quad s \neq 0, 1.$$

Proof We only have to show the eigen-equation. We have

$$E(z, s) = \pi^{-s} \Gamma(s) \zeta(2s) \tilde{E}(z, s)$$

with $\tilde{E}(z, s) = \sum_{\Gamma_\infty \backslash \Gamma} \operatorname{Im}(\gamma z)^s$. So it suffices to show the eigen-equation for $\tilde{E}$. We have

$$\Delta(y^s) = -y^2 \left(\frac{\partial^2}{\partial x^2} + \frac{\partial^2}{\partial y^2} \right) y^s = s(1 - s)y^s.$$

By invariance of the Laplace operator we get

$$\Delta \operatorname{Im}(\gamma z)^s = s(1 - s) \operatorname{Im}(\gamma z)^s$$

for every $\gamma \in \Gamma$. By means of Lemma 1.2.1 one shows that for $\operatorname{Re}(s) > 3$ the series $\sum_{\Gamma_\infty \backslash \Gamma} \frac{\partial}{\partial y} \operatorname{Im}(z)^s$ and $\sum_{\Gamma_\infty \backslash \Gamma} \frac{\partial^2}{\partial y^2} \operatorname{Im}(z)^s$ as well as the x-derivatives converge locally uniformly, so we may differentiate the Eisenstein series term-wise. This implies the claim for $\tilde{E}$ and therefore also for E in the domain $\operatorname{Re}(s) > 3$. For arbitrary $s \in \mathbb{C}$ the Fourier expansion shows that $\Delta E(z, s) - s(1 - s)E(z, s)$ is a meromorphic function in s, which for $\operatorname{Re}(s) > 3$ is constantly equal to zero. By the identity theorem it is zero everywhere.   □

The differential equation can also be expressed in the form

$$\Delta E\left(z, \nu + \frac{1}{2}\right) = \left(\frac{1}{4} - \nu^2\right) E\left(z, \nu + \frac{1}{2}\right).$$

Let f be an arbitrary Maaß form for the group $\Gamma(1)$. Because of $f(z + 1) = f(z)$ the function f has a Fourier expansion

$$f(x + iy) = \sum_{r=-\infty}^{\infty} a_r(y) e^{2\pi i r x}.$$

Lemma 2.8.4 *Let $\lambda \in \mathbb{C}$ be the Laplace eigenvalue of the Maaß form f. There is a $\nu \in \mathbb{C}$, which is unique up to sign, such that $\lambda = \frac{1}{4} - \nu^2$. The Fourier coefficients of f are*

$$a_r(y) = a_r \sqrt{y} K_\nu(2\pi |r| y)$$

if $r \neq 0$, where $a_r \in \mathbb{C}$ depends only on r. For $r = 0$ one has

$$a_0(y) = a_0 y^{\frac{1}{2} - \nu} + b_0 y^{\frac{1}{2} + \nu}$$

for some $a_0, b_0 \in \mathbb{C}$.

Proof We have $\Delta f = (\frac{1}{4} - \nu^2) f(z)$. The definition of $a_r(y)$,

$$a_r(y) = \int_0^1 f(x + iy) e^{-2\pi i r x} \, dx,$$

implies

$$\left(\frac{1}{4}-\nu^2\right)a_r(y)=\int_0^1 \Delta f(x+iy)e^{-2\pi irx}\,dx$$

$$=-y^2\int_0^1\frac{\partial^2 f}{\partial y^2}+\frac{\partial^2 f}{\partial x^2}(x+iy)e^{-2\pi irx}\,dx$$

$$=-y^2\frac{\partial^2}{\partial y^2}a_r(y)-y^2\int_0^1 f(x+iy)\left(-4\pi^2r^2\right)e^{-2\pi irx}\,dx$$

$$=-y^2\frac{\partial^2}{\partial y^2}a_r(y)+4\pi^2r^2y^2a_r(y).$$

So there is a differential equation of second order,

$$y^2\frac{\partial^2}{\partial y^2}a_r(y)+\left(\frac{1}{4}-\nu^2-4\pi r^2y^2\right)a_r(y)=0.$$

The rth Fourier coefficient of the Eisenstein series is a solution of this differential equation. Therefore the function

$$a_r(y)=\sqrt{y}K_\nu\left(2\pi|r|y\right)$$

solves this linear differential equation. A second solution is given by

$$b_r(y)=\sqrt{y}I_\nu\left(2\pi|r|y\right),$$

where I_ν is the I-Bessel function [AS64]. As the differential equation is linear of order 2, every solution is a linear combination of these two basis solutions. A proof of this classical fact can be found for example in [Rob10]. Further, the I-Bessel function grows exponentially, whereas the K-Bessel function decreases exponentially [AS64]. According to the definition of a Maaß form, the function $a_r(y)$ can only grow moderately, and the claim follows.   □

Let $\iota:\mathbb{H}\to\mathbb{H}$ be the anti-holomorphic map $\iota(z)=-\bar{z}$, so $\iota(x+iy)=-x+iy$. Then $\iota\circ\iota=\mathrm{Id}_\mathbb{H}$ and one finds that ι commutes with the hyperbolic Laplacian Δ, when ι acts on functions f of the upper half plane by $\iota(f)(z)\overset{\mathrm{def}}{=}f(\iota(z))$. Therefore ι maps the λ-eigenspace into itself for every $\lambda\in\mathbb{C}$. By $\iota^2=\mathrm{Id}$ the map ι itself has at most the eigenvalues ±1. A Maaß form f is called an *even Maaß form* if $\iota(f)=f$ and an *odd Maaß form* if $\iota(f)=-f$. By

$$f=\frac{1}{2}(f+\iota(f))+\frac{1}{2}(f-\iota(f))$$

every Maaß form is the sum of an even and an odd Maaß form.

Theorem 2.8.5 *Let*

$$f(z) = \sum_{r \neq 0} a_r \sqrt{y} K_\nu \left(2\pi |r| y\right) e^{2\pi i r x}$$

be a Maaß cusp form and let

$$L(s, f) = \sum_{n=1}^{\infty} a_n n^{-s}$$

be the corresponding L-series. The series $L(s, f)$ converges for $\mathrm{Re}(s) > 3/2$ and extends to an entire function on $\mathbb{C}$. Let f be even or odd and let $\varepsilon = 0$ if f is even and $\varepsilon = 1$ if f is odd. Let $\Delta f = (\frac{1}{4} - \nu^2) f$. Then with

$$\Lambda(s, f) = \pi^{-s} \Gamma\left(\frac{s - \varepsilon + \nu}{2}\right) \Gamma\left(\frac{s - \varepsilon - \nu}{2}\right) L(s, f),$$

one has the functional equation

$$\Lambda(s, f) = (-1)^\varepsilon \Lambda(1 - s, f).$$

Proof Note that $a_{-r} = (-1)^\varepsilon a_r$ holds. The convergence is clear by the following lemma.

Lemma 2.8.6 *We have $a_n = O(n^{1/2})$.*

Proof There are $C, N > 0$ such that for $y > 1$ the inequality $|f(x + iy)| \leq C y^N$ holds. If $y < 1/2$ and if $w \in D$ is conjugate to z modulo $\Gamma(1)$, then $\mathrm{Im}(w) \leq \frac{1}{y}$. So suppose $y < 1/2$. Then it follows that $|f(x + iy)| \leq C y^{-N}$. So that for $y < 1/2$ one has

$$\left|a_r\right| \sqrt{y} \left|K_s\left(2\pi |r| y\right)\right| \leq \int_0^1 \left|f(x + iy)\right| dx \leq C y^{-N}.$$

With $y = 1/|r|$ we get from this

$$|a_r| \leq C r^{N+\frac{1}{2}} \left|K_s(2\pi)\right|^{-1}.$$

As the K-Bessel function is rapidly decreasing and f is a cusp form, we conclude that f is bounded on D and therefore on $\mathbb{H}$. This argument can be repeated with $N = 0$. The claim is proven.   □

Lemma 2.8.7 *The integral*

$$\int_0^\infty K_\nu(y) y^s \frac{dy}{y} = 2^{s-2} \Gamma\left(\frac{s + \nu}{2}\right) \Gamma\left(\frac{s - \nu}{2}\right)$$

converges absolutely if $\mathrm{Re}(s) > |\mathrm{Re}(\nu)|$.

Proof Plugging in the definition of K_ν, the left-hand side becomes

$$\frac{1}{2}\int_0^\infty\int_0^\infty e^{-(t+t^{-1})y/2}t^\nu y^s\frac{dy\,dt}{y\,t}.$$

We use the change of variables rule with the diffeomorphism $\phi:(0,\infty)\times(0,\infty)\to(0,\infty)\times(0,\infty)$ given by

$$\phi(t,y)=\left(\frac{1}{2}ty,\frac{1}{2}t^{-1}y\right)=(u,v).$$

Then $y=2\sqrt{uv}$ and $t=\sqrt{u/v}$. The Jacobian matrix of ϕ is

$$D\phi(t,y)=\frac{1}{2}\begin{pmatrix} y & t \\ -\frac{y}{t^2} & \frac{1}{t} \end{pmatrix}.$$

Its determinant equals $\det D\phi=\frac{y}{2t}$. By the change of variables rule the integral equals

$$2^{s-1}\int_0^\infty\int_0^\infty e^{-u-v}v^{(s-\nu)/2}u^{(s+\nu)/2}\frac{du\,dv}{u\,v}.$$

The claim follows. $\qquad\square$

We now prove the theorem in the case when f is even. Then

$$\int_0^\infty f(iy)y^{s-1/2}\frac{dy}{y}=\frac{1}{2}\Lambda(s,f).$$

By Lemma 2.8.6 we infer that $f(iy)$ is rapidly decreasing for $y\to\infty$. Because of $f(iy)=f(i\frac{1}{y})$ the claim follows similar to Theorem 2.4.5.

If f is odd, put

$$g(z)=\frac{1}{4\pi i}\frac{\partial f}{\partial x}(z)=\sum_{n=1}^\infty a_n n\sqrt{y}K_\nu(2\pi ny)\cos(2\pi nx).$$

Then

$$\int_0^\infty g(iy)y^{s+1/2}\frac{dy}{y}=\Lambda(s,f).$$

Because of $g(iy)=-\frac{1}{y^2}g(\frac{i}{y})$ the claim follows in this case as well. $\qquad\square$

More generally, for every $k\in\mathbb{Z}$ we introduce the operator

$$\Delta_k=-y^2\left(\frac{\partial^2}{\partial x^2}+\frac{\partial^2}{\partial y^2}\right)+iky\frac{\partial}{\partial x}.$$

A computation shows that

$$\Delta_k=-L_{k+2}R_k-\frac{k}{2}\left(1+\frac{k}{2}\right)=-R_{k-2}L_k+\frac{k}{2}\left(1-\frac{k}{2}\right),$$

where

$$R_k=iy\frac{\partial}{\partial x}+y\frac{\partial}{\partial y}+\frac{k}{2},\qquad L_k=-iy\frac{\partial}{\partial x}+y\frac{\partial}{\partial y}-\frac{k}{2}.$$

Definition 2.8.8 For $f \in C^\infty(\mathbb{H})$ and $g = \left(\begin{smallmatrix} * & * \\ c & d \end{smallmatrix} \right) \in G = \mathrm{SL}_2(\mathbb{R})$ let

$$f\|_k g(z) = \left(\frac{cz+d}{|cz+d|} \right)^{-k} f(gz) = \left(\frac{c\bar{z}+d}{|cz+d|} \right)^{k} f(gz).$$

Lemma 2.8.9 *With $f \in C^\infty(\mathbb{H})$ and $g \in G = \mathrm{SL}_2(\mathbb{R})$ we have*

$$(R_k f)\|_{k+2} g = R_k(f\|_k g), \qquad (L_k f)\|_{k-2} g = L_k(f\|_k g)$$

and

$$(\Delta_k f)\|_k g = \Delta_k(f\|_k g).$$

Proof A direct computation verifies the first two identities. The third then follows. Alternatively, one waits until the next section, where a Lie-theoretic and more structural proof is given. □

Differential operators are naturally defined on infinite-dimensional spaces like $C^\infty(\mathbb{R})$. These are not Hilbert spaces, but one can define differential operators on dense subspaces of natural Hilbert spaces. This motivates the next definition.

Definition 2.8.10 Let H be a Hilbert space. By an *operator* on H we mean a pair (D_T, T), where $D_T \subset H$ is a linear subspace of H and $T : D_T \to H$ is a linear map. The space D_T is the *domain* of the operator. The operator is said to be *densely defined* if D_T is dense in H. The operator is called a *closed operator* if its graph $G(T) = \{(h, T(h)) : h \in D_T\}$ is a closed subset of $H \times H$.

An operator T is called *symmetric* if

$$\langle T(v), w \rangle = \langle v, T(w) \rangle$$

holds for all $v, w \in D_T$.

Given a densely defined operator T on H we define its *adjoint operator* T^* as follows. Firstly the domain D_{T^*} is defined to be the set of all $v \in H$, for which the map $w \mapsto \langle Tw, v \rangle$ is a bounded linear map on D_T. As D_T is dense, this map extends uniquely to a continuous linear map on H. By the Riesz Representation Theorem there exists a uniquely determined vector $T^* v \in H$, such that $\langle Tw, v \rangle = \langle w, T^* v \rangle$ holds for every $w \in D_T$. It is easy to see that the so-defined map $T^* : D_{T^*} \to H$ is linear. If the domain D_{T^*} is dense, one can show that the adjoint operator T^* is closed.

An operator T is called *self-adjoint* if $D_{T^*} = D_T$ and $T^* = T$. We have

$$T \text{ self-adjoint} \quad \Rightarrow \quad T \text{ closed and symmetric},$$

but the converse is false in general, as the following example shows.

Example 2.8.11 Let $H = L^2([0,1])$ and let D_T be the set of all continuous functions f on $[0,1]$ of the form $f(x) = \int_0^x f'(t)\, dt = \langle f', \mathbf{1}_{[0,x]} \rangle$ for some

$f' \in L^2(0, 1)$ with $f' \perp \mathbf{1}_{[0,1]}$. Then f is uniquely determined by f. For every $f \in D_T$ one has $f(0) = 0 = f(1)$. Let T be the operator with domain D_T given by

$$T(f) = f'.$$

Since $C([0, 1])$ is dense in H, for every $f \in D_T$ there is a sequence of continuously differentiable functions f_j with $f_j \to f$ and $Tf_j \to Tf$. Using integration by parts we get

$$\langle Tf, g \rangle = \langle f, Tg \rangle$$

for all $f, g \in D_T$. This means that T is indeed symmetric. It is also closed, since for every sequence $f_j \in D_T$ with $f_j \to f$ and $Tf_j \to g$ we have $f \in D_T$ and $g = Tf$. It remains to show that $T^* \neq T$. The constant function 1, for example, lies in D_{T^*}, but not in D_T. Furthermore, the adjoint operator T^* is not symmetric.

If H is finite-dimensional, then every densely defined operator T is defined on all of H, as the only dense subspace is H itself.

We recall from linear algebra:

Theorem 2.8.12 (Spectral theorem) *Let $T : H \to H$ be a self-adjoint operator on a finite-dimensional Hilbert space H. Then H is a direct sum of eigenspaces,*

$$H = \bigoplus_{\lambda \in \mathbb{R}} \mathrm{Eig}(T, \lambda),$$

where

$$\mathrm{Eig}(T, \lambda) = \{v \in H : T(v) = \lambda v\}.$$

The proof is part of a linear algebra lecture. If H is infinite-dimensional, there is also a spectral theorem for self-adjoint operators. However, the space is not a direct sum in general, but a so-called direct integral of eigenspaces. We shall come back to this later.

Definition 2.8.13 The *support* of a function $f : X \to \mathbb{C}$ on a topological space X is the closure of the set $\{x \in X : f(x) \neq 0\}$. By $C_c(X)$ we denote the set of all continuous functions of compact support.

As usual, we denote by $L^2(\mathbb{H})$ the space of all measurable functions $f : \mathbb{H} \to \mathbb{C}$ such that $\int_{\mathbb{H}} |f(z)|^2 \, d\mu(z) < \infty$ modulo the subspace of all functions vanishing outside a set of measure zero. The measure μ is the invariant measure $\frac{dx\, dy}{y^2}$. The space $D = C_c^\infty(\mathbb{H})$ of all infinitely differentiable functions on $\mathbb{H}$ of compact support is a dense subspace on which the operator Δ_k is defined.

Proposition 2.8.14 *The operator Δ_k with domain $C_c^\infty(\mathbb{H})$ is a symmetric operator on the Hilbert space $H = L^2(\mathbb{H})$.*

Proof Let

$$\Delta^e = \frac{\partial^2}{\partial x^2} + \frac{\partial^2}{\partial y^2}$$

be the Euclidean Laplace operator and let d denote the exterior differential, which maps n-differential forms to $(n+1)$-forms. For $f, g \in C_c^\infty(\mathbb{H})$ we have

$$d\left(g\left(\frac{\partial f}{\partial x}dy - \frac{\partial f}{\partial y}dx\right) - f\left(\frac{\partial g}{\partial x}dy - \frac{\partial g}{\partial y}dx\right)\right) = \left(g\Delta^e f - f\Delta^e g\right)dx \wedge dy.$$

By Stokes's integral theorem we conclude

$$\int_{\mathbb{H}}\left(\overline{g}\Delta^e f - f\Delta^e \overline{g}\right)dx \wedge dy = 0,$$

so

$$\int_{\mathbb{H}}\overline{g}\Delta^e f \, dx \wedge dy = \int_{\mathbb{H}} f\Delta^e \overline{g}\, dx \wedge dy.$$

Write $T = \frac{i}{y}\frac{\partial}{\partial x}$. Integration by parts yields

$$\int_{\mathbb{H}}\left((Tf)\overline{g} - f(\overline{Tg})\right)dx \wedge dy = i\int_{\mathbb{H}}\frac{1}{y}\left(\frac{\partial f}{\partial x}\overline{g} + f\overline{\frac{\partial g}{\partial x}}\right)dx \wedge dy$$

$$= i\int_{\mathbb{H}} d\left(\frac{1}{y} f\overline{g}dy\right) = \int_{\partial\Omega}\frac{1}{y} f\overline{g}dy = 0,$$

where Ω is any relatively compact open subset of $\mathbb{H}$ with smooth boundary, containing the support of $f\overline{g}$. So

$$\int_{\mathbb{H}}(Tf)\overline{g}\, dx \wedge dy = \int_{\mathbb{H}} f(\overline{Tg})\, dx \wedge dy.$$

One has

$$\langle \Delta_k f, g\rangle = \int_{\mathbb{H}}(\Delta_k f)\overline{g}\frac{dx \wedge dy}{y^2} = \int_{\mathbb{H}}\left(-\Delta^e f + kTf\right)\overline{g}\, dx \wedge dy.$$

Hence the operator Δ_k is symmetric. $\square$

Pick a discrete subgroup Γ of $SL_2(\mathbb{R})$. By invariance, the operator Δ_k preserves the set of all smooth functions f on $\mathbb{H}$, which satisfy $f\|_k\gamma = f$ for every $\gamma \in \Gamma$. Write $C^\infty(\Gamma\backslash\mathbb{H}, k)$ for the vector space of all these functions and $L^2(\Gamma\backslash\mathbb{H}, k)$ for the space of all measurable functions f on $\mathbb{H}$ with $f\|_k\gamma = f$ for every $\gamma \in \Gamma$ and

$$\|f^2\| \overset{\text{def}}{=} \int_{\Gamma\backslash\mathbb{H}}|f(z)|^2\frac{dx\, dy}{y^2} < \infty,$$

modulo the subspace of functions with $\|f\| = 0$. Note that the integral is well-defined, since the function $|f(z)|^2$ is invariant under Γ. Then $L^2(\Gamma\backslash\mathbb{H}, k)$ is a Hilbert space with inner product

$$\langle f, g\rangle = \int_{\Gamma\backslash\mathbb{H}} f(z)\overline{g(z)}\frac{dx \wedge dy}{y^2}.$$

For the rest of this section we assume that the topological space $\Gamma\backslash\mathbb{H}$ is *compact*. This is equivalent to the quotient $\Gamma\backslash\mathrm{SL}_2(\mathbb{R})$ being compact.

In that case one calls Γ a *cocompact subgroup* of $\mathrm{SL}_2(\mathbb{R})$. A subgroup of $\mathrm{SL}_2(\mathbb{Z})$ is never cocompact, because $\mathrm{SL}_2(\mathbb{Z})$ is not cocompact itself. Do cocompact groups exist at all? Yes they do, and we will show this, using some facts of complex analysis, topology and elementary number theory.

- We start with a concrete example. Pick two rationals $0 < p, q \in \mathbb{Q}$. The matrices

$$i = \begin{pmatrix} \sqrt{p} & \\ & -\sqrt{p} \end{pmatrix}, \qquad j = \begin{pmatrix} & \sqrt{q} \\ \sqrt{q} & \end{pmatrix}$$

generate a $\mathbb{Q}$-subalgebra M of $\mathrm{M}_2(\mathbb{R})$ with the relations

$$i^2 = p, \qquad j^2 = q, \qquad ij = -ji.$$

These relations imply that the vectors $1, i, j, ij$ form a basis of M over $\mathbb{Q}$, so M has dimension four over $\mathbb{Q}$. The algebra M is a special case of a *quaternion algebra*.

We now insist that p and q are prime numbers and that q is not quadratic modulo p, i.e. we assume that $q \not\equiv k^2 \bmod p$ for every number k modulo p. In that case one can show (Exercise 2.21), that M is a *division algebra*, which means that every $m \neq 0$ in M is invertible. The set

$$M_{\mathbb{Z}} = \mathbb{Z}1 \oplus \mathbb{Z}i \oplus \mathbb{Z}j \oplus \mathbb{Z}ij$$

is a subring. Let

$$\Gamma = \{\gamma \in M_{\mathbb{Z}} : \det(\gamma) = 1\}.$$

One can show that Γ is a discrete subgroup of $\mathrm{SL}_2(\mathbb{R})$, such that $\Gamma\backslash\mathbb{H}$ is compact.

- Let X be a Riemann surface of genus $g \geq 0$. Let $\tilde{X}$ be its universal covering and Γ its fundamental group, which we consider as a group of biholomorphic maps on $\tilde{X}$. Then there is a natural identification $\Gamma\backslash\tilde{X} \cong X$. The Riemann surface $\tilde{X}$ is simply connected and Γ acts on $\tilde{X}$ without fixed points. By the Riemann mapping theorem, there are the following possibilities:

(a) $\tilde{X} \cong \mathbb{P}_1(\mathbb{C}) = \widehat{\mathbb{C}}$ the Riemann number sphere,
(b) $\tilde{X} \cong \mathbb{C}$,
(c) $\tilde{X} \cong \mathbb{H}$.

In case (a) every biholomorphic map $\gamma : \tilde{X} \to \tilde{X}$ is a linear fractional $\gamma(z) = \frac{az+b}{cz+d}$ and every such transformation has at least one fixed point in $\widehat{\mathbb{C}}$, which means that $\Gamma = \{1\}$ and $X = \tilde{X} = \widehat{\mathbb{C}}$, so $g = 0$.

Case (b): A biholomorphic map on $\mathbb{C}$ is a linear fractional γ with $\gamma(\infty) = \infty$, so $\gamma(z) = az + b$. If $a \neq 1$, then γ has a fixed point given by $z_0 = b/(1-a)$. So Γ consists only of transformations of the form $\gamma(z) = z + b$. The set of all $b \in \mathbb{C}$ with $(z \mapsto z+b) \in \Gamma$ then is a lattice and X is topologically isomorphic to $\mathbb{R}^2/\mathbb{Z}^2$, so $g = 1$.

In case (c) the group Γ is a discrete cocompact subgroup of $\mathrm{SL}_2(\mathbb{R})/\pm 1$, as the latter is the group of all biholomorphic maps on $\mathbb{H}$. Every X as in (c) therefore gives a Γ as we need it. This still doesn't prove existence, but one can show that there are uncountably many such Γ, even modulo conjugation.

Definition 2.8.15 A *torsion element* of a group Γ is an element of finite order. A group Γ is said to be *torsion-free* if the neutral element 1 is the only torsion element.

Now let $\Gamma \subset SL_2(\mathbb{R})$ be a discrete cocompact subgroup. One can show that Γ always contains a torsion-free subgroup of finite index. Hence we do not lose too much if we restrict our attention to torsion-free groups Γ. The upper half plane $\mathbb{H}$ has a natural orientation $(\frac{\partial}{\partial x}, \frac{\partial}{\partial y})$. If you don't know the notion of an *orientation* on a manifold or Stokes's theorem, you may for example consult [Lee03]. You may, on the other hand, understand what follows also if you consider the next proposition as a definition of the set $C^\infty(\Gamma\backslash\mathbb{H})$.

Proposition 2.8.16 *If the group* $\Gamma \subset SL_2(\mathbb{R})$ *is discrete and torsion-free, then the topological space* $\Gamma\backslash\mathbb{H}$ *carries exactly one structure of a smooth manifold such that the map* $\mathbb{H} \to \Gamma\backslash\mathbb{H}$ *is smooth. In that case one has*

$$C^\infty(\Gamma\backslash\mathbb{H}) = C^\infty(\mathbb{H})^\Gamma.$$

The natural orientation on $\mathbb{H}$ *induces an orientation on* $\Gamma\backslash\mathbb{H}$, *so that* $\Gamma\backslash\mathbb{H}$ *is an oriented smooth manifold.*

Proof (Sketch) As Γ is torsion-free, one can show that the group Γ acts *discontinuously* on $\mathbb{H}$, which means that for every $z \in \mathbb{H}$ there exists an open neighborhood U, such that for every $\gamma \in \Gamma$ one has: $\gamma U \cap U \neq \emptyset \Rightarrow \gamma = 1$. This implies that the projection $p : \mathbb{H} \to \Gamma\backslash\mathbb{H}$ maps the open neighborhood U homeomorphically onto its image $p(U)$, so that $p|_U$ is a chart. The set of all these charts is an atlas for $\Gamma\backslash\mathbb{H}$. Since Γ acts by orientation-preserving maps, the orientation descends to the quotient $\Gamma\backslash\mathbb{H}$. $\square$

The smooth manifold $\Gamma\backslash\mathbb{H}$ being oriented, one can integrate differential forms. If ω is a differential form on $\Gamma\backslash\mathbb{H}$ and if $p : \mathbb{H} \to \Gamma\backslash\mathbb{H}$ is the canonical projection, then the pullback form $p^*\omega$ is a Γ-invariant form on $\mathbb{H}$.

Lemma 2.8.17 *Let* ω *be a* 1*-form on* $\Gamma\backslash\mathbb{H}$. *Then*

$$\int_{\Gamma\backslash\mathbb{H}} d\omega = 0.$$

Proof This follows from the theorem of Stokes, since $\Gamma\backslash\mathbb{H}$ is a compact manifold without boundary. $\square$

Definition 2.8.18 Let $C^\infty(\Gamma\backslash\mathbb{H}, k)$ denote the set of all smooth functions f on $\mathbb{H}$ with $f\|_k\gamma = f$ for every $\gamma \in \Gamma$.

Lemma 2.8.19

(a) *If $f \in C^\infty(\Gamma\backslash\mathbb{H}, k)$ and $g \in C^\infty(\Gamma\backslash\mathbb{H}, k')$, then $fg \in C^\infty(\Gamma\backslash\mathbb{H}, k + k')$.*
(b) *If $f \in C^\infty(\Gamma\backslash\mathbb{H}, k)$, then $\overline{f} \in C^\infty(\Gamma\backslash\mathbb{H}, -k)$.*
(c) *$C^\infty(\Gamma\backslash\mathbb{H}, 0) = C^\infty(\Gamma\backslash\mathbb{H})$.*

Proof A smooth function f on $\mathbb{H}$ lies in $C^\infty(\Gamma\backslash\mathbb{H}, k)$ if and only if for every $\gamma = \left(\begin{smallmatrix} * & * \\ c & d \end{smallmatrix}\right) \in \Gamma$ one has

$$f(\gamma z) = \left(\frac{cz + d}{|cz + d|}\right)^k f(z).$$

The claims follow.   □

Proposition 2.8.20 *The operator Δ_k with domain $C^\infty(\Gamma\backslash\mathbb{H}, k)$ is a symmetric operator on the Hilbert space $L^2(\Gamma\backslash\mathbb{H}, k)$.*

Proof Similar to the proof of Proposition 2.8.14.   □

The Spectral Problem of Δ_k Is it possible to decompose the Hilbert space $L^2(\Gamma\backslash\mathbb{H}, k)$ into a direct sum of eigenspaces? If this is the case, we say that Δ_k has a pure *eigenvalue spectrum*. In this case every $\phi \in L^2(\Gamma\backslash\mathbb{H}, k)$ can be written as a L^2-convergent sum

$$\phi = \sum_{\lambda \in \mathbb{R}} \phi_\lambda,$$

with $\Delta_k \phi_\lambda = \lambda \phi_\lambda$.

If Γ is not cocompact, one will not have such a sum decomposition. Instead there is a so-called *direct integral* of eigenspaces. This is generally true for self-adjoint operators. We will not properly define a direct integral here, but we give an example of such a spectral decomposition.

Example 2.8.21 Let V be the Hilbert space $L^2(\mathbb{R})$ and let $D = -\frac{\partial^2}{\partial x^2}$ with domain $C_c^\infty(\mathbb{R})$. Then D is symmetric and one can show that D has a self-adjoint extension.

The operator D has no eigenfunction in $L^2(\mathbb{R})$. For $y \in \mathbb{R}$ the function $e_y(x) = e^{2\pi i x y}$ is an eigenfunction for the eigenvalue $4\pi^2 y^2$, but this function does not belong to the space $L^2(\mathbb{R})$. Nevertheless, according to the theory of Fourier transformation, every $\phi \in L^2(\mathbb{R})$ can be written as an L^2-convergent integral

$$\phi = \int_{\mathbb{R}} \hat{\phi}(y) e_y \, dy.$$

2.9 Exercises and Remarks

Exercise 2.1 Show that for $\left(\begin{smallmatrix} a & b \\ c & d \end{smallmatrix}\right) \in GL_2(\mathbb{C})$ and $z \in \mathbb{C}$ the expressions $az + b$ and $cz + d$ cannot both be zero.

Exercise 2.2 Find all $\gamma \in \Gamma = SL_2(\mathbb{Z})$, which commute with

(a) $S = \begin{pmatrix} & -1 \\ 1 & \end{pmatrix}$,

(b) $T = \begin{pmatrix} 1 & 1 \\ & 1 \end{pmatrix}$,

(c) ST.

Exercise 2.3 Which point in the fundamental domain D is Γ-conjugate to

(a) $6 + \frac{1}{2}i$,

(b) $\frac{8+6i}{3+2i}$?

Exercise 2.4 Let $\Gamma = SL_2(\mathbb{Z})$ and let $N \in \mathbb{N}$. Show that the set $\Gamma_0(N)$ of all matrices $\begin{pmatrix} a & b \\ c & d \end{pmatrix} \in \Gamma$ with $c \equiv 0 \bmod N$ is a subgroup of Γ.
(Hint: consider the reduction map $SL_2(\mathbb{Z}) \to SL_2(\mathbb{Z}/N\mathbb{Z})$.)

Exercise 2.5 (Bruhat decomposition) Let $G = SL_2(\mathbb{R})$ and let B be the subgroup of upper triangular matrices. Show that

$$G = B \cup BSB, \qquad S = \begin{pmatrix} & -1 \\ 1 & \end{pmatrix},$$

where the union is disjoint.

Exercise 2.6 Show that the group $SL_2(\mathbb{R})$ is generated by all elements of the form $\begin{pmatrix} a & \\ & 1/a \end{pmatrix}$ with $a \in \mathbb{R}^\times$, $\begin{pmatrix} 1 & x \\ & 1 \end{pmatrix}$ with $x \in \mathbb{R}$ and $S = \begin{pmatrix} & -1 \\ 1 & \end{pmatrix}$.

Exercise 2.7 Carry out the proof of Lemma 2.8.1.

Exercise 2.8 Show, without using differential forms, that the measure $\frac{dx\,dy}{y^2}$ is invariant under the action of $SL_2(\mathbb{R})$.
(Hint: use the change of variables rule.)

Exercise 2.9 Show that $\overline{D}$ has finite measure under $\frac{dx\,dy}{y^2}$.

Exercise 2.10 Show that for every $g \in SL_2(\mathbb{R})$ with $g \neq \pm 1$ one has

$$|tr(g)| < 2 \quad \Leftrightarrow \quad g \text{ has a fixed point in } \mathbb{H}.$$

Exercise 2.11 The Ramanujan τ-function is defined by the Fourier expansion

$$\Delta(z) = (2\pi)^{12} \sum_{n=1}^{\infty} \tau(n)q^n, \qquad q = e^{2\pi iz}.$$

Show $\tau(n) = 8000((\sigma_3 \star \sigma_3) \star \sigma_3)(n) - 147(\sigma_5 \star \sigma_5)(n)$, where $f \star g$ is the Cauchy product of two sequences:

$$f \star g(n) = \sum_{k=0}^{n} f(k)g(n-k).$$

Here we put $\sigma_a(n) = \sum_{d|n} d^a$ for $n \geq 1$ and $\sigma_3(0) = \frac{1}{240}$ as well as $\sigma_5(0) = -\frac{1}{504}$.

Exercise 2.12 (Jacobi product formula) Show that for $0 < |q| < 1$ and $\tau \in \mathbb{C}^\times$ one has

$$\sum_{n=-\infty}^{\infty} q^{n^2} \tau^n = \prod_{n=1}^{\infty} (1 - q^{2n})(1 + q^{2n-1}\tau)(1 + q^{2n-1}\tau^{-1}).$$

This can be done in the following steps.

Let $\vartheta(z, w) = \sum_{n=-\infty}^{\infty} q^{n^2} \tau^n$, where $z \in \mathbb{H}$, $w \in \mathbb{C}$ and $q = e^{2\pi i z}$, $\tau = e^{2\pi i w}$. Let

$$P(z, w) = \prod_{n=1}^{\infty} (1 + q^{2n-1}\tau)(1 + q^{2n-1}\tau^{-1}).$$

(a) Show: $\vartheta(z, w + 2z) = (q\tau)^{-1}\vartheta(z, w)$ and $P(z, w + 2z) = (q\tau)^{-1}P(z, w)$.
(b) Show that for fixed z the function $f(w) = \vartheta(z, w)/P(z, w)$ is constant.
   (Hint: show that f is entire and periodic for the lattice $\Lambda(1, 2z)$.)
(c) Show that for the function $\phi(q) = \vartheta(z, w)/P(z, w)$ one has

$$\phi(q) = \prod_{n=1}^{\infty} (1 - q^{2n}).$$

(Hint: show that $\vartheta(4z, 1/2) = \vartheta(z, 1/4)$ and

$$P\left(4z, \frac{1}{2}\right) \Big/ P\left(z, \frac{1}{4}\right) = \prod_{n=1}^{\infty} (1 - q^{4n-2})(1 - q^{8n-4}).$$

Therefore $\phi(q) = \frac{P(4z, \frac{1}{2})}{P(z, \frac{1}{4})}\phi(q^4)$. Now show that $\phi(q) \to 1$ for $q \to 0$.)

Exercise 2.13 Show that the L-series $L(f, s) = \sum_{n \geq 1} a_n n^{-s}$ also possesses an analytic continuation if $f \in M_{2k}$, $f(z) = \sum_{n \geq 0} a_n q^n$ is not a cusp form. It is not necessarily entire, but meromorphic on $\mathbb{C}$. Where are the poles?

Exercise 2.14 Let $f \in M_k$ with $k \geq 4$. Assume that f is not a cusp form. Show that f is a normalized Hecke eigenform if and only if

$$f = \frac{(k-1)!}{2(2\pi i)^k} G_k.$$

Exercise 2.15 For $f, g \in M_{2k}$ let

$$\langle f, g \rangle_{\text{Pet}} = \int_{\Gamma \backslash \mathbb{H}} f(z)\overline{g(z)} y^{2k} \frac{dx\,dy}{y^2}.$$

Show that the integrand is invariant under Γ and that the integral converges if at least one of the functions f, g is a cusp form. Show that for $k \geq 2$ the Eisenstein series G_{2k} is perpendicular to all cusp forms.

Exercise 2.16 Show that the map $\Gamma(1) \to SL_2(\mathbb{Z}/N\mathbb{Z})$ is surjective.
(Hint: use the Elementary Divisor Theorem to reduce to the case of a diagonal matrix of the form $\left(\begin{smallmatrix} a & \\ & an \end{smallmatrix}\right)$. Vary n modulo N and consider matrices of the form $\left(\begin{smallmatrix} a & Nx \\ N & an \end{smallmatrix}\right)$. Recall that a and N are coprime.)

Exercise 2.17 Let $\Gamma \subset \Gamma(1)$ be a congruence subgroup and let Σ be a normal subgroup of finite index in Γ. Show that the finite group Γ/Σ acts on $\mathcal{M}_k(\Sigma)$ by $f \mapsto f|\gamma$. Show that this action is unitary with respect to the Petersson inner product.

Exercise 2.18 Let $\Gamma_0(N)$ be the group of all $\left(\begin{smallmatrix} a & b \\ c & d \end{smallmatrix}\right) \in \Gamma(1)$ with $c \equiv 0 \bmod(N)$ and let $\Gamma_1(N)$ be the subgroup of all $\left(\begin{smallmatrix} a & b \\ c & d \end{smallmatrix}\right) \in \Gamma_0(N)$ with $a \equiv d \equiv 1 \bmod(N)$. Let χ be a *Dirichlet character modulo N*, i.e. a group homomorphism $\chi : (\mathbb{Z}/N\mathbb{Z})^\times \to \mathbb{C}^\times$. Let $S_k(\Gamma_0(N), \chi)$ be the set of all $f \in S_k(\Gamma_1(N))$ with $f|\gamma = \chi(d)f$ for every $\gamma = \left(\begin{smallmatrix} a & b \\ c & d \end{smallmatrix}\right) \in \Gamma_0(N)$. Show

$$S_k\big(\Gamma_1(N)\big) = \bigoplus_\chi S_k\big(\Gamma_0(N), \chi\big),$$

where the sum is orthogonal with respect to the Petersson inner product.

Exercise 2.19 Let $f \in \mathcal{M}_k(\Gamma)$ for a congruence subgroup Γ. Show that there is a $\alpha \in GL_2(\mathbb{Q})^+$ and a $N \in \mathbb{N}$, such that $f|\alpha \in \mathcal{M}_k(\Gamma_1(N))$.

Let S be the finite set of all primes which divide N and let $\mathbb{Z}_S$ be the localization of $\mathbb{Z}$ in S, i.e. the set of all rational numbers a/b, where the denominator b is coprime to N. Then $N\mathbb{Z}_S$ is an ideal of $\mathbb{Z}_S$ and $\mathbb{Z}_S/N\mathbb{Z}_S \cong \mathbb{Z}/N\mathbb{Z}$. Let $G_0(N)$ be the subgroup of $GL_2(\mathbb{Z}_S)$ consisting of all matrices $\left(\begin{smallmatrix} a & b \\ c & d \end{smallmatrix}\right)$ with positive determinant such that $c \in N\mathbb{Z}_S$. Show that a set of representatives of $\Gamma_0(N)\backslash G_0(N)/\Gamma_0(N)$ is given by the set of all matrices $\left(\begin{smallmatrix} an & \\ & a \end{smallmatrix}\right)$, where $a \in \mathbb{Z}_S$ is positive and $n \in \mathbb{N}$ is coprime to N.

Exercise 2.20 Let f be a continuous function on an open set $D \subset \mathbb{C}^2$. Suppose that for every $z_0 \in \mathbb{C}$ the function $w \mapsto f(z_0, w)$ is holomorphic where it is defined, and that for every $w_0 \in \mathbb{C}$ the function $z \mapsto f(z, w_0)$ is holomorphic where it is defined. So f is holomorphic in each argument separately. Show that f is representable as a power series in both arguments simultaneously. This means that for every $(z_0, w_0) \in D$ there is an open neighborhood in which

$$f(z, w) = \sum_{n=0}^\infty \sum_{m=0}^\infty a_{m,n}(z - z_0)^n (w - w_0)^m$$

holds. Here $a_{m,n}$ are complex numbers and the double series converges absolutely. Conclude that f is a smooth function.
(Hint: it suffices to assume $(0,0) \in D$ and to show the power series expansion around that point. Let K, L be two discs around zero in $\mathbb{C}$ such that $K \times L \subset D$.

Let z be in the interior of K and w in the interior of L. Apply Cauchy's integral formula in both arguments to get

$$f(z, w) = \frac{1}{2\pi i} \int_{\partial K} \frac{f(\xi, w)}{\xi - z} \, d\xi = \frac{1}{-4\pi^2} \int_{\partial K} \int_{\partial L} \frac{f(\xi, \zeta)}{(\xi - z)(\zeta - w)} \, d\zeta \, d\xi.$$

Write

$$\frac{f(\xi, \zeta)}{(\xi - z)(\zeta - w)} = \frac{1}{\xi \zeta} \frac{f(\xi, \zeta)}{(1 - z/\xi)(1 - w/\zeta)} = \frac{1}{\xi \zeta} f(\xi, \zeta) \sum_{n=0}^{\infty} \sum_{m=0}^{\infty} \frac{z^n}{\xi^n} \frac{w^m}{\zeta^m}.)$$

Exercise 2.21 Let $0 < p, q \in \mathbb{Q}$. The matrices

$$i = \begin{pmatrix} \sqrt{p} & \\ & -\sqrt{p} \end{pmatrix}, \qquad j = \begin{pmatrix} & \sqrt{q} \\ \sqrt{q} & \end{pmatrix}$$

generate a $\mathbb{Q}$-subalgebra M of $M_2(\mathbb{R})$ satisfying the relations

$$i^2 = p, \qquad j^2 = q, \qquad ij = -ji.$$

These relations imply that the vectors $1, i, j, ij$ for a basis of M over $\mathbb{Q}$, so M is four-dimensional. Such an algebra is called a *quaternion algebra*. Show that M is a division algebra if p and q are prime numbers such that q is not a quadratic remainder modulo p.

Remarks A *homothety* on $\mathbb{C}$ is a map of the form $z \mapsto \lambda z$, where $\lambda \in \mathbb{C}^{\times}$. The bijection given in Theorem 2.1.5, $\Gamma \backslash \mathbb{H} \to \text{LATT}/\mathbb{C}^{\times}$, shows that $\Gamma \backslash \mathbb{H}$ is the *moduli space* of the lattices modulo homothethies. Generally a moduli space is a mathematical object, whose points classify other mathematical objects. If you want to learn about moduli spaces, you should read [HM98] and [KM85].

The j-function is a bijection from $\Gamma \backslash \mathbb{H}$ to $\mathbb{C}$. If one adds the Γ-orbit of the point ∞, one gets a bijection to $\widehat{\mathbb{C}} = \mathbb{C} \cup \{\infty\} = \mathbb{P}^1(\mathbb{C})$. More generally one compactifies $\Gamma \backslash \mathbb{H}$ for a congruence subgroup Γ by adding the cusps of a fundamental domain. The so-defined compact space has the structure of an algebraic curve which can be realized in some projective space.

Instead of congruence subgroups, one can also look at arbitrary subgroups of finite index in $\text{SL}_2(\mathbb{Z})$ or even more general at discrete subgroups Γ of $G = \text{SL}_2(\mathbb{R})$ of finite covolume; see [Iwa02]. In this book we will concentrate on congruence groups, as they are most important to number theory.

Non-holomorphic Eisenstein series give the continuous contribution in the spectral decomposition of the Maaß wave forms; see [Iwa02]. In the proof of this, the Rankin–Selberg method is crucial. In this book, we mentioned this method also for another reason. The Rankin–Selberg convolution is the first example of an automorphic L-function, which does not belong to the group GL_2, but rather to GL_4. This is seen by the order of the polynomials in the Euler product. The *Langlands conjectures* imply roughly that every L-function, that shows up in number theory, is automorphic. This can only hold if one considers automorphic L-functions from all groups GL_n; see [BCdS$^+$03].

Chapter 3
Representations of $SL_2(\mathbb{R})$

In this chapter we will show that cusp forms can be viewed as representation vectors for Lie groups. This opens the way for the application of representation theory to automorphic forms.

3.1 Haar Measures and Decompositions

The group $G = SL_2(\mathbb{R})$ acts on the upper half plane by linear fractionals $\left(\begin{smallmatrix} a & b \\ c & d \end{smallmatrix}\right) z = \frac{az+b}{cz+d}$. Let $g = \left(\begin{smallmatrix} a & b \\ c & d \end{smallmatrix}\right) \in G$ be in the stabilizer of $i \in \mathbb{H}$, i.e. $gi = i$. Then one gets $\frac{ai+b}{ci+d} = i$ or $ai + b = -c + di$, which gives $a = d$ and $b = -c$ by comparison of real and imaginary parts. It follows that the stabilizer of $i \in \mathbb{H}$ is the group

$$K = SO(2) = \left\{ \begin{pmatrix} a & -b \\ b & a \end{pmatrix} : a, b \in \mathbb{R}, \ a^2 + b^2 = 1 \right\}.$$

This is the group of all matrices of the form

$$\begin{pmatrix} \cos\varphi & -\sin\varphi \\ \sin\varphi & \cos\varphi \end{pmatrix} \quad \text{for } \varphi \in \mathbb{R}.$$

The action of G on $\mathbb{H}$ is transitive (i.e. there is only one orbit), because for $z = x + iy \in \mathbb{H}$ one has

$$z = \begin{pmatrix} \sqrt{y} & \frac{x}{\sqrt{y}} \\ 0 & \frac{1}{\sqrt{y}} \end{pmatrix} i.$$

So the map

$$G/K \to \mathbb{H},$$

$$gK \mapsto gi$$

is a bijection which identifies the upper half plane with the quotient G/K.

A. Deitmar, *Automorphic Forms*, Universitext,
DOI 10.1007/978-1-4471-4435-9_3, © Springer-Verlag London 2013

The group G is a subset of $M_2(\mathbb{R}) \cong \mathbb{R}^4$, and thus inherits a natural topology. A sequence $\begin{pmatrix} a_n & b_n \\ c_n & d_n \end{pmatrix}$ converges in G to an element $\begin{pmatrix} a & b \\ c & d \end{pmatrix}$ if and only if $a_n \to a$, $b_n \to b$, $c_n \to c$ and $d_n \to d$. The action on $\mathbb{H}$ is continuous in that the map

$$G \times \mathbb{H} \to \mathbb{H}$$

$$(g, z) \mapsto gz$$

is a continuous map. (See Exercise 3.2.)

Theorem 3.1.1 (Iwasawa decomposition) *Let A be the group of all diagonal matrices in G with positive entries. Let N be the group of all matrices of the form $\begin{pmatrix} 1 & s \\ 0 & 1 \end{pmatrix}$ with $s \in \mathbb{R}$. Then $G = ANK$. More precisely, the map*

$$\psi : A \times N \times K \to G,$$

$$(a, n, k) \mapsto ank$$

is a homeomorphism.

Proof Let $g \in G$ and let $gi = x + yi$. With

$$a = \begin{pmatrix} \sqrt{y} & \\ & 1/\sqrt{y} \end{pmatrix} \quad \text{and} \quad n = \begin{pmatrix} 1 & x/y \\ & 1 \end{pmatrix},$$

one has $gi = ani$ and so $g^{-1}an$ lies in K, i.e. there is $k \in K$ with $g = ank$. By means of the formula $gi = \frac{ai+b}{ci+d}$ we can determine the inverse map as follows. Let

$$\phi : G \to A \times N \times K$$

be given by $\phi(g) = (\underline{a}(g), \underline{n}(g), \underline{k}(g))$, where

$$\underline{a}\begin{pmatrix} a & b \\ c & d \end{pmatrix} = \begin{pmatrix} \frac{1}{\sqrt{c^2+d^2}} & \\ & \sqrt{c^2+d^2} \end{pmatrix},$$

$$\underline{n}\begin{pmatrix} a & b \\ c & d \end{pmatrix} = \begin{pmatrix} 1 & ac+bd \\ & 1 \end{pmatrix},$$

$$\underline{k}\begin{pmatrix} a & b \\ c & d \end{pmatrix} = \frac{1}{\sqrt{c^2+d^2}} \begin{pmatrix} d & -c \\ c & d \end{pmatrix}.$$

It is an easy computation to show that $\phi\psi = \mathrm{Id}$ and $\psi\phi = \mathrm{Id}$. $\square$

The notation $\underline{a}(g)$, $\underline{n}(g)$ and $\underline{k}(g)$ will be used in the sequel. For $x, t, \theta \in \mathbb{R}$ we write

$$a_t \overset{\text{def}}{=} \begin{pmatrix} e^t & \\ & e^{-t} \end{pmatrix} \in A$$

$$n_x \overset{\text{def}}{=} \begin{pmatrix} 1 & x \\ & 1 \end{pmatrix} \in N$$

$$k_\theta \overset{\text{def}}{=} \begin{pmatrix} \cos 2\pi\theta & -\sin 2\pi\theta \\ \sin 2\pi\theta & \cos 2\pi\theta \end{pmatrix} \in K.$$

Definition 3.1.2 Let $k \in \mathbb{N}_0$. A function $f : G \to \mathbb{C}$ is said to be k-times *continuously differentiable* if the map $\mathbb{R}^3 \to \mathbb{C}$, given by

$$(t, x, \theta) \mapsto f(a_t n_x k_\theta)$$

is k-times continuously differentiable. The map is called *smooth* if it is infinitely often continuously differentiable. The set of smooth functions on G is written as $C^\infty(G)$. The set of smooth functions of compact support is written as $C_c^\infty(G)$.

   The group G is a locally compact Hausdorff space, since G is a closed subset of $\mathbb{R}^4$. Further, G is a *topological group*, i.e. the group operations,

$$\begin{array}{cc} G \times G \to G & G \to G \\ (x, y) \mapsto xy, & x \mapsto x^{-1}, \end{array}$$

are continuous maps. A topological group, which is locally compact and Hausdorff, is called a *locally compact group*.

Examples 3.1.3

- An arbitrary group G becomes a locally compact group, if we equip it with the discrete topology (i.e. every set is open). In that case, one speaks of a *discrete group*.
- The groups $\mathrm{GL}_2(\mathbb{R})$ and $\mathrm{SL}_2(\mathbb{R})$ are locally compact groups when equipped with the subspace topology of $\mathbb{R}^4$.
- We will construct further examples using p-adic numbers and adeles later.

Lemma 3.1.4 *Let G be a locally compact group. Every function $f \in C_c(G)$ is uniformly continuous, which means that for every $\varepsilon > 0$ there exists a neighborhood U of the unit element of G, such that for $x, y \in G$ with $x^{-1}y \in U$ or $yx^{-1} \in U$ one has $|f(x) - f(y)| < \varepsilon$.*

Proof We show only the case $x^{-1}y \in U$, as the other case is analogous. Let S be the support of f. Choose $\varepsilon > 0$ and a compact neighborhood V of the unit in G. Since f is continuous, for given $x \in G$ there exists an open unit neighborhood $V_x \subset V$ such that $y \in xV_x \Rightarrow |f(x) - f(y)| < \varepsilon/2$. As the group multiplication is continuous, there is an open unit neighborhood U_x such that $U_x^2 \subset V_x$. The sets xU_x, with $x \in SV$, form an open covering of the compact set SV. So there exist $x_1, \ldots, x_n \in KV$ such that $SV \subset x_1U_1 \cup \cdots \cup x_nU_n$, where we have written U_j for U_{x_j}. Let $U = U_1 \cap \cdots \cap U_n$. Then U is an open unit neighborhood. Let $x, y \in G$ with $x^{-1}y \in U$. If $x \notin SV$, then $y \notin S$, since $x \in yU^{-1} = yU \subset yV$. In that case we conclude $f(x) =$

$f(y) = 0$. So let $x \in SV$. Then there is j with $x \in x_j U_j$, so $y \in x_j U_j U \subset x_j V_j$. We get

$$\left| f(x) - f(y) \right| \le \left| f(x) - f(x_j) \right| + \left| f(x_j) - f(y) \right| < \frac{\varepsilon}{2} + \frac{\varepsilon}{2} = \varepsilon$$

as claimed.   □

Definition 3.1.5 A measure μ, defined on the Borel σ-algebra of a locally compact Hausdorff space X, is called a *Radon measure* if the following conditions are met:

(a) $\mu(K) < \infty$ for every compact subset K;
(b) $\mu(A) = \inf_{U \supset A} \mu(U)$, where the infimum runs over all open subsets with $U \supset A$ and $A \subset X$ is measurable; this property is called *outer regularity*;
(c) $\mu(A) = \sup_{K \subset A} \mu(K)$, where the supremum runs over all compact subsets $K \subset X$, which lie in A, and A is open or of finite measure; this property is called *inner regularity*.

Examples 3.1.6

• If X is discrete, i.e. every set is open, then the counting measure

$$\mu(A) = \begin{cases} n & \text{if } A \text{ is finite with } n \text{ elements,} \\ \infty & \text{otherwise,} \end{cases}$$

is a Radon measure.
• If $X = \mathbb{R}$, then the Lebesgue measure is a Radon measure.

Proposition 3.1.7 *Let μ be a Radon measure on the locally compact Hausdorff space X. For every $1 \le p \le \infty$ the space $C_c(X)$ is dense in $L^p(X)$.*

Proof Let $f \in L^p(X)$. We want to write f as a limit of a sequence in $C_c(X)$. We can decompose f into real and imaginary parts and these again into positive and negative parts. If we can write all these as limits of $C_c(X)$ functions, then we can do so for f. So it suffices to assume $f \ge 0$. One can write f as the point-wise limit of a monotonically increasing sequence of Lebesgue step functions. It therefore suffices to assume that f is a Lebesgue step-function. By linearity it suffices to consider the case $f = \mathbf{1}_A$ for a set A of finite measure. By outer regularity there exists a sequence U_n of open sets, $U_n \supset U_{n+1} \supset A$, such that $\mu(U_n) \to \mu(A)$, i.e. $\|\mathbf{1}_A - \mathbf{1}_{U_n}\|_p \to 0$. So we can assume that A is open. By inner regularity there exists a sequence of compact sets $K_n \subset K_{n+1} \subset A$ such that $\|\mathbf{1}_A - \mathbf{1}_{K_n}\|_p \to 0$. By the lemma of Urysohn, A.3.2, there is, for each n, a function $\varphi_n \in C_c(X)$ with $\mathbf{1}_{K_n} \le \varphi_n \le \mathbf{1}_A$. Then the sequence φ_n converges to $\mathbf{1}_A$ in the L^p-norm.   □

If μ is a Radon measure on X, then the integral $I = I_\mu : \varphi \mapsto \int_X \varphi(x) \, d\mu(x)$ is a linear map $C_c(X) \to \mathbb{C}$, which is positive in the sense that $\varphi \ge 0 \Rightarrow I(\varphi) \ge 0$. The *Representation Theorem of Riesz* says that the map $\mu \mapsto I_\mu$ is a bijection between the set of all Radon measures and the set of all positive linear functionals on $C_c(X)$.

Theorem 3.1.8 *Let G be a locally compact group. Then there exists a Radon measure $\mu \neq 0$ on the Borel σ-algebra, which is left-invariant, i.e. one has $\mu(xA) = \mu(A)$ for every $x \in G$ and every measurable set $A \subset G$. This measure μ is uniquely determined up to scaling by positive numbers. It is called a* Haar measure *of G.*

Every open set has positive Haar measure, and every compact set has finite Haar measure. The Haar measure is finite if and only if the group is compact.

Proof One finds a proof for instance in each of the books [DE09, Str06]. □

Note that a Radon measure μ on a locally compact group G is a Haar measure if and only if

$$\int_G f(y)\,d\mu(y) = \int_G f(xy)\,d\mu(y)$$

holds for every $f \in L^1(G)$ and every $x \in G$.

Examples 3.1.9

- If G is a discrete group, then the counting measure is a Haar measure.
- For the group $G = (\mathbb{R}, +)$, the Lebesgue measure dx is a Radon measure.
- A Haar measure on the group $G = (\mathbb{R}^\times, \times)$ is given by $\frac{dx}{x}$.
- A Haar measure for the group $G = GL_2(\mathbb{R})$ is given by

$$\frac{dx\,dy\,dz\,dw}{|xw - yz|^2},$$

where the coordinates are the entries of the matrix $\left(\begin{smallmatrix} x & y \\ z & w \end{smallmatrix}\right) \in G$. This is easily shown by change of variables.

Convention For simplicity we will write dx instead of $d\mu(x)$, when integrating over a Haar measure μ, where we always assume a fixed choice of Haar measure. So we write

$$\int_G f(x)\,dx$$

instead of $\int_G f(x)\,d\mu(x)$. The volume $\mu(A)$ of a measurable set $A \subset G$ will be denoted as

$$\mathrm{vol}(A) \quad \text{or} \quad \mathrm{vol}_{dx}(A).$$

Definition 3.1.10 Let G be a locally compact group with Haar measure dx. For $f, g \in L^1(G)$ the *convolution* is defined by

$$f * g(x) = \int_G f(y)g(y^{-1}x)\,dy.$$

Proposition 3.1.11 *Let G be a locally compact group with Haar measure dx. For* $f, g \in L^1(G)$ *the convolution integral* $f * g(x)$ *exists for x outside a set of measure zero and the so-defined function* $f * g$ *lies in* $L^1(G)$. *More precisely, one has* $\|f * g\|_1 \le \|f\|_1 \|g\|_1$. *Equipped with the convolution product, the space* $L^1(G)$ *is an algebra over* ℂ, *i.e. for all* $f, g, h \in L^1(G)$ *one has*

$$(f * g) * h = f * (g * h), \quad f * (g + h) = f * g + f * h, \quad (f + g) * h = f * h + g * h$$

and for every $\lambda \in ℂ$ *the equalities*

$$\lambda(f * g) = (\lambda f) * g = f * (\lambda g)$$

hold.

Proof The proposition follows from simple applications of the invariance of Haar measure. As an example, we will prove the first assertion. Let $f, g \in L^1(G)$. Then we compute, formally at first,

$$\|f * g\|_1 = \int_G |f * g(x)| \, dx = \int_G \left| \int_G f(y) g(y^{-1} x) \, dy \right| dx$$

$$\le \int_G \int_G |f(y) g(y^{-1} x)| \, dy \, dx = \int_G \int_G |f(y) g(y^{-1} x)| \, dx \, dy$$

$$= \int_G \int_G |f(y) g(x)| \, dx \, dy = \|g\|_1 \|f\|_1.$$

The existence of the integrals on the right-hand side implies their existence on the left. The existence of $f * g(x)$ outside a set of measure zero follows from Fubini's theorem, A.2.3. □

3.1.1 The Modular Function

A Haar measure μ needn't be right-invariant as well, i.e. in general one has $\mu(Ax) \ne \mu(A)$. For given $x \in G$ let

$$\mu_x(A) = \mu(Ax).$$

Then μ_x is a left-invariant measure, as

$$\mu_x(yA) = \mu(yAx) = \mu(Ax) = \mu_x(A).$$

The uniqueness of Haar measures implies that there is a unique number $\Delta(x) > 0$ with $\mu_x = \Delta(x)\mu$. The resulting function $\Delta = \Delta_G : G \to (0, \infty)$ is called the *modular function* of G. The modular function is a group homomorphism of G to the multiplicative group $\mathbb{R}^\times_{>0}$, since

$$\Delta(xy)\mu(A) = \mu(Axy) = \Delta(y)\mu(Ax) = \Delta(y)\Delta(x)\mu(A).$$

Further one shows that Δ is a continuous function (see [DE09], Chap. 1).

Example 3.1.12 Let B denote the group of all real matrices of the form $\begin{pmatrix} 1 & x \\ 0 & y \end{pmatrix}$ with $y \neq 0$. Then the modular function of B is $\Delta\begin{pmatrix} 1 & x \\ 0 & y \end{pmatrix} = |y|$ (see Exercise 3.4).

Definition 3.1.13 Let the set X be equipped with a σ-algebra. Assume the group G acts on X by measurable maps. A measure μ on X is called an *invariant measure* if for every measurable set $A \subset X$ and every $g \in G$ one has $\mu(gA) = \mu(A)$. We will examine the case when X is the coset space G/H of a subgroup H of G.

Lemma 3.1.14 *Let G be a locally compact group and let H be a closed subgroup. Then H is again a locally compact group and the quotient space G/H, equipped with the quotient topology, is a locally compact Hausdorff space.*

Proof [DE09].   □

Theorem 3.1.15

(a) *Let $H \subset G$ be a closed subgroup of the locally compact group G. On the locally compact space G/H there exists a non-trivial, G-invariant Radon measure if and only if*

$$\Delta_G(h) = \Delta_H(h)$$

holds for every $h \in H$. If it exists, the invariant measure is unique up to scaling. Given Haar measures on G and H it can be normalized such that for every $f \in L^1(G)$ the integral formula

$$\int_G f(x)\,dx = \int_{G/H} \int_H f(yh)\,dh\,dy$$

holds.

(b) *For $y \in G$ and $f \in L^1(G)$ one has*

$$\int_G f(xy)\,dx = \Delta(y^{-1}) \int_G f(x)\,dx.$$

(c) *The equation*

$$\int_G f(x^{-1})\,\Delta(x^{-1})\,dx = \int_G f(x)\,dx$$

holds for every $f \in L^1(G)$.

(d) *If $H \subset G$ is a closed subgroup and $K \subset G$ a compact subgroup such that $G = HK$, then one can normalize the Haar measures of G, H, K in such a way that for every integrable function f the identity*

$$\int_G f(x)\,dx = \int_H \int_K f(hk)\,dk\,dh$$

holds.

Proof [DE09, Str06].   □

Example 3.1.16 The upper half plane $\mathbb{H}$ can be identified with the quotient $G/K = SL_2(\mathbb{R})/SO(2)$. The invariant measure $d\mu = \frac{dx\,dy}{y^2}$ is a measure as in the theorem. Therefore the G-invariance determines the measure μ uniquely up to scaling.

A locally compact group G is called *unimodular* if $\Delta \equiv 1$.

Examples 3.1.17

- If G is abelian, then G is unimodular.
- If G is compact, then G is unimodular, since the image $\Delta(G)$ is a compact subgroup of $\mathbb{R}^\times_{>0}$, so $\Delta(G) = \{1\}$.
- If G is a discrete group, then G is unimodular, since the cardinality of a set $A \subset G$ equals the cardinality of Ax for every $x \in G$.

Proposition 3.1.18 *The group $G = SL_2(\mathbb{R})$ is unimodular.*

Proof Let $\phi : G \to \mathbb{R}^\times_+$ be a continuous group homomorphism. We show that ϕ is trivial. Firstly, we have that $\phi(K)$ is a compact subgroup of $(0, \infty)$, but the latter group has only one compact subgroup, the trivial one. Hence it follows that $\phi(K) = 1$. The restriction of ϕ to A is a continuous group homomorphism, so there exists a real number $x \in \mathbb{R}$ such that $\phi(a_t) = e^{tx}$ for every $t \in \mathbb{R}$. Let $w = \left(\begin{smallmatrix} & -1 \\ 1 & \end{smallmatrix}\right)$. Then $wa_t w^{-1} = a_{-t}$, and therefore $e^{tx} = \phi(a_t) = \phi(wa_t w^{-1}) = e^{-tx}$ for every $t \in \mathbb{R}$. This means $x = 0$ and so $\phi(A) = 1$. Analogously we have $\phi(n_x) = e^{rx}$ for some $r \in \mathbb{R}$. By $a_t n_x a_t^{-1} = n_{e^{2t}x}$ it follows that $e^{rs} = e^{re^{2t}s}$ for every $t \in \mathbb{R}$, so $r = 0$ and $\phi(N) = 1$. By the Iwasawa decomposition we infer that $\phi(G) = \phi(ANK) = \phi(A)\phi(N)\phi(K) = 1$.   □

Let $g \in G = SL_2(\mathbb{R})$. We write $\underline{t}(g)$ for the unique $t \in \mathbb{R}$ with $\underline{a}(g) = a_t$, i.e. we have

$$\underline{a}(g) = a_{\underline{t}(g)}.$$

Theorem 3.1.19 (Iwasawa integral formula) *Let $G = SL_2(\mathbb{R})$ and $A, N, K \subset G$ as in Theorem 3.1.1. For any given Haar measures on three of the groups G, A, N, K there is a unique Haar measure on the fourth, such that for every $f \in L^1(G)$ the equality*

$$\int_G f(x)\,dx = \int_A \int_N \int_K f(ank)\,dk\,dn\,da$$

holds.

In the sequel, we will choose fixed Haar measures as follows. On the compact group K we choose the unique Haar measure of volume 1. On A we choose the measure $2dt$, where $t = \underline{t}(a)$, and on N we choose $\int_{\mathbb{R}} f(n_s)\,ds$. The factor of 2 in the measure on A was put in to grant compatibility with the measure $\frac{dx\,dy}{y^2}$ on the upper half plane.

Proof of the Theorem Let $B = AN$ be the subgroup of all upper triangular matrices with positive diagonal entries. It is easy to show that $db = da\,dn$ is a Haar measure on B and that B is not unimodular. Indeed, one has $\Delta_B(a_t) = e^{-2t}$, as the equation $a_t n_x a_s n_y = a_{t+s} n_{y+e^{-2s}x}$ yields. Let $\underline{b} : G \to B$ be the projection $\underline{b}(g) = \underline{a}(g)\underline{n}(g)$. The map $B \to G/K \cong \mathbb{H}$, sending b to bK, is a B-equivariant homeomorphism. Every G-invariant measure on $G/K \cong \mathbb{H}$ gives a Haar measure on B and the uniqueness of these measures implies that every B-invariant measure on G/K already is G-invariant. The formula $\int_G f(x)\,dx = \int_{G/K} \int_K f(xk)\,dk\,dx$ gives $\int_G f(x)\,dx = \int_B \int_K f(bk)\,dk\,db$. Since $db = da\,dn$, the integral formula follows. $\square$

3.2 Representations

We define the notion of a continuous representation of a topological group on a Banach space V.

Definition 3.2.1 Let G be a topological group. A *representation* of G is a pair (π, V) consisting of a Banach space V and a group homomorphism $\pi : G \to \mathrm{GL}(V)$, such that the map

$$G \times V \to V, \quad (g, v) \mapsto \pi(g)v$$

is continuous.

Examples 3.2.2

- Let $\chi : G \to \mathbb{C}^\times$ be a continuous group homomorphism, i.e. a so-called *quasi-character*. One can consider χ as a representation, since naturally, $\mathbb{C}^\times \cong \mathrm{GL}(\mathbb{C})$.
- Let $G = \mathrm{SL}_2(\mathbb{R})$. This group has a canonical representation on the Banach space $\mathbb{C}^2$, given by matrix multiplication.

Now assume that V is even a Hilbert space. A representation π of the group G on V is called a *unitary representation* if $\pi(g)$ is unitary for every $g \in G$. This means that π is unitary if and only if for every $g \in G$ and all $v, w \in V$ one has $\langle \pi(g)v, \pi(g)w \rangle = \langle v, w \rangle$.

Lemma 3.2.3 *A representation π of a topological group G on a Hilbert space V is unitary if and only if $\pi(g^{-1}) = \pi(g)^*$ holds for every $g \in G$.*

Proof An operator T is unitary if and only if it is invertible and satisfies $T^{-1} = T^*$. A representation π is a group homomorphism, so it satisfies $\pi(g^{-1}) = \pi(g)^{-1}$ for every $g \in G$. These two assertions give the claim.   □

Example 3.2.4 Let $\chi : G \to \mathbb{C}^\times$ be a quasi-character. As a representation, χ is unitary if and only if its image lies in the compact torus $\mathbb{T} = \{z \in \mathbb{C} : |z| = 1\}$. In this case we say that χ is a *character*.

We will next consider an important example. Let G be a locally compact group and let $x \in G$. On the Hilbert space $L^2(G)$, we define an operator L_x by

$$L_x\varphi(y) = \varphi(x^{-1}y), \quad \varphi \in L^2(G).$$

Lemma 3.2.5 *The map $x \mapsto L_x$ is a unitary representation of the locally compact group G. It is called the* left regular representation.

Proof We first show that L_x is a unitary operator. This is a consequence of the left-invariance of the Haar measure:

$$\langle L_x\varphi, L_x\psi \rangle = \int_G L_x\varphi(y)\overline{L_x\psi(y)}\,dy = \int_G \varphi(x^{-1}y)\overline{\psi(x^{-1}y)}\,dy$$

$$= \int_G \varphi(y)\overline{\psi(y)}\,dy = \langle \varphi, \psi \rangle.$$

Next we show that $x \mapsto L_x$ is a group homomorphism. This is immediate by

$$L_{xy}\varphi(z) = \varphi((xy)^{-1}z) = \varphi(y^{-1}x^{-1}z) = (L_y\varphi)(x^{-1}z) = L_zL_y\varphi(z).$$

It remains to show continuity of the map $\Phi : G \times L^2(G) \to L^2(G)$; $(x, \varphi) \mapsto L_x\varphi$. For this let $\varphi_0 \in L^2(G)$ be given. An open neighborhood of φ_0 in $L^2(G)$ is given by the set $B_r(\varphi_0)$ of all $\varphi \in L^2(G)$ with $\|\varphi - \varphi_0\| < r$, where $r > 0$ and $\|\cdot\|$ is the L^2-norm. We have to show that $S = \Phi^{-1}(B_r(\varphi_0))$ is an open subset of the product $G \times L^2(G)$. Let $(x, \varphi) \in S$, so $\|L_x\varphi - \varphi_0\| < r$. The claim asserts that there are open neighborhoods U of x and V of φ with $U \times V \subset S$. To prove this, we estimate for $y \in G$ and $\psi \in L^2(G)$,

$$\|L_y\psi - \varphi_0\| \le \|L_x\varphi - L_y\psi\| + \|L_x\varphi - \varphi_0\|$$

$$\le \|L_x\varphi - L_y\varphi\| + \|L_y\varphi - L_y\psi\| + \|L_x\varphi - \varphi_0\|$$

$$= \|(L_x - L_y)\varphi\| + \|\varphi - \psi\| + \|L_x\varphi - \varphi_0\|.$$

The last summand is strictly smaller that r. Set

$$\varepsilon \overset{\text{def}}{=} r - \|L_x\varphi - \varphi_0\| > 0.$$

The claim follows, if we can show that there are neighborhoods U of x and V of φ, such that for $(y, \psi) \in U \times V$ the estimate

$$\|(L_x - L_y)\varphi\| + \|\varphi - \psi\| < \varepsilon$$

holds. So let V be the set of all $\psi \in L^2(G)$ with $\|\varphi - \psi\| < \varepsilon/2$. We show that there exists an open neighborhood U of x, such that for every $y \in U$ one has $\|(L_x - L_y)\varphi\| < \varepsilon/2$. By Proposition 3.1.7 there exists a $g \in C_c(G)$ with $\|\varphi - g\| < \varepsilon/8$. It follows that

$$\begin{aligned}
\|(L_x - L_y)\varphi\| &\leq \|(L_x - L_y)g\| + \|(L_x - L_y)(g - \varphi)\| \\
&\leq \|(L_x - L_y)g\| + \|L_x(g - \varphi)\| + \|L_y(g - \varphi)\| \\
&= \|(L_x - L_y)g\| + 2\|g - \varphi\| < \|(L_x - L_y)g\| + \varepsilon/4.
\end{aligned}$$

Let $C \subset G$ be a compact set such that $g \equiv 0$ outside C. We can assume that C has positive Haar measure $\mathrm{vol}(C) > 0$. By uniform continuity of g, i.e. Lemma 3.1.4, there exists an open neighborhood U of x, such that for all $y \in U$ and all $t \in G$ the inequality

$$\left|g\left(x^{-1}t\right) - g\left(y^{-1}t\right)\right| < \frac{\varepsilon}{4\sqrt{2\mathrm{vol}(C)}}$$

holds. So

$$\left|g\left(x^{-1}t\right) - g\left(y^{-1}t\right)\right|^2 < \frac{\varepsilon^2}{32\mathrm{vol}(C)}.$$

We integrate this over $t \in G$ to get for $y \in U$,

$$\begin{aligned}
\|(L_x - L_y)g\|^2 &= \int_G \left|g\left(x^{-1}t\right) - g\left(y^{-1}t\right)\right|^2 dt \\
&= \int_{xC \cup yC} \left|g\left(x^{-1}t\right) - g\left(y^{-1}t\right)\right|^2 dt \\
&< \left(\mathrm{vol}(xC) + \mathrm{vol}(yC)\right) \frac{\varepsilon^2}{16\mathrm{vol}(C)} = \frac{\varepsilon^2}{16}.
\end{aligned}$$

Taking square roots, we get $\|(L_x - L_y)g\| < \varepsilon/4$ and the lemma is proven. $\qquad\square$

We define the *right regular representation* $x \mapsto R_x$ on the Hilbert space $L^2(G)$ by

$$R_x\varphi(y) = \sqrt{\Delta(x)}\varphi(yx), \qquad \varphi \in L^2(G),$$

where Δ is the modular function of G. Analogous to the right regular case, one shows that R is a unitary representation as well.

Definition 3.2.6 Two representations (π, V_π) and (η, V_η) of a topological group G are called *equivalent representations* if there exists a linear operator $T : V_\pi \to V_\eta$, such that

- T is continuous, invertible, and the inverse T^{-1} is continuous as well, and further
- $T\pi(g) = \eta(g)T$ holds for every $g \in G$.

The second property can also be written as $\pi(g) = T^{-1}\eta(g)T$. Every such operator T is called an *intertwining operator* between the two representations π and η. If V_π and V_η are Hilbert spaces, and there is a unitary intertwining operator, then the representations are said to be *unitarily equivalent*.

Let (π_1, V_1) and (π_2, V_2) be two unitary representations. On the direct sum $V = V_1 \oplus V_2$ there is a *direct sum representation* $\pi = \pi_1 \oplus \pi_2$. More generally, one can define direct sum representations of infinitely many summands as follows.

Definition 3.2.7 Let I be an index set and for every $i \in I$ let a Hilbert space V_i be given. The *Hilbert direct sum*

$$V = \widehat{\bigoplus_{i \in I}} V_i$$

is the set of all $v \in \prod_{i \in I} V_i$ such that

$$\|v^2\| \overset{\text{def}}{=} \sum_{i \in I} \|v_i\|^2 < \infty.$$

The algebraic direct sum $\bigoplus_{i \in I} V_i$ of all $v \in \prod_{i \in I} V_i$ such that $v_i = 0$ for almost all $i \in I$ is a subset of the Hilbert direct sum.

Since the set I may be uncountable, it is not immediately clear how the sum $\sum_{i \in I} \|v_i\|^2$ is defined. It is defined to be the integral with respect to the counting measure on I of the function $i \mapsto \|v_i\|^2$. By definition of the integral one gets

$$\sum_{i \in I} \|v_i\|^2 = \sup_{\substack{E \subset I \\ E \text{ finite}}} \sum_{i \in E} \|v_i\|^2,$$

where the supremum is extended over all finite subsets E of I. The reader not familiar with integration theory may take the right-hand side of this equality as the definition of the left-hand side. The condition $\sum_{i \in I} \|v_i\|^2 < \infty$ then implies that there exists a countable subset $J \subset I$ such that $\|v_i\| = 0$ for all $i \in I \setminus J$ and the sum $\sum_{i \in J} \|v_i\|^2$ converges in any order (see Exercise 3.8).

Note that by the definition it is not clear that V is a vector space, i.e. it is not clear why with $v, w \in V$ their sum $v + w$ lies in V.

Lemma 3.2.8 *The direct Hilbert sum* $V = \widehat{\bigoplus_{i \in I}} V_i$ *is a sub vector space of* $\prod_i V_i$. *It is a Hilbert space with inner product*

$$\langle v, w \rangle \overset{\text{def}}{=} \sum_{i \in I} \langle v_i, w_i \rangle_i.$$

The algebraic direct sum is a dense subspace.

If for every $i \in I$ *there is given a unitary representation* π_i *of the topological group* G *on the space* V_i, *then*

$$\pi(g)\left(\sum_{i \in I} v_i\right) \overset{\text{def}}{=} \sum_{i \in I} \pi_i(g) v_i$$

defines a unitary representation of G *on* V, *called the* direct sum representation.

Notation In the sequel, we will often leave out the hat in the direct sum. So we write $\bigoplus_{i \in I} V_i$ when it is clear from the context that we really mean $\widehat{\bigoplus}_{i \in I} V_i$.

Proof Let $v, w \in V$. We shall show that the sum $v + w$ lies in V, if v does. Also, we shall show that the sum $\sum_{i \in I} \langle v_i, w_i \rangle_i$ converges absolutely (see Exercise 3.8).

If we equip the set I with the counting measure defined on the σ-algebra of all subsets of I, the corresponding L^2-space is the space $\ell^2(I)$ of all functions $\varphi : I \to \mathbb{C}$ with $\|\varphi\|^2 = \sum_{i \in I} |\varphi(i)|^2 < \infty$. This is a Hilbert space with the inner product $\langle \varphi, \psi \rangle = \sum_{i \in I} \varphi(i) \overline{\psi(i)}$. For $v \in V$, the map $\varphi_v(i) = \|v_i\|$ is in $\ell^2(I)$ and one has $\|\varphi_v\| = \|v\|$. We now apply the Cauchy–Schwarz inequality first for the Hilbert spaces V_i and then for $\ell^2(I)$ to get

$$\sum_{i \in I} |\langle v_i, w_i \rangle| \leq \sum_{i \in I} \|v_i\| \|w_i\| = |\langle \varphi_v, \varphi_w \rangle| \leq \|\varphi_v\| \|\varphi_w\| = \|v\| \|w\| < \infty.$$

This implies the claimed absolute convergence of the sum $\langle v, w \rangle$ and by

$$|\langle v, w \rangle| = \left|\sum_{i \in I} \langle v_i, w_i \rangle_i\right| \leq \sum_{i \in I} |\langle v_i, w_i \rangle_i|$$

we infer the Cauchy–Schwarz inequality for this inner product, so that for $v, w \in V$ we have

$$\left|\|v + w\|^2\right| = |\langle v + w, v + w \rangle| \leq \langle v, v \rangle + \langle w, w \rangle + |\langle v, w \rangle| + |\langle w, v \rangle|$$

$$\leq \|v\|^2 + \|w\|^2 + 2\|v\| \|w\| = \left(\|v\| + \|w\|\right)^2 < \infty,$$

so that finally we get that V indeed is a sub vector space of $\prod_i V_i$. The last claim is trivial. $\qquad\square$

Example 3.2.9 Let $G = \mathbb{R}/\mathbb{Z}$. The left regular representation L of G lives on the Hilbert space $V = L^2(\mathbb{R}/\mathbb{Z})$. By the theory of Fourier series, L is isomorphic to the direct sum representation on $\widehat{V} = \bigoplus_{k \in \mathbb{Z}} \mathbb{C} v_k$, where G acts on $\mathbb{C} v_k$ by the character $\chi_{-k}(x) = e^{-2\pi i x}$, i.e. we have $\pi(t)v = \chi_{-k}(t)v$ if v lies in the one-dimensional space $\mathbb{C} v_k$.

Definition 3.2.10 A representation (π, V_π) is called a *subrepresentation* of a representation (η, V_η), if V_π is a closed, G-stable linear subspace of V_η and $\pi(g)$ is the restriction of $\eta(g)$ to V_π. So every G-stable, closed linear subspace gives a subrepresentation of η.

A representation (π, V_π) is called *irreducible* if $V_\pi \neq 0$ and it has no proper subrepresentations, i.e. if for every closed linear subspace $U \subset V_\pi$, which is G-stable, one has $U = 0$ or $U = V_\pi$.

Example 3.2.11 Let $G = SL_2(\mathbb{C})$ and $V = \mathbb{C}^2$. Let π be the standard representation of G on V given by matrix multiplication. Then π is irreducible. For a proof let $e_1 = (1, 0)^t$ be the first standard basis vector. For $g \in G$ the vector $\pi(g)e_1$ equals the first column of g. Therefore, for every $v \in V \smallsetminus \{0\}$ there exists a $g \in G$ with $v = \pi(g)e_1$. This implies irreducibility as follows. Let $0 \neq U \subset V$ be a G-stable linear subspace. We have to show that $U = V$. Let $0 \neq v \in V$ be arbitrary. Choose a $0 \neq u \in U$. Then there are $g, h \in G$ such that $\pi(g)e_1 = u$ and $\pi(h)e_1 = v$. Then $\pi(g^{-1})u = e_1$ and so $\pi(hg^{-1})u = \pi(h)e_1 = v$. As U is stable under the G-action, we conclude $v \in U$. Since v is arbitrary, $U = V$.

3.3 Modular Forms as Representation Vectors

Lemma 3.3.1 *Let G be a locally compact group and let $\Gamma \subset G$ be a discrete subgroup. Then Γ is closed in G.*

Proof Let g be in the closure $\overline{\Gamma}$ of Γ. We want to show that $g \in \Gamma$, so we assume the contrary. Since $g \in \overline{\Gamma}$, every neighborhood of g contains an element of Γ. If U is a neighborhood of the unit, then gU is a neighborhood of g, and hence contains an element γ_U of Γ. Since $g \neq \gamma_U$, we get $g^{-1}\gamma_U \neq 1$, so, by the Hausdorff property, there is a neighborhood $V \subset U$ of the unit such that $\gamma_U \notin gV$. As gV again contains an element γ_V of Γ, we infer that $\gamma_V \neq \gamma_U$, so we get that for every unit neighborhood U there exist two different elements $\gamma_U \neq \tau_U$ of Γ with $\gamma_U, \tau_U \in gU$.

As Γ is discrete, there exists a unit neighborhood $V \subset G$ such that $V \cap \Gamma = \{1\}$. Since the multiplication of G is a continuous map, there exists a unit neighborhood U with $U^2 \subset V$, where

$$U^2 = \{uu' : u, u' \in U\}.$$

Since $U^{-1} = \{u^{-1} : u \in U\}$ again is a unit neighborhood, we can, by replacing U with $U \cap U^{-1}$, assume that $U = U^{-1}$. This means in particular that for any two $u_1, u_2 \in U$ we have $u_1^{-1}u_2 \in V$. Consider $\gamma_U, \tau_U \in gU$. Then $g^{-1}\gamma_U, g^{-1}\tau_U \in U$; hence

$$\gamma_U^{-1}\tau_U = \left(g^{-1}\gamma_U\right)^{-1}g^{-1}\tau_U \in V.$$

Since $\gamma_U^{-1}\tau_U \in \Gamma$ this implies $\gamma_U^{-1}\tau_U = 1$, in contradiction with $\gamma_U \neq \tau_U$. Hence the assumption is false, so Γ is indeed closed. $\qquad\square$

The group $K = SO(2)$ is isomorphic with $\mathbb{T} = \{z \in \mathbb{C} : |z| = 1\}$ via $\begin{pmatrix} a & -b \\ b & a \end{pmatrix} \mapsto$ $a + ib$. Any character of K is of the form

$$\varepsilon_\nu \begin{pmatrix} a & -b \\ b & a \end{pmatrix} = (a + ib)^\nu, \quad \nu \in \mathbb{Z}.$$

Let $\Gamma \subset G = SL_2(\mathbb{R})$ be a congruence subgroup. The topology of G makes Γ a discrete subgroup. By Lemma 3.3.1 it is a closed subgroup. As Γ is discrete, the counting measure is a Haar measure, so Γ is unimodular. Since G is unimodular, too, Theorem 3.1.15 implies that there exists a non-trivial G-invariant Radon measure on G/Γ. It is now convenient to switch between left and right cosets, since we will use left cosets with respect to the subgroup K. The map

$$G/\Gamma \to \Gamma\backslash G,$$
$$g\Gamma \mapsto \Gamma g^{-1}$$

is a homeomorphism sending the left translation action of G to the right translation action. It transports the invariant Radon measure on G/Γ to a right G-invariant Radon measure on $\Gamma\backslash G$. Let $L^2(\Gamma\backslash G)$ be the corresponding L^2-space. This is a Hilbert space with an action of G by right translations:

$$R_g\varphi(x) = \varphi(xg).$$

Lemma 3.3.2 *The representation R of G on the Hilbert space $L^2(\Gamma\backslash G)$ is unitary.*

Proof We have to show that for $g \in G$ the operator $R_g : L^2(\Gamma\backslash G) \to L^2(\Gamma\backslash G)$ is unitary, i.e. that

$$\langle R_g\varphi, R_g\psi\rangle = \langle \varphi, \psi\rangle$$

for all $\varphi, \psi \in L^2(\Gamma\backslash G)$. We compute

$$\langle R_g\varphi, R_g\psi\rangle = \int_{\Gamma\backslash G} \varphi(xg)\overline{\psi(xg)}\,dx = \int_{\Gamma\backslash G} \varphi(x)\overline{\psi(x)}\,dx = \langle \varphi, \psi\rangle,$$

where we have used the G-invariance of the measure. Further we have to show that the representation is continuous. This is proven analogously to the proof of Lemma 3.2.5. $\qquad\square$

For $\nu \in \mathbb{Z}$ recall the character $\varepsilon_\nu : K \to \mathbb{T}$ given by

$$\varepsilon_\nu(k_\theta) = e^{2\pi i\nu\theta}.$$

Let $L^2(\Gamma\backslash G)^1$ be the space of all $\varphi \in L^2(\Gamma\backslash G)$ which are continuously differentiable; the latter means that φ as a function on G is continuously differentiable in the sense of Definition 3.1.2.

Lemma 3.3.3

(a) *The space $L^2(\Gamma\backslash G)^1$ is dense in the Hilbert space $L^2(\Gamma\backslash G)$.*

(b) *The space $L^2(\Gamma\backslash G)$ is a direct Hilbert sum*

$$L^2(\Gamma\backslash G) = \bigoplus_{v\in\mathbb{Z}} L^2(\Gamma\backslash G)(\varepsilon_v),$$

where

$$L^2(\Gamma\backslash G)(\varepsilon_v) = \{\varphi \in L^2(\Gamma\backslash G) : \varphi(xu) = \varepsilon_v(u)\varphi(x)\ \forall u \in K\}.$$

Proof (a) It suffices to show that the orthogonal space $W = L^2(\Gamma\backslash G)^{1,\perp}$ is zero. Let $\psi \in L^2(\Gamma\backslash G)$, let $f \in C_c^\infty(G)$ and define

$$R(f)\psi(x) = \int_G f(y)\psi(xy)\,dy = \int_G f(x^{-1}y)\psi(y)\,dy.$$

Then $R(f)\psi$ is continuously differentiable, as the second representation shows. The first representation shows that $R(f)\psi$ is still Γ-invariant on the left. To show that $R(f)\psi$ lies in $L^2(\Gamma\backslash G)^1$ it suffices to show that its L^2-norm is finite. Let $C \subset G$ be the support of f and let $M = \int_C |f(y)|^2\,dy$. We use the Hölder inequality and the Fubini theorem to get

$$\|R(f)\psi\|^2 = \int_{\Gamma\backslash G} |R(f)\psi(x)|^2\,dx = \int_{\Gamma\backslash G} \left|\int_G f(y)\psi(xy)\,dy\right|^2 dx$$

$$= \int_{\Gamma\backslash G} \left|\int_C f(y)\psi(xy)\,dy\right|^2 dx$$

$$\leq \int_{\Gamma\backslash G} \int_C |f(y)|^2\,dy \int_C |\psi(xy)|^2\,dy\,dx$$

$$= M \int_C \int_{\Gamma\backslash G} |\psi(xy)|^2\,dx\,dy = M\,\text{vol}(C) \int_{\Gamma\backslash G} |\psi(x)|^2\,dx < \infty.$$

So we see that $R(f)$ is an operator on $L^2(\Gamma\backslash G)$ with image inside $L^2(\Gamma\backslash G)^1$. Now let $\psi \in W$, i.e. we have $\langle\psi, \varphi\rangle = 0$ for every $\varphi \in L^2(\Gamma\backslash G)^1$. We want to show that $\psi = 0$. We first show that $R(f)\psi \in W$ again. To see this, let $\varphi \in L^2(\Gamma\backslash G)^1$ and compute

$$\langle R(f)\psi, \varphi\rangle = \int_{\Gamma\backslash G} R(f)\psi(x)\overline{\varphi(x)}\,dx$$

$$= \int_{\Gamma\backslash G} \int_G f(y)\psi(xy)\,dy\,\overline{\varphi(x)}\,dx$$

$$= \int_G \int_{\Gamma\backslash G} f(y)\psi(xy)\overline{\varphi(x)}\,dx\,dy$$

$$= \int_G \int_{\Gamma\backslash G} f(y)\psi(x)\overline{\varphi(xy^{-1})}\,dx\,dy$$

$$= \int_{\Gamma\backslash G} \psi(x)\overline{\int_G f(y^{-1})\varphi(xy)\,dy}\,dx = \langle\psi, R(f^*)\varphi\rangle,$$

where $f^*(y) = \overline{f(y^{-1})}$. Now $R(f^*)\varphi$ lies in $L^2(\Gamma\backslash G)^1$ as we have seen, so the latter product is zero. It follows that $R(f)\psi$ lies in $L^2(\Gamma\backslash G)^1$ and its orthogonal space; hence it is zero. Being a continuous function, it must vanish everywhere, so $R(f)\psi = 0$ holds for every $f \in C_c^\infty(G)$. We finally deduce $\psi = 0$ from this. As ψ is continuous at x, for given $\varepsilon > 0$ there exists a unit neighborhood U such that for every $y \in U$ one has $|\psi(xy) - \psi(x)| < \varepsilon$. Let f be a function in $C_c^\infty(G)$ with support in U, such that $f \geq 0$ and $\int_G f(x)\,dx = 1$. Such a function exists for every U by Exercise 3.6. Then

$$\left|R(f)\psi(x) - \psi(x)\right| = \left|\int_U f(y)\big(\psi(xy) - \psi(x)\big)\,dx\right|$$

$$\leq \int_U f(y)\big|\psi(xy) - \psi(x)\big|\,dx < \varepsilon.$$

Since $R(f)\psi = 0$ for all f, we infer that $\psi = 0$ as claimed.

For part (b) we first show that $L^2(\Gamma\backslash G)(\varepsilon_\nu) \perp L^2(\Gamma\backslash G)(\varepsilon_\mu)$ for $\nu \neq \mu$. For this let $\varphi \in L^2(\Gamma\backslash G)(\varepsilon_\nu)$ and $\psi \in L^2(\Gamma\backslash G)(\varepsilon_\mu)$. Then

$$\langle\varphi, \psi\rangle = \int_{\Gamma\backslash G} \varphi(x)\overline{\psi(x)}\,dx = \int_K \int_{\Gamma\backslash G} \varphi(xu)\overline{\psi(xu)}\,dx\,du$$

$$= \int_{\Gamma\backslash G} \int_K \varphi(xu)\overline{\psi(xu)}\,du\,dx$$

$$= \int_{\Gamma\backslash G} \int_K \varepsilon_\nu(u)\varepsilon_{-\mu}(u)\varphi(x)\overline{\psi(x)}\,du\,dx$$

$$= \underbrace{\int_K \varepsilon_{\nu-\mu}(u)\,du}_{=0} \int_{\Gamma\backslash G} \varphi(x)\overline{\psi(x)}\,dx.$$

So the sum is indeed orthogonal.

Let $\varphi \in L^2(\Gamma\backslash G)$ be continuously differentiable, which means that φ as a function on G is continuously differentiable in the sense of Definition 3.1.2. For given $x \in G$ the group $K \cong \mathbb{T}$ acts on $xK \subset G$. By the theory of Fourier series we have

$$\varphi(xk_\theta) = \sum_{\nu\in\mathbb{Z}} \varphi_\nu(x)e^{2\pi i\nu\theta},$$

where the series converges absolutely. The function $\varphi_v(x)$ is given by .

$$\varphi_v(x) = \int_0^1 \varphi(xk_\theta)e^{-2\pi i v\theta}\,d\theta.$$

This implies that $\varphi_v \in L^2(\Gamma\backslash G)(\varepsilon_v)$ and that we have the desired decomposition for the subspace $L^2(\Gamma\backslash G)^1$ of all continuously differentiable functions. As this space is dense in L^2 by part (a), the claim follows.   □

Let $f \in S_k(\Gamma)$ for a congruence subgroup $\Gamma \subset \Gamma(1)$. Then $f(\gamma z) = (cz+d)^k \times f(z)$ holds for every $\gamma = \left(\begin{smallmatrix} * & * \\ c & d \end{smallmatrix}\right) \in \Gamma$. For $g \in G$ set

$$\phi_f(g) = (cz+d)^{-k}f(gi),$$

if $g = \left(\begin{smallmatrix} * & * \\ c & d \end{smallmatrix}\right)$. For $g = \left(\begin{smallmatrix} * & * \\ c & d \end{smallmatrix}\right) \in G$ and $z \in \mathbb{H}$ set $\mu(g,z) = (cz+d)^k$. A computation shows

$$\mu(gh,z) = \mu(g,hz)\mu(h,z)$$

for all $g,h \in G$ and all $z \in \mathbb{H}$. For $u \in SO(2)$ one also has

$$\mu(u,i) = \varepsilon_k(u).$$

Proposition 3.3.4 *The map $f \mapsto \phi_f$ is an isometric injection of the Hilbert space $S_k(\Gamma)$ into the space $L^2(\Gamma\backslash G)(\varepsilon_{-k})$.*

Proof We show that ϕ_f is invariant under Γ. For this let $\gamma \in \Gamma$ and $g \in G$. Then

$$\phi_f(\gamma g) = \mu(\gamma g,i)^{-1}g(\gamma gi) = \mu(\gamma,gi)^{-1}\mu(g,i)^{-1}\mu(\gamma,gi)g(gi) = \phi_f(g).$$

We see that this map is an isometry by noting that

$$\langle \phi_f, \phi_g\rangle = \int_{\Gamma\backslash G} \phi_f(x)\overline{\phi_g(x)}\,dx = \int_{\Gamma\backslash\mathbb{H}} \underbrace{\mu(x,i)^{-1}\overline{\mu(x,i)}^{-1}}_{=\mathrm{Im}(xi)^k} f(xi)\overline{g(xi)}\,dx$$

$$= \int_{\Gamma\backslash\mathbb{H}} \mathrm{Im}(z)^k f(z)\overline{g(z)}\,d\mu(z) = \langle f,g\rangle_{\mathrm{Pet}}.$$

The image lies in $L^2(\Gamma\backslash G)(\varepsilon_{-k})$, as for $u \in K$,

$$\phi_f(xu) = \mu(xu,i)^{-1}f(xui) = \mu(x,i)^{-1}\mu(u,i)^{-1}f(xi) = \varepsilon_{-k}(u)\phi_f(x). \quad □$$

Let $G = SL_2(\mathbb{R})$. By Lemma 3.3.2, the representation of G on $L^2(\Gamma\backslash G)$ is unitary.

Definition 3.3.5 An *automorphic form* is a function φ in $L^2(\Gamma\backslash G)$. Later we will extend this notion to include functions of the adele valued group which are invariant under $GL_2(\mathbb{Q})$.

The word *automorphic* indicates the invariance $\varphi(\gamma x) = \varphi(x)$ under the Γ-left-translation. It derives from the Greek word for *of the same form*, meaning unchanged under the transform. Felix Klein formed this notion in his paper *Zur Theorie der Laméschen Functionen* which appeared in the year 1890.

The Automorphic Spectral Problem Can the unitary representation $(R, L^2(\Gamma \backslash G))$ be decomposed into a direct sum of irreducible subrepresentations? We will show that a positive solution of this problem solves the spectral problem for each of the operators Δ_k. This problem admits a positive solution if and only if Γ is cocompact. If Γ is not cocompact, there will also be the so-called continuous spectrum.

3.4 The Exponential Map

Let $n \in \mathbb{N}$. The following assertions will be used subsequently only in the case $n = 2$. But as the proofs are the same for any dimension n, we prove them in general.

On the real vector space $M_n(\mathbb{R})$ of all real $n \times n$ matrices we consider the Euclidean norm

$$\|A\| = \sqrt{\sum_{i,j=1}^{n} A_{i,j}^2},$$

where we have written the entries of a matrix A as $A_{i,j}$. For matrix multiplication we have $\|AB\| \leq \|A\| \|B\|$, since by means of the Cauchy–Schwarz inequality we get

$$\|AB\|^2 = \sum_{i,j} \left(\sum_k A_{i,k} B_{k,j} \right)^2 \leq \sum_{i,j} \left(\sum_k A_{i,k}^2 \right) \left(\sum_l B_{l,j}^2 \right) = \|A\|^2 \|B\|^2.$$

A series $\sum_{j=0}^{\infty} A_j$ of matrices in $M_n(\mathbb{R})$ is said to be *absolutely convergent* if $\sum_j \|A_j\| < \infty$. In this case, the series converges in $M_n(\mathbb{R})$ and the limit doesn't change under reordering of the series.

For a given real $n \times n$ matrix $X \in M_n(\mathbb{R})$, the *exponential series*

$$\exp(X) = \sum_{\nu=0}^{\infty} \frac{1}{\nu!} X^\nu$$

converges absolutely in $M_n(\mathbb{R})$, as follows from the estimate $\|X^\nu\| \leq \|X\|^\nu$ and the absolute convergence of the $\mathbb{R}$-valued exponential series.

Proposition 3.4.1

(a) *If the matrices $A, B \in M_n(\mathbb{R})$ commute, i.e. $AB = BA$, then*

$$\exp(A + B) = \exp(A) \exp(B).$$

(b) *For every $A \in M_n(\mathbb{R})$ the matrix $\exp(A)$ is invertible and we have $\exp(A)^{-1} = \exp(-A)$.*

(c) *The exponential map is a smooth map $M_n(\mathbb{R}) \cong \mathbb{R}^{n^2} \to M_n(\mathbb{R})$, the image of which lies in $GL_n(\mathbb{R})$, and whose differential at zero $D\exp(0) : \mathbb{R}^{n^2} \to \mathbb{R}^{n^2}$ is an invertible linear map.*

(d) *In the group $GL_n(\mathbb{R})$ there is a unit neighborhood U, which does not contain any subgroup of $GL_n(\mathbb{R})$ other than the trivial one.*

Proof The equality $AB = BA$ implies $(A + B)^\nu = \sum_{k=0}^\nu \binom{\nu}{k} A^k B^{\nu-k}$ for every $\nu \in \mathbb{N}$. Therefore,

$$\exp(A + B) = \sum_{\nu=0}^\infty \frac{1}{\nu!} \sum_{k=0}^\nu \binom{\nu}{k} A^k B^{\nu-k}$$

$$= \sum_{\nu=0}^\infty \sum_{k=0}^\nu \frac{1}{k!} \frac{1}{(\nu-k)!} A^k B^{\nu-k} = \exp(A)\exp(B).$$

This proves (a). Part (b) follows from part (a), as $\exp(A)\exp(-A) = I$. For part (c) note that the entries of the matrix $\exp(A)$ are convergent power series in the entries of A. So the exponential map is infinitely differentiable. We compute the directional derivative in the direction $X \in M_2(\mathbb{R})$ as follows

$$\lim_{t \downarrow 0} \frac{1}{t}(\exp(tX) - \exp(0)) = \lim_{t \downarrow 0} \sum_{n=1}^\infty \frac{t^{n-1}}{n!} X^n = X.$$

This means $D\exp(0) = \mathrm{Id}$, so the claim follows.

(d) Since the Jacobi matrix of exp is invertible, there exists an open neighborhood $\tilde{V} \subset M_n(\mathbb{R})$ of zero in $M_n(\mathbb{R})$, such that $\tilde{V}$ is mapped diffeomorphically to an open neighborhood V of the unit matrix in $GL_n(\mathbb{R})$ by the exponential map. We may choose $\tilde{V}$ to be bounded. Let $\tilde{U} = \frac{1}{2}\tilde{V}$ and $U = \exp(\tilde{U})$. Then U contains no non-trivial subgroup. To see this let $a = \exp(X) \in U$ with $X \in \tilde{U}$ and assume $X \neq 0$. Then there exists $\nu \in \mathbb{N}$ with $\nu X \in \tilde{V} \setminus \tilde{U}$. Assuming that $a^\nu = \exp(\nu X)$ lies in U, there exists $Y \in \tilde{U}$ with $\exp(Y) = a^\nu = \exp(\nu X)$, so the two elements Y and νX of $\tilde{V}$ have the same image under the exponential map, and hence they are equal, contradicting $\nu X \notin \tilde{U}$. $\qquad\square$

Every $X \in M_n(\mathbb{R})$ defines, by taking right derivatives, a differential operator of first order $\tilde{R}_X$ on $G = GL_n(\mathbb{R})$ by

$$\tilde{R}_X f(x) = \frac{d}{dt}\bigg|_{t=0} f(x \exp(tX)).$$

Analogously, there is a differential operator $\tilde{L}_X$ given by left derivatives,

$$\tilde{L}_X f(x) = \frac{d}{dt}\bigg|_{t=0} f(\exp(-tX)x).$$

Let (π, V_π) be a representation of the group $G = \mathrm{GL}_n(\mathbb{R})$. A vector $v \in V_\pi$ is called a *smooth vector* if the map

$$x \mapsto \pi(x)v$$

as a map from the open set $\mathrm{GL}_n(\mathbb{R}) \subset M_n(\mathbb{R}) \cong \mathbb{R}^{n^2}$ into the Banach space V_π is infinitely differentiable. The set V_π^∞ of all smooth vectors in V_π is a linear subspace of the vector space V_π.

Let $v \in V_\pi$ be a smooth vector and let $X \in M_n(\mathbb{R})$. Then the map $t \mapsto \pi(\exp(tX))v$ is differentiable, so the limit

$$\tilde{\pi}(X)v \stackrel{\mathrm{def}}{=} \lim_{t \to 0} \frac{1}{t}\big(\pi\big(\exp(tX)\big)v - v\big) = \frac{d}{dt}\Big|_{t=0} \pi\big(\exp(tX)\big)v$$

exists.

Lemma 3.4.2

(a) *The operator $\tilde{R}_X$, $X \in M_n(\mathbb{R})$ is left-invariant, i.e. $L_x \tilde{R}_X L_{x^{-1}} = \tilde{R}_X$, where $x \in G = \mathrm{GL}_n(\mathbb{R})$ and $L_x f(y) = f(x^{-1}y)$ for every smooth function f on G. Analogously, the operator $\tilde{L}_X$ is right-invariant, i.e. $R_x \tilde{L}_X R_{x^{-1}} = \tilde{L}_X$ with $R_x f(y) = f(yx)$.*

(b) *A continuous intertwining operator $T : V_\pi \to V_\eta$ between two representations of $G = \mathrm{GL}_n(\mathbb{R})$ maps smooth vectors to smooth vectors, i.e. $T(V_\pi^\infty) \subset V_\eta^\infty$.*

(c) *For $v \in V_\pi$ and $f \in C_c^\infty(G)$, the vector $\pi(f)v$ is smooth. The space of smooth vectors is dense in V_π. For $X \in M_n(\mathbb{R})$ we have*

$$\tilde{\pi}(X)\pi(f)v = \pi(\tilde{L}_X f)v.$$

(d) *Let $f \in C_c^\infty(G)$ and let φ be a locally square-integrable function on G, that is, for every $x \in G$ there exists a neighborhood U of x, such that $\varphi|_U \in L^2(U)$. Then the integral*

$$R(f)\varphi(x) = \int_G f(y)\varphi(xy)\,dy$$

exists for every given $x \in G$ and defines a smooth function $R(f)\varphi$. For every $X \in M_n(\mathbb{R})$ one has

$$\tilde{R}_X\big(R(f)\varphi\big) = R(\tilde{L}_X f)\varphi.$$

Proof (a) Let f be a differentiable function on $\mathrm{GL}_n(\mathbb{R})$. We have to show that $\tilde{R}_X L_x f = L_x \tilde{R}_X f$, where $L_x f(y) = f(x^{-1}y)$. For $y \in \mathrm{GL}_n(\mathbb{R})$ we compute

$$\tilde{R}_X L_x f(y) = \frac{d}{dt}\Big|_{t=0} L_x f\big(y\exp(tX)\big)$$

$$= \frac{d}{dt}\Big|_{t=0} f\big(x^{-1}y\exp(tX)\big) = \tilde{R}_X f\big(x^{-1}y\big) = L_x \tilde{R}_X f(y).$$

(b) If $f : \mathbb{R}^N \to V$ is a smooth map into a Banach space and if $T : V \to W$ is a continuous linear map between Banach spaces, then $T \circ f : \mathbb{R}^N \to W$ is smooth.

Now for (c). Since the differential of the exponential function is surjective, a vector v is smooth if and only if for every $x \in G$ the map $X \mapsto \pi(\exp(X)x)v$ as a map $M_n(\mathbb{R}) \to V_\pi$ is smooth. So the smoothness, as well as the claimed formula, follow from

$$\pi\left(\exp(tX)\right)\pi(f)v = \int_G f(x)\pi\left(\exp(tX)x\right)v\,dx = \int_G f\left(\exp(-tX)x\right)\pi(x)v\,dx.$$

Finally we have to show that smooth vectors are dense. This follows if we show that vectors of the form $\pi(f)v$ for $f \in C_c^\infty(G)$ are dense. Let $\varepsilon > 0$ and $v \in V_\pi$. By continuity of π there exists a neighborhood U of the unit in G such that $\|\pi(x)v - v\| < \varepsilon$ for every $x \in U$. Let $f \in C_c^\infty(G)$ be supported inside U, satisfy $f \geq 0$ and $\int_G f(x)\,dx = 1$; see Exercise 3.6 for the existence of such a function. For $v \in V_\pi$ we get

$$\|\pi(f)v - v\| = \left\|\int_G f(x)\left(\pi(x)v - v\right)dx\right\| \leq \int_G f(x)\|\pi(x)v - v\|\,dx < \varepsilon.$$

So the set of all vectors of the form $\pi(f)v$ for $v \in V_\pi$ and $f \in C_c^\infty(G)$ is indeed dense in V_π.

(d) follows similar to (c).   □

3.5 Exercises and Remarks

Exercise 3.1 Prove Proposition 3.1.11.

Exercise 3.2 Show that the map SL$_2$($\mathbb{R}$) $\times$ $\mathbb{H} \to \mathbb{H}$, given by $(g, z) \mapsto gz$ is continuous.

Exercise 3.3 Let $K = SO(2)$ and let D be the set of all diagonal matrices in $G = SL_2(\mathbb{R})$ or in $G = GL_2(\mathbb{R})$. Show in both cases that $G = KDK$.

Exercise 3.4 Let B be the group of all real matrices of the form $\left(\begin{smallmatrix}1 & x \\ 0 & y\end{smallmatrix}\right)$ with $y \neq 0$. Show that the modular function is $\Delta\left(\begin{smallmatrix}1 & x \\ 0 & y\end{smallmatrix}\right) = |y|$.

Exercise 3.5 Let μ be a Haar measure of the locally compact group G. Let $\emptyset \neq U \subset G$ be an open subset. Show that $\mu(U) > 0$.

Exercise 3.6

(a) Show that the function

$$f(t) = \begin{cases} 0 & \text{if } t \le 0, \\ e^{-t} & \text{if } t > 0, \end{cases}$$

is infinitely differentiable on $\mathbb{R}$.

(b) Show that for every neighborhood U of zero in $\mathbb{R}$ there exists a function $f \ge 0$ on $\mathbb{R}$ which is infinitely differentiable, has support in U and satisfies $\int_{\mathbb{R}} f(x)\,dx = 1$.

(c) Let $G = \mathrm{SL}_2(\mathbb{R})$. Show that for every unit neighborhood U there exists a function $f \in C_c^\infty(G)$ such that $f \ge 0$, the integral $\int_G f(x)\,dx$ is equal to 1 and f has support inside U. Show the same for $G = \mathrm{GL}_n(\mathbb{R})$.

Exercise 3.7

(a) Show that the map $f \mapsto \phi_f$ with

$$\phi_f(g) = (ci + d)^{-k} f(gi)$$

defines a bijection between the set $C^\infty(\mathbb{H})$ of all smooth maps on $\mathbb{H}$ and the set V_k of all smooth functions ϕ on $\mathrm{SL}_2(\mathbb{R})$ with $\phi(xu) = \varepsilon_{-k}(u)$ for every $u \in \mathrm{SO}(2)$.

(Hint: the inverse map is given by $\phi \mapsto f_\phi$ with $f_\phi(x + iy) = \sqrt{y}^{-k} \times \phi\left(\begin{smallmatrix} \sqrt{y} & x/\sqrt{y} \\ & 1/\sqrt{y} \end{smallmatrix}\right)$.)

(b) Let $H = \left(\begin{smallmatrix} 1 & \\ & -1 \end{smallmatrix}\right)$, $E = \left(\begin{smallmatrix} 0 & 1 \\ 0 & 0 \end{smallmatrix}\right)$ and $F = \left(\begin{smallmatrix} 0 & 0 \\ 1 & 0 \end{smallmatrix}\right)$. For $\phi \in V_k$ define the differential operator

$$D\phi(g) = \frac{d}{dt}\Big|_{t=0} \phi\big(g\exp(tH)\big) - i\phi\big(g\exp(tE)\big) - i\phi\big(g\exp(tF)\big).$$

Show that $f_{D\phi} = -2iy\bar\partial f_\phi$, where $\bar\partial = \frac{\partial}{\partial x} + i\frac{\partial}{\partial y}$. Conclude that f is holomorphic if and only if f is continuously differentiable and $D\phi_f = 0$.

Exercise 3.8 Let I be an index set and for each $i \in I$ let $a_i \in \mathbb{C}$ be given. We define

$$\sum_{i \in I} |a_i| = \sup_{E \subset I \text{ finite}} \sum_{i \in E} |a_i|.$$

Suppose that $\sum_{i \in I} |a_i| < \infty$. Show that there is a countable subset $C \subset E$ such that $a_i = 0$ if $i \notin C$ and that the sum

$$\sum_{i \in C} a_i$$

converges in any order, always yielding the same complex number. We call this number the sum $\sum_{i \in I} a_i$ and say that this sum *converges absolutely*.

Exercise 3.9 Assume that the series $L(s) = \sum_{n=1}^{\infty} a_n n^{-s}$ converges in $s = s_0 \in \mathbb{C}$. Show that the series converges locally uniformly absolutely in $\mathrm{Re}(s) > \mathrm{Re}(s_0) + 1$.

Exercise 3.10 Let $L(a,s) = \sum_{n=1}^{\infty} a_n n^{-s}$ and $L(b,s) = \sum_{n=1}^{\infty} b_n n^{-s}$ both convergent for some $s \in \mathbb{C}$. Assume $L(a, s_\nu) = L(b, s_\nu)$ for a sequence $s_\nu \in \mathbb{C}$ with $\mathrm{Re}(s_\nu) \to \infty$. Show that $a_n = b_n$ for every $n \in \mathbb{N}$.

Exercise 3.11 A sequence (a_n) of complex numbers is called *weakly multiplicative* if $a_{mn} = a_n a_m$ for every coprime pair of natural numbers (m, n). The sequence is called *strongly multiplicative* or just *multiplicative* if this holds for every pair of natural numbers. Let $L(s) = \sum_{n=1}^{\infty} a_n n^{-s}$ be absolutely convergent for $\mathrm{Re}(s) > \sigma_0$. Show:

- The sequence (a_n) is weakly multiplicative if and only if for every $s \in \mathbb{C}$ with $\mathrm{Re}(s) > \sigma_0$ one has

$$L(s) = \prod_p \sum_{k=0}^{\infty} a_{p^k} p^{ks}.$$

- The sequence (a_n) is strongly multiplicative if and only if for every $s \in \mathbb{C}$ with $\mathrm{Re}(s) > \sigma_0$ one has

$$L(s) = \prod_p \frac{1}{1 - a_p p^{-s}}.$$

Remarks We have encountered classical modular forms $\mathcal{M}_k(\Gamma)$ and cusp forms $S_k(\Gamma)$, and we have seen how to attach L-functions $L(f, s)$ to these.

The L-functions have Euler products if the forms are Hecke eigenforms. Finally we saw how to embed $S_k(\Gamma)$ into the space $L^2(\Gamma\backslash G)$, where the latter space carries a unitary representation of $G = \mathrm{SL}_2(\mathbb{R})$. The spectral problem of decomposition of this representation into irreducibles is a central problem in the theory of automorphic forms.

We shall show that sometimes, $f \in S_k(\Gamma) \subset L^2(\Gamma\backslash G)$ generates an irreducible subrepresentation π. We want to define $L(f, s)$ using π, but there are data missing to do that. These data are encoded in the Hecke action.

Let $\Gamma \subset \mathrm{SL}_2(\mathbb{Z})$ be a congruence subgroup. Recall that the Hecke action comes about by the operation of the group $G_{\mathbb{Q}} = \mathrm{GL}_2(\mathbb{Q})$. For $\alpha \in G_{\mathbb{Q}}$ the group $\Gamma \cap \alpha \Gamma \alpha^{-1}$ is again a congruence subgroup. the Hecke operator T_α can be defined by the action of α: $f \mapsto f|\alpha$, which maps $S_k(\Gamma)$ to $S_k(\Gamma \cap \alpha \Gamma \alpha^{-1})$, followed by a sum over $\Gamma/(\Gamma \cap \alpha \Gamma \alpha^{-1})$, the latter mapping to $S_k(\Gamma)$ again. This summation is just the orthogonal projection onto the subspace $S_k(\Gamma)$ in the Hilbert space $S_k(\Gamma \cap \alpha \Gamma \alpha^{-1})$. By projecting, one loses information. If we leave out the projection, the Hecke operator does not map the space $S_k(\Gamma)$ to itself. We solve this problem by enlarging the space to

$$S_k = \bigcup_\Gamma S_k(\Gamma),$$

where Γ runs through all congruence subgroups of Γ. This is a vector space with a linear action of the group $G_{\mathbb{Q}}$. The space

$$\bigcup_{\Gamma} L^2(\Gamma \backslash G)$$

inherits an inner product, if one normalizes the inner products on the space $L^2(\Gamma \backslash G)$ as follows:

$$\langle f, g \rangle = \frac{1}{[\overline{\Gamma} : \overline{\Gamma}]} \int_{\Gamma \backslash G} f(x)\overline{g(x)}\,dx.$$

Let H be the completion of the resulting pre-Hilbert space. One can define Hecke operators on this space. However, in the current description, this space is not easy to understand.

Here comes the punch line: There exists a natural ring $\mathbb{A}$, the *ring of adeles*, such that as $SL_2(\mathbb{R})$-module,

$$H \cong L^2\big(SL_2(\mathbb{Q}) \backslash SL_2(\mathbb{A})\big).$$

Given the definition of H, this assertion seems ridiculous. More precisely, one has $\mathbb{A} = \mathbb{A}_{\mathrm{fin}} \times \mathbb{R}$, where $\mathbb{A}_{\mathrm{fin}}$ is called the *ring of finite adeles*. The group $SL_2(\mathbb{A}) = SL_2(\mathbb{A}_{\mathrm{fin}}) \times SL_2(\mathbb{R})$ acts by right translations on H. The action of $SL_2(\mathbb{R})$ is the same as before. It turns out that the action of $SL_2(\mathbb{A}_{\mathrm{fin}})$ is, in a sense to be made precise, equivalent to the action of the Hecke algebra! Consequently, a decomposition of H into $SL_2(\mathbb{A})$-irreducibles gives Hecke eigenforms with Euler products. So the L-function should be expressible in representation-theoretic terms of the group $SL_2(\mathbb{A})$.

Chapter 4
p-Adic Numbers

In this chapter we introduce p-adic numbers, the siblings of the real numbers. These live in a bizarre universe, in which every point of a disk is its center and two disks of the same radius are either equal or disjoint. The p-adic numbers are, as the reals, a completion of the rationals $\mathbb{Q}$. One can show that the reals and the p-adic numbers give all completions of the rationals.

The set of real numbers is the completion of $\mathbb{Q}$ with respect to the usual absolute value

$$|x|_\infty = \begin{cases} x & \text{if } x \geq 0, \\ -x & \text{if } x < 0. \end{cases}$$

We will see that there are other absolute values, which give other completions. In order to make this precise, we first have to make clear, what exactly we mean by 'absolute value'.

4.1 Absolute Values

Definition 4.1.1 By an *absolute value* on the field K we mean a map $|\cdot| : K \to [0, \infty)$ such that for all $a, b \in K$ we have

- $|a| = 0 \Leftrightarrow a = 0$,   definiteness
- $|ab| = |a||b|$,   multiplicativity
- $|a + b| \leq |a| + |b|$.   triangle inequality

Remark Every absolute value satisfies $|-1| = 1$, since firstly $|1| = |1 \cdot 1| = |1|^2$ so $|1| = 1$ and secondly $|-1|^2 = |(-1)^2| = |1| = 1$ and therefore $|-1| = 1$.

Examples 4.1.2

- For $K = \mathbb{Q}$ the usual absolute value $|\cdot|_\infty$ is an example.

A. Deitmar, *Automorphic Forms*, Universitext,
DOI 10.1007/978-1-4471-4435-9_4, © Springer-Verlag London 2013

- The *trivial absolute value* exists for every field and is given by

$$|x|_{\text{triv}} = \begin{cases} 0 & \text{if } x = 0, \\ 1 & \text{if } x \neq 0. \end{cases}$$

Next we give a more elaborate example for the field $K = \mathbb{Q}$ of rational numbers. For a fixed prime number p, every rational number $r \neq 0$ can be written as

$$r = p^k \frac{m}{n},$$

where $m, n \in \mathbb{Z}$ are coprime to p. The number $k \in \mathbb{Z}$ is uniquely determined by r. We define the *p-adic absolute value* by

$$|r|_p = \left| p^k \frac{m}{n} \right|_p \stackrel{\text{def}}{=} p^{-k}.$$

We complete this definition by

$$|0|_p \stackrel{\text{def}}{=} 0.$$

Lemma 4.1.3 *Let p be a prime number. Then $|\cdot|_p$ is an absolute value on $\mathbb{Q}$, which satisfies the strong triangle inequality*

$$|x + y|_p \leq \max(|x|_p, |y|_p).$$

Here we have equality, if $|x|_p \neq |y|_p$.

Proof The definiteness is clear by definition. For multiplicativity write $x = p^k \frac{m}{n}$ and $y = p^{k'} \frac{m'}{n'}$, where m, n, m', n' are coprime to p. Then

$$xy = p^{k+k'} \frac{mm'}{nn'},$$

which implies multiplicativity, i.e. $|xy|_p = |x|_p |y|_p$. For the strong triangle inequality we can assume $k \leq k'$. Then

$$x + y = p^k \left(\frac{m}{n} + p^{k'-k} \frac{m'}{n'} \right) = p^k \frac{mn' + p^{k'-k} nm'}{nn'}.$$

If $|x|_p \neq |y|_p$, i.e. $k' - k > 0$, then the number $mn' + p^{k'-k} nm'$ is coprime to p and then we have $|x + y| = p^{-k} = \max(|x|_p, |y|_p)$. If, on the other hand, $|x|_p = |y|_p$, then the numerator $mn' + p^{k'-k} nm' = mn' + nm'$ is of the form $p^l N$, where $l \geq 0$ and N is coprime to p. This means that $|x + y|_p = |p^{k+l} \frac{N}{nn'}|_p = p^{-k-l} \leq \max(|x|_p, |y|_p)$. $\qquad\square$

Proposition 4.1.4 *For every $x \in \mathbb{Q}^\times$ we have the product formula*

$$\prod_{p \leq \infty} |x|_p = 1.$$

The product extends over all prime numbers and $p = \infty$. For any given $x \in \mathbb{Q}^\times$, almost all factors of the product are equal to one.

Here we have used the phrase

almost all, which means all but finitely many.

Proof Write x as a coprime fraction and write the numerator and denominator as a product of prime powers. Then

$$x = \pm p_1^{k_1} \cdots p_n^{k_n}$$

for pairwise different primes $p_1, \ldots, p_n$ and $k_1, \ldots, k_n \in \mathbb{Z}$. Then $|x|_p = 1$ if p is a prime that does not coincide with any of the p_j. Hence the product indeed has only finitely many factors $\neq 1$. Further we have,

$$|x|_{p_j} = p^{-k_j} \quad \text{and} \quad |x|_\infty = p_1^{k_1} \cdots p_n^{k_n}.$$

Therefore,

$$\prod_{p \leq \infty} |x|_p = \left(\prod_{j=1}^{n} p^{-k_j} \right) \cdot p_1^{k_1} \cdots p_n^{k_n} = 1.$$

$\square$

Remark One can show that for every non-trivial absolute value $|\cdot|$ there exists exactly one $p \leq \infty$ and exactly one $a > 0$, such that $|x| = |x|_p^a$ for every $x \in \mathbb{Q}$.

4.2 $\mathbb{Q}_p$ as Completion of $\mathbb{Q}$

We now give the first construction of the set $\mathbb{Q}_p$ of p-adic numbers. This set will be the completion of $\mathbb{Q}$ by the p-adic metric defined as follows.

Let $x \mapsto |x|$ be an absolute value on $\mathbb{Q}$. Then

$$d(x, y) \stackrel{\text{def}}{=} |x - y|$$

is a *metric* on $\mathbb{Q}$, i.e. one has

- $d(x, y) = 0 \Leftrightarrow x = y$,   definiteness
- $d(x, y) = d(y, x)$,   symmetry
- $d(x, y) \leq d(x, z) + d(z, y)$.   triangle inequality

A *metric space* is a pair (X, d) consisting of a non-empty set X and a metric d on X. A sequence $(x_n)_{n\in\mathbb{N}}$ in X is called *convergent* to $x \in X$, if the sequence of real numbers $(d(x_n, x))_{n\in\mathbb{N}}$ tends to zero. In this case, x is uniquely determined and is called the *limit* of the sequence. A *Cauchy sequence* in the metric space X is a sequence (x_n) in X such that for every $\varepsilon > 0$ there is $n_0 \in \mathbb{N}$ such that $d(x_m, x_n) < \varepsilon$ holds for all $m, n \geq n_0$. It is easy to see that every convergent sequence is a Cauchy sequence. If the converse holds as well, i.e. if every Cauchy sequence converges, we say that the metric space X is *complete*.

A map $\phi : X \to Y$ between metric spaces is called an *isometry* if for all $x, x' \in X$ the equality

$$d\big(\phi(x), \phi(x')\big) = d\big(x, x'\big)$$

holds true. An isometry is injective. If it is also surjective, then its inverse map is an isometry as well. In this case φ is called an *isometric isomorphism* of metric spaces. A *completion* of a metric space X is a complete metric space Y together with an isometry $\phi : X \to Y$, such that the image $\phi(X)$ is dense in Y, which means that every $y \in Y$ is the limit of a sequence in $\phi(X)$.

Theorem 4.2.1 *Every metric space X has a completion Y. The completion is uniquely determined up to isometric isomorphy.*

Proof One constructs a completion as the set of all Cauchy sequences in X modulo the equivalence relation $\sim$, where

$$(x_n) \sim (y_n) \quad \Leftrightarrow \quad d(x_n, y_n) \text{ tends to zero.}$$

The map $X \to Y$ is given by sending an element $x \in X$ to the class of the constant sequence $x_n = x$. Details can be found for instance in [Dei05], Chap. 6. □

Proposition 4.2.2 *Let $p \leq \infty$. Then $\mathbb{Q}$ is not complete in the metric d_p, induced by $|\cdot|_p$.*

We denote the resulting completion by $\mathbb{Q}_p$. The addition and multiplication of $\mathbb{Q}$ extend in a unique way to continuous maps $\mathbb{Q}_p \times \mathbb{Q}_p \to \mathbb{Q}_p$. With these operations, $\mathbb{Q}_p$ is a field, called the field of p-adic numbers. The absolute value $|\cdot|_p$ extends to a unique continuous map on $\mathbb{Q}_p$, which is an absolute value.

Proof We assume the proposition known for $p = \infty$, where $\mathbb{Q}_p = \mathbb{R}$. So let $p < \infty$ and write $|\cdot| = |\cdot|_p$. The incompleteness of $\mathbb{Q}$ is an easy consequence of a different characterization of $\mathbb{Q}_p$ which we will see in the next section.

Now for the extension of the operations. Let $x, y \in \mathbb{Q}_p$. As $\mathbb{Q}$ is dense in $\mathbb{Q}_p$, there are sequences (x_n) and (y_n), which converge in $\mathbb{Q}_p$ to x and y, respectively. These are Cauchy sequences in $\mathbb{Q}$. Therefore $(x_n + y_n)$ is a Cauchy sequence in $\mathbb{Q}$, too. Hence $(x_n + y_n)$ converges in $\mathbb{Q}_p$ to an element z. If (x'_n) and (y'_n) are a second set of sequences in $\mathbb{Q}$, converging to x and y as well, the sequence $(x'_n + y'_n)$

converges in $\mathbb{Q}$ to a limit, say z'. We show that $z = z'$. Then this element z depends only on x and y and we may call it $x + y$, thus extending the addition to $\mathbb{Q}_p$. To see that $z = z'$ let $\varepsilon > 0$. Then there exists $n \in \mathbb{N}$ with

- $|z - (x_n + y_n)| < \varepsilon/3$,
- $|z' - (x'_n + y'_n)| < \varepsilon/3$, and
- $|(x_n + y_n) - (x'_n + y'_n)| \leq |x_n - x'_n| + |y_n - y'_n| < \varepsilon/3$.

Then

$$|z - z'| \leq |z - (x_n + y_n)| + |z' - (x_n + y_n)|$$
$$\leq |z - (x_n + y_n)| + |z' - (x'_n + y'_n)| + |(x'_n + y'_n) - (x_n + y_n)| < \varepsilon.$$

As ε was arbitrary, we get $|z - z'| = 0$, so $z = z'$ and we have extended the addition to a continuous map. Analogously one extends multiplication to $\mathbb{Q}_p$. It is a straightforward verification to show that $\mathbb{Q}_p$ is a field with these operations. The absolute value $|\cdot|$ of $\mathbb{Q}$ extends to $\mathbb{Q}_p$ by the definition $|x| \overset{\text{def}}{=} \lim_{j \to \infty} |x_j|$, if $x = \lim_j X_j$. One checks that $|x|$ does not depend on the choice of the sequence (x_j) and that $|\cdot|$ is an absolute value. $\square$

Note that the strong triangle inequality $|x + y| \leq \max(|x|, |y|)$ holds on $\mathbb{Q}_p$ as well. It has astonishing consequences, such as that the set

$$\mathbb{Z}_p = \left\{ x \in \mathbb{Q}_p : |x|_p \leq 1 \right\}$$

is a subring of $\mathbb{Q}_p$. This ring is called the *ring of p-adic integers*.

4.3 Power Series

Fix a prime number p. In this section we give a second construction of the p-adic numbers. Every integer $n \geq 0$ can be written in the *base p expansion*,

$$n = \sum_{j=0}^{N} a_j p^j,$$

with uniquely determined coefficients $a_j \in \{0, 1, \ldots, p-1\}$. The sum of n and a second number $m = \sum_{i=0}^{M} b_i p^i$ is

$$n + m = \sum_{j=0}^{\max(M,N)+1} c_j p^j,$$

where each c_j depends only on $a_0, \ldots, a_j$ and $b_0, \ldots, b_j$. More precisely, one calculates the coefficients c_j as follows. First one sets $c'_j = a_j + b_j$. Then $0 \leq c'_j \leq 2p - 2$

and it may happen that $c'_j \geq p$. One finds the smallest index j, for which this happens, replaces c'_j by the remainder modulo p and increases c'_{j+1} by one. One repeats these steps until all coefficients are $\leq p - 1$. Further,

$$nm = \sum_{j=0}^{M+N+1} d_j p^j,$$

where d_j depends only on $a_0, \ldots, a_j$ and $b_0, \ldots, b_j$. These properties of addition and multiplication make it possible to extend these operations to the set Z of formal series of the form

$$\sum_{j=0}^{\infty} a_j p^j$$

with $0 \leq a_j < p$. By a *formal series* we mean the sequence of coefficients $(a_0, a_1, \ldots)$. The series is only a convenient way to denote the sequence.

Lemma 4.3.1 *With these operations the set Z is a ring. An element $x = \sum_{j=0}^{\infty} a_j p^j$ is invertible in Z if and only if $a_0 \neq 0$.*

Proof In order to show that Z is a ring, we only need to prove the existence of additive inverses. So let $x = \sum_{j=0}^{\infty} a_j p^j$. We have to show that there exists $y = \sum_{j=0}^{\infty} b_j p^j$ such that $x + y = 0$. One constructs the coefficients b_j inductively. In the case $a_0 = 0$ one sets $b_0 = 0$. Otherwise one sets $b_0 = p - a_0$. Suppose now that $b_0, \ldots, b_n$ have already been constructed such that for $y_n = \sum_{j=0}^{n} b_j p^j$ one has

$$x + y_n = \sum_{j=n+1}^{\infty} c_j p^j, \quad 0 \leq c_j < p.$$

If $c_{n+1} = 0$, one sets $b_{n+1} = 0$. Otherwise one sets $b_{n+1} = p - c_{n+1}$. In this way one gets an element $y = \sum_{j=0}^{\infty} b_j p^j$ which satisfies $x + y = 0$.

We now prove the second assertion. If $x = \sum_{j=0}^{\infty} a_j p^j$ is invertible, then $a_0 \neq 0$, as otherwise the series xy has vanishing zeroth term for every $y \in Z$. For the converse direction let $x = \sum_{j=0}^{\infty} a_j p^j$ with $a_0 \neq 0$. We construct a multiplicative inverse $y = \sum_{j=0}^{\infty} b_j p^j$ by giving the coefficients b_j successively. Since $\mathbb{F}_p = \mathbb{Z}/p\mathbb{Z}$ is a field, there is exactly one $1 \leq b_0 < p$ such that $a_0 b_0 \equiv 1 \bmod p$. Next let $0 \leq b_0, \ldots, b_n < p$ already be constructed such that

$$\left(\sum_{0 \leq j} a_j p^j \right) \left(\sum_{0 \leq j \leq n} b_j p^j \right) \equiv 1 \bmod p^{n+1}.$$

Then there is exactly one $0 \leq b_{n+1} < p$ such that

$$\left(\sum_{0\le j} a_j p^j\right)\left(\sum_{0\le j\le n+1} b_j p^j\right) \equiv 1 \bmod p^{n+2}.$$

The element $y = \sum_{j=0}^{\infty} b_j p^j$ thus constructed, satisfies the equation $xy = 1$. $\Box$

Lemma 4.3.2 *Let (a_j) be a sequence in $\{0, 1, \ldots, p - 1\}$. Then the series $\sum_{j=0}^{\infty} a_j p^j$ converges in $\mathbb{Q}_p$. The resulting map $\psi : Z \to \mathbb{Q}_p$ induces a ring iso-morphism $Z \cong \mathbb{Z}_p$.*

Proof Let $x_n = \sum_{j=0}^{n} a_j p^j$. We have to show that (x_n) is a Cauchy sequence in $\mathbb{Q}_p$. For $m \ge n \ge n_0$ one has

$$|x_m - x_n| = \left| \sum_{j=n+1}^{m} a_j p^j \right| \le \max_{n < j \le m} |a_j|_p |p^j|_p \le p^{-n_0}.$$

It follows that the sequence is Cauchy, so ψ is well defined. The map ψ is clearly a ring homomorphism. It remains to show bijectivity of $\phi : Z \to \mathbb{Z}_p$.

Injectivity: Let $x = \sum_{j=0}^{\infty} a_j p^j \ne 0$. Then there is a minimal j_0 with $a_{j_0} \ne 0$. We get

$$|\psi(x)| = \left| a_{j_0} p^{j_0} + \sum_{j=j_0+1}^{\infty} a_j p^j \right| = p^{-j_0},$$

since $|\sum_{j=j_0+1}^{\infty} a_j p^j| \le \max_{j > j_0} |a_j| p^{-j} < p^{-j_0}$. So $\psi(x) \ne 0$ and ψ is injective.

Surjectivity: We define an absolute value on Z by

$$|z| = \left| \phi(z) \right|_p.$$

We claim that Z is complete with this absolute value. Let (z_j) be a Cauchy sequence in Z. Then for every $k \in \mathbb{N}$ there exists a $j_0(k) \in \mathbb{N}$, such that for all $i, j \ge j_0(k)$ we have $|z_i - z_j| \le p^{-k}$, which means that $\psi(z_i) - \psi(z_j) \in p^k \mathbb{Z}_p$, so $z_i - z_j \in p^k Z$. We conclude that the coefficients of the power series z_i and z_j coincide up to the index $k - 1$. Hence there are coefficients a_ν for $\nu = 0, 1, 2, \ldots$, such that for every $k \in \mathbb{N}$ and every $j \ge j_0(k)$ one has $z_j \equiv \sum_{\nu=0}^{k-1} a_\nu p^\nu \bmod p^k Z$. Set

$$z = \sum_{\nu=0}^{\infty} a_\nu p^\nu \in Z.$$

It follows that the sequence (z_j) converges to z, so Z is indeed complete. It now suffices to show that $\psi(Z)$ contains a dense subset of $\mathbb{Z}_p$. But such a dense subset is the set of all rational numbers in $\mathbb{Z}_p$, i.e. the set of all $q = \pm p^k \frac{m}{n}$ with $k \ge 0$ and m, n coprime to p. Since Z is a ring, it suffices to show that $\frac{1}{n} \in Z$, if $n \in \mathbb{N}$ is coprime to p. Since n is coprime to p, the zeroth coefficient of its base p expansion is $\ne 0$, which means that n is invertible in Z. $\Box$

In this way we identify $\mathbb{Z}_p$ with the set of all power series in p. Since $\mathbb{Z}_p$ is the set of all $z \in \mathbb{Q}_p$ with $|z| \le 1$, the set $p^{-j}\mathbb{Z}_p$ is the set of all $z \in \mathbb{Q}_p$ with $|z| \le p^j$. So we have

$$\mathbb{Q}_p = \bigcup_{j=0}^{\infty} p^{-j}\mathbb{Z}_p.$$

Putting things together, $\mathbb{Q}_p$ can be viewed as the set of all Laurent series in p, with only finitely many negative powers of p occurring, i.e.

$$\mathbb{Q}_p = \left\{ \sum_{j=-N}^{\infty} a_j p^j : N \in \mathbb{N}, \ 0 \le a_j < p \right\}.$$

This description implies that the set $\mathbb{Q}_p$ is uncountable. In particular we have $\mathbb{Q} \ne \mathbb{Q}_p$ and so $\mathbb{Q}$ is not complete in the p-adic metric.

4.4 Haar Measures

The absolute value $|\cdot|_p$ defines a metric, which in turn defines a topology on $\mathbb{Q}_p$. We show that the groups $(\mathbb{Q}_p, +)$ and $(\mathbb{Q}_p^{\times}, \times)$ are locally compact groups with this topology. We then determine the Haar measures of these groups.

Lemma 4.4.1 *The additive and multiplicative groups of $\mathbb{Q}_p$, i.e. the groups $(\mathbb{Q}_p, +)$ and $(\mathbb{Q}^{\times}, \times)$, are locally compact groups. $\mathbb{Z}_p$ is a compact open subgroup of $(\mathbb{Q}_p, +)$ and $\mathbb{Z}_p^*$ is a compact open subgroup of $\mathbb{Q}_p^{\times}$.*

Proof The topology is given by a metric; hence it is Hausdorff. Next we need to show that $(\mathbb{Q}_p, +)$ is a topological group. We have to show that the maps

$$\begin{array}{cc} \mathbb{Q}_p \times \mathbb{Q}_p \to \mathbb{Q}_p & \mathbb{Q}_p \to \mathbb{Q}_p \\ (x, y) \mapsto x + y & x \mapsto (-x) \end{array}$$

are continuous. Let $x_j \to x$ and $y_j \to y$ be convergent sequences in $\mathbb{Q}_p$. We have to show that $x_j + y_j$ converges to $x + y$. By the triangle inequality we have

$$\left|(x_j + y_j) - (x + y)\right| = \left|(x_j - x) + (y_j - y)\right| \le |x_j - x| + |y_j - y|.$$

As the right-hand side converges to zero, the left-hand side does, and the claim follows. The continuity of the negation is even simpler as $|-x| = |x|$ holds for every x.

We finally have to show that every point in $\mathbb{Q}_p$ has a compact neighborhood. It suffices to do this for the point zero, because if $a \in \mathbb{Q}_p$ and U is a compact zero neighborhood, then $U + a$ is a compact neighborhood of a.

The subgroup $\mathbb{Z}_p$ is the closed ball of radius 1. Since the function $x \mapsto |x|_p$ only takes values in $\{p^k : k \in \mathbb{Z}\}$, the set $\mathbb{Z}_p$ is also the open ball of radius α for any $1 < \alpha < p$, so

$$\mathbb{Z}_p = \{x \in \mathbb{Q}_p : |x| < \alpha\}.$$

Therefore, it is an open neighborhood of zero, i.e. an open subgroup. We show that it is also compact. In a metric space, a set K is compact if and only if every sequence in K has an accumulation point in K. So let x_j be a sequence in $\mathbb{Z}_p$. We have to show that it has an accumulation point in $\mathbb{Z}_p$. Consider the subgroup $p\mathbb{Z}_p$. For every $k \in \mathbb{N}$ the group homomorphism

$$\mathbb{Z}_p \to \mathbb{Z}/p^k\mathbb{Z}, \quad \sum_{n=0}^{\infty} a_n p^n \mapsto \sum_{v=0}^{k-1} a_v p^v$$

is surjective and has kernel $p^k\mathbb{Z}_p$. So the index $[\mathbb{Z}_p : p^k\mathbb{Z}_p]$ equals $|\mathbb{Z}/p^k\mathbb{Z}| = p^k$. Consider first the case $k = 1$. In the disjoint coset decomposition $\mathbb{Z}_p = \bigcup_{i=1}^{p}(a_i + p\mathbb{Z}_p)$ there exists a coset which contains x_j for infinitely many $j \in \mathbb{N}$. Of these infinitely many, there are infinitely many for which the x_j lie in the same class modulo $p^2\mathbb{Z}_p$ and so on. This descending sequence of cosets is of the form

$$a + p^k\mathbb{Z}_p = \{x \in \mathbb{Z}_p : |x - a| \leq p^{-k}\} = \bar{B}_{p^{-k}}(a),$$

so they are closed balls in the p-adic metric whose radii tend to zero. The intersection of these balls contains an element by completeness. So there is an element $x \in \mathbb{Z}_p$ such that for every $n \in \mathbb{N}$ one has $x_j \equiv x \bmod p^n$ for infinitely many j. This means that x is an accumulation point.

The multiplicative group $\mathbb{Q}_p^\times$ is an open subset of $\mathbb{Q}_p$ and therefore Hausdorff and locally compact. The proofs of continuity of the group operations and the assertions about $\mathbb{Z}_p^\times$ are similar. We leave these to the reader as an exercise. $\qquad\square$

Note that $\mathbb{Z}_p^\times$ is the set of all $x \in \mathbb{Q}_p$ with $|x|_p = 1$.

Let μ denote the Haar measure on the group $(\mathbb{Q}_p, +)$, which maps the compact open subgroup $\mathbb{Z}_p$ to one, i.e.

$$\mu(\mathbb{Z}_p) = 1.$$

By invariance of μ we have for every measurable $A \subset \mathbb{Q}_p$ and every $x \in \mathbb{Q}_p$,

$$\mu(x + A) = \mu(A).$$

Lemma 4.4.2 *For every measurable subset $A \subset \mathbb{Q}_p$ and every $x \in \mathbb{Q}_p$ one has*

$$\mu(xA) = |x|_p \mu(A).$$

In particular, for every integrable function f and every $x \neq 0$:

$$\int_{\mathbb{Q}_p} f(x^{-1}y)\, d\mu(y) = |x|_p \int_{\mathbb{Q}_p} f(y)\, d\mu(y).$$

Proof Let $x \in \mathbb{Q}_p \setminus \{0\}$. The measure μ_x, defined by

$$\mu_x(A) = \mu(xA),$$

is again a Haar measure, as is easy to see. By uniqueness of Haar measures, there is $M(x) > 0$ with $\mu_x = M(x)\mu$. We have to show that $M(x) = |x|_p$. It suffices to show that $\mu(x\mathbb{Z}_p) = |x|_p$. Suppose $|x|_p = p^{-k}$; then $x = p^k y$ for some $y \in \mathbb{Z}_p^\times$, and therefore $x\mathbb{Z}_p = p^k\mathbb{Z}_p$, so it suffices to show $\mu(p^k\mathbb{Z}_p) = p^{-k}$. Start with the case $k \geq 0$. Then $[\mathbb{Z}_p : p^k\mathbb{Z}_p] = p^k$, so there is a disjoint decomposition of $\mathbb{Z}_p$,

$$\mathbb{Z}_p = \bigcup_{j=1}^{p^k} (x_j + p^k\mathbb{Z}_p).$$

By invariance of Haar measure it follows that

$$1 = \mu(\mathbb{Z}_p) = \sum_{j=1}^{p^k} \mu(x_j + p^k\mathbb{Z}_p) = p^k \mu(p^k\mathbb{Z}_p),$$

implying the assertion. If $k < 0$, one uses $[p^k\mathbb{Z}_p : \mathbb{Z}_p] = p^{-k}$ and proceeds similarly.   □

According to our convention, we shall write integration with respect to the Haar measure μ simply as dx, so we write

$$\int_{\mathbb{Q}_p} f(x)\, d\mu(x) = \int_{\mathbb{Q}_p} f(x)\, dx.$$

Proposition 4.4.3 *The measure* $\frac{dx}{|x|_p}$ *is a Haar measure on the multiplicative group* $(\mathbb{Q}_p^\times, \times)$.

Proof Let f be an integrable function and $y \in \mathbb{Q}_p^\times$. Then

$$\int_{\mathbb{Q}_p^\times} f(y^{-1}x)\frac{dx}{|x|_p} = |y|_p^{-1} \int_{\mathbb{Q}_p^\times} f(y^{-1}x)\frac{1}{|y^{-1}x|_p}\, dx = \int_{\mathbb{Q}_p^\times} f(x)\frac{dx}{|x|_p}$$

by Lemma 4.4.2.   □

The subgroup $\mathbb{Z}_p^\times$ of $\mathbb{Q}_p^\times$ is the kernel of the group homomorphism

$$\mathbb{Q}_p^\times \to \mathbb{Z}$$

$$x \mapsto \frac{\log(|x|_p)}{\log p}.$$

We can therefore write $\mathbb{Q}_p^\times$ as the disjoint union $\mathbb{Q}_p^\times = \bigcup_{k \in \mathbb{Z}} p^k \mathbb{Z}_p^\times$. Now

$$\mathrm{vol}_{\frac{dx}{|x|}}\left(p^k \mathbb{Z}_p^\times\right) = \mathrm{vol}_{\frac{dx}{|x|}}\left(\mathbb{Z}_p^\times\right).$$

It is therefore interesting to know the measure $\mathrm{vol}_{\frac{dx}{|x|}}\left(\mathbb{Z}_p^\times\right)$. One has

$$\mathrm{vol}_{\frac{dx}{|x|_p}}\left(\mathbb{Z}_p^\times\right) = \int_{\mathbb{Z}_p^\times} \frac{dx}{|x|} = \int_{\mathbb{Z}_p^\times} dx = \mathrm{vol}_{dx}\left(\mathbb{Z}_p^\times\right).$$

Considering the power series representation of $\mathbb{Z}_p$, we order the elements of $\mathbb{Z}_p^\times$ by their first coefficient to get a disjoint decomposition

$$\mathbb{Z}_p^\times = \bigcup_{\substack{a \bmod p \\ a \not\equiv 0 \bmod p}} (a + p\mathbb{Z}_p).$$

This means that the subgroup $1_p\mathbb{Z}_p$ of $\mathbb{Z}_p^\times$ has index $p - 1$, so

$$\mathrm{vol}_{dx}\left(\mathbb{Z}_p^\times\right) = (p - 1)\,\mathrm{vol}_{dx}\left(p\mathbb{Z}_p\right) = \frac{p - 1}{p}.$$

Definition 4.4.4 We define the *normalized multiplicative measure* on $\mathbb{Q}_p$ by

$$d^\times x = \frac{p}{p - 1} \frac{dx}{|x|_p}.$$

This Haar measure is uniquely determined by the property

$$\mathrm{vol}_{d^\times x}\left(\mathbb{Z}_p^\times\right) = 1.$$

To get used to integration with these measures, we compute the integral $\int_{\mathbb{Z}_p \smallsetminus \{0\}} |x|_p^s \, d^\times x$ for $s \in \mathbb{C}$ with $\mathrm{Re}(s) > 0$. We decompose $\mathbb{Z}_p \smallsetminus \{0\}$ into a disjoint union of the sets $p^k \mathbb{Z}_p^\times$ for $k \geq 0$. We get

$$\int_{\mathbb{Z}_p \smallsetminus \{0\}} |x|_p^s \, d^\times x = \sum_{k=0}^{\infty} p^{-ks} \underbrace{\int_{\mathbb{Z}_p^\times} d^\times x}_{=1} = \frac{1}{1 - p^{-s}}.$$

Remarkably, this is the pth Euler factor of the Riemann zeta function.

4.5 Direct and Projective Limits

In this section we give yet another description of the p-adic numbers. We will not be using this description much, but the techniques to do so will be useful in later chapters.

Direct and projective limits can be defined for all algebraic structures like groups, rings, vector spaces, modules and so on. In this section we construct these limits exemplarily for groups and rings. We shall show that $\mathbb{Z}_p$ can be written as a projective limit of finite rings. Later, in Sect. 7.5, we shall make use of the notion of direct limits of vector spaces.

We first recall the notion of a *partial order* on a set I. This is a relation $\leq$ such that for all $a, b, c \in I$ one has

- $a \leq a$,   reflexivity
- $a \leq b$ and $b \leq a$ implies $a = b$, and   antisymmetry
- $a \leq b$ and $b \leq c$ implies $a \leq c$.   transitivity

Definition 4.5.1 A *directed set* is a tuple $(I, \leq)$ consisting of a non-empty set I and a partial order $\leq$ on I, such that any two elements of I have a common upper bound, which means that for any two $a, b \in I$ there exists an element $c \in I$ with

$$a \leq c \quad \text{and} \quad b \leq c.$$

Examples 4.5.2

- The set $\mathbb{N}$ of natural numbers is an example with the natural order $\leq$. In this case the order is even *linear*, which means that any two elements on $\mathbb{N}$ can be compared. Every linear order is directed.
- Let Ω be an infinite set and let I be the set of all finite subsets of Ω, ordered by inclusion, so $A \leq B \Leftrightarrow A \subset B$. Then I is directed, as for $A, B \in I$ the union $C = A \cup B$ is an upper bound.

Definition 4.5.3 A *direct system* of groups consists of the following data

- a directed set $(I, \leq)$,
- a family $(A_i)_{i \in I}$ of groups and
- a family of group homomorphisms

$$\varphi_i^j : A_i \to A_j, \quad \text{if } i \leq j,$$

such that the following axioms are satisfied:

$$\varphi_i^i = \mathrm{Id}_{A_i} \quad \text{and} \quad \phi_j^k \circ \varphi_i^j = \varphi_i^k, \text{ if } i \leq j \leq k.$$

Examples 4.5.4

- Let A be a group and let $(A_i)_{i \in I}$ be a family of pairwise distinct subgroups, such that for any two indices $i, j \in I$ there exists an index $k \in I$, such that $A_i, A_j \subset A_k$. Then the A_i form a direct system, if on I one installs the partial order

$$i \leq j \quad \Leftrightarrow \quad A_i \subset A_j,$$

and if for group homomorphisms φ_i^j one takes the inclusions.

- Let X be a topological space, fix $x_0 \in X$ and let I be the set of all neighborhoods of x_0 in X. For $U \in I$ let A_U be the group $C(U)$ of all continuous functions from U to $\mathbb{C}$. We order I by the *inverse inclusion*, i.e. $U \leq V \Leftrightarrow U \supset V$. The *restriction homomorphisms*

$$\varphi_U^V : C(U) \to C(V), \quad \varphi_U^V(f) = f|_V$$

form a direct system.

Analogously one introduces a *direct system of rings*, by insisting that all A_i be rings and the φ_i^j be ring homomorphisms.

Definition 4.5.5 Let $((A_i)_{i \in I}, (\varphi_i^j)_{i \leq j})$ be a direct system of groups. The direct limit of the system is the set

$$\varinjlim_{i \in I} A_i \overset{\text{def}}{=} \coprod_{i \in I} A_i \, / \sim,$$

where $\coprod$ denotes the disjoint union and $\sim$ the following equivalence relation: For $a \in A_i$ and $b \in A_j$ we say $a \sim b$, if there is $k \in I$ with $k \geq i, j$ and $\varphi_i^k(a) = \varphi_j^k(b)$.

On the set $A = \varinjlim A_i$ we define a group multiplication as follows. Let $a \in A_i$ and $b \in A_j$ and let $[a]$ and $[b]$ denote their equivalence classes in A. Then there is $k \in I$ with $k \geq i$ and $k \geq j$. We define $[a][b]$ to be the equivalence class of the element $\varphi_i^k(a)\varphi_j^k(b)$ in A_k, so $[a][b] = [\varphi_i^k(a)\varphi_j^k(b)]$.

Proposition 4.5.6 *The multiplication is well defined and makes the set A a group. This group is called the direct limit of the system (A_i, φ_j^i). For every $i \in I$ the map*

$$\psi_i : A_i \hookrightarrow \coprod_{j \in I} A_j \to \coprod_{j \in I} A_j / \sim$$

is a group homomorphism.

The direct limit has the following universal property: Let Z be a group and for every $i \in I$ let a group homomorphism $\alpha_i : A_i \to Z$ be given, such that $\alpha_i = \alpha_j \circ \varphi_i^j$ holds if $i \leq j$. Then there exists exactly one group homomorphism $\alpha : A \to Z$ with $\alpha_i = \alpha \circ \psi_i$ for every $i \in I$.

In other words: if for any two indices $i \leq j$ the diagram

$$
\begin{array}{ccc}
A_i & \overset{\varphi_i^j}{\longrightarrow} & A_j \\
{\scriptstyle \alpha_j} \downarrow & \swarrow {\scriptstyle \alpha_j} & \\
Z & &
\end{array}
$$

commutes, then there exists exactly one $\alpha : A \to Z$, *such that for every i the diagram*

commutes.

Proof To show well-definedness, we need to show that the product is independent of the choice of k. If k' is another element of I with $k' \geq i, j$, there exists a common upper bound l for k and k', so $l \geq k, k'$. We show that the construction gives the same element with l as with k. Then we apply the same argument to k' and l. Note that by definition for every $c \in A_k$ one has $[c] = [\varphi_k^l(c)]$. As φ_k^l is a group homomorphism, it follows that

$$\left[\varphi_i^k(a)\varphi_j^k(b)\right] = \left[\varphi_k^l\left(\varphi_i^k(a)\varphi_j^k(b)\right)\right] = \left[\varphi_k^l\left(\varphi_i^k(a)\right)\varphi_k^l\left(\varphi_j^k(b)\right)\right] = \left[\varphi_i^l(a)\varphi_j^l(b)\right].$$

This proves well-definedness. The rest is left as an exercise to the reader.   □

If (A_i) is a direct system of rings, then A is a ring with the same universal property for ring homomorphisms.

Example 4.5.7 *In the case of the direct system* $(C(U))_U$, *where U runs through all neighborhoods of a point in a topological space, one calls the elements of* $\varinjlim C(U)$ *germs of continuous functions.*

There is a dual construction to the direct limit, called the projective limit. Since our most prominent example is a projective limit of rings, we will formulate the construction for rings. For groups, vector spaces and so on, one simply replaces the word *ring* by the word *group*, etc.

Definition 4.5.8 A *projective system* of rings consists of the following data

- a directed set $(I, \leq)$,
- a family $(A_i)_{i \in I}$ of rings and
- a family of ring homomorphisms

$$\pi_i^j : A_j \to A_i, \quad \text{if } i \leq j,$$

such that the following axioms are met:

$$\pi_i^i = \text{Id}_{A_i} \quad \text{and} \quad \pi_i^j \circ \pi_j^k = \pi_i^k, \text{ if } i \leq j \leq k.$$

Note that, in comparison to a direct system, the homomorphisms now run in the opposite direction.

Example 4.5.9 *Let p be a prime number. Let $I = \mathbb{N}$ with the usual order. For $n \in \mathbb{N}$ let $A_n = \mathbb{Z}/p^n\mathbb{Z}$ and for $m \geq n$ let $\pi_n^m : \mathbb{Z}/p^m\mathbb{Z} \to \mathbb{Z}/p^n\mathbb{Z}$ be the canonical projection. Then (A_n, π_n^m) form a projective system of rings.*

Definition 4.5.10 Let (A_i, π_i^j) be a projective system of rings. The *projective limit* of the system is the set

$$A = \varprojlim A_i$$

of all $a \in \prod_{i \in I} A_i$ such that $a_i = \pi_i^j(a_j)$ holds for every pair $i \leq j$ in I.

Proposition 4.5.11 *The projective limit A of the system (A_i) is a subring of the product $\prod_{i \in I} A_i$. Let $\mathrm{pr}_i : A \to A_i$ be the map given by the projection to the ith coordinate. Then pr_i is a ring homomorphism. The projective limit has the following universal property: If Z is a ring with ring homomorphisms $\alpha_i : Z \to A_i$, such that $\alpha_i = \pi_i^j \circ \alpha_j$ holds for all $i \leq j$ in I, then there exists exactly one ring homomorphism $\alpha : Z \to A$, such that $\alpha_i = \mathrm{pr}_i \circ \alpha$ for every $i \in I$. In other words: If for all $i \leq j$ the diagram*

commutes, then there is a unique ring homomorphism α such that for every $i \in I$ the diagram

$$
\begin{array}{ccc}
A & \xrightarrow{\mathrm{pr}_i} & A_i \\
\alpha \uparrow & \nearrow \alpha_i & \\
Z & &
\end{array}
$$

commutes.

Proof The proof is left to the reader.   □

The next theorem says that the p-adic integers can be constructed as a projective limit.

Theorem 4.5.12 *Let p be a prime number. Then the ring $\varprojlim \mathbb{Z}/p^n\mathbb{Z}$ is isomorphic with $\mathbb{Z}_p$.*

Proof We view $\mathbb{Z}_p$ as the ring of power series $a = \sum_{j=0}^{\infty} a_j p^j$ with $a_j \in \{0, \ldots, p-1\}$. We define a map $\phi : \mathbb{Z}_p \to \varprojlim \mathbb{Z}/p^n\mathbb{Z}$, by setting the nth coordinate $\phi(a)_n$ of $\phi(a)$ to $\sum_{j=0}^{n-1} a_j p^j \mod p^n$. This defines a ring homomorphism, which is easily seen to be bijective.   □

4.6 Exercises

Exercise 4.1 For $a \in \mathbb{Q}_p$ and $r > 0$ let $B_r(a)$ be the open ball

$$B_r(a) = \left\{ x \in \mathbb{Q}_p : |a - x|_p < r \right\}.$$

Show:

(a) If $b \in B_r(a)$, then $B_r(a) = B_r(b)$.
(b) Two open balls are either disjoint or one is contained in the other.

Exercise 4.2 Show that there is a canonical ring isomorphism

$$\mathbb{Q}_p \cong \mathbb{Q} \otimes_\mathbb{Z} \mathbb{Z}_p.$$

Exercise 4.3 Show that

$$\sum_{j=-N}^{\infty} a_j p^j \mapsto \sum_{j=-N}^{\infty} a_j p^{-j}, \quad 0 \le a_j < p,$$

defines a continuous map $\mathbb{Q}_p \to \mathbb{R}$. Is this a ring homomorphism? Describe its image.

Exercise 4.4 Let $\mathbb{T} = \{z \in \mathbb{C} : |z| = 1\}$ denote the circle group and let $\chi : \mathbb{Z}_p \to \mathbb{T}$ be a continuous group homomorphism, i.e. $\chi(a+b) = \chi(a)\chi(b)$.

Show that there exists $k \in \mathbb{N}$ with $\chi(p^k \mathbb{Z}_p) = 1$. It follows that χ factors through the finite group $\mathbb{Z}_p/p^k\mathbb{Z}_p \cong \mathbb{Z}/p^k\mathbb{Z}$, so the image of χ is finite.
(Hint: let $U = \{z \in \mathbb{T} : \mathrm{Re}(z) > 0\}$. Then U is an open neighborhood of the unit, so $\chi^{-1}(U)$ is an open neighborhood of zero.)

Exercise 4.5 Let $e_p : \mathbb{Q}_p \to \mathbb{T}$ be defined by

$$e_p\left(\sum_{j=-N}^{\infty} a_j p^j \right) = \exp\left(2\pi i \sum_{j=-N}^{-1} a_j p^j \right),$$

where $a_j \in \mathbb{Z}$ with $0 \le a_j < p$. Show that e_p is a continuous group homomorphism.

Exercise 4.6 Let $\chi : \mathbb{Q}_p \to \mathbb{T}$ be a continuous group homomorphism. Show that there exists exactly one $a \in \mathbb{Q}_p$ with $\chi(x) = e_p(ax)$.
(Hint: reduce to the case $\chi(\mathbb{Z}_p) = 1$ and consider $\chi(1/p^k)$ for $k \in \mathbb{N}$.)

Exercise 4.7 Let p be a prime number and let $\mathcal{O}$ be the polynomial ring $\mathbb{F}_p[t]$. As one can perform division with remainder, the ring $\mathcal{O}$ is a factorial principal domain. The prime ideals of $\mathcal{O}$ are the principal ideals of the form 0 or (η), where $\eta \neq 0$ is an irreducible polynomial in $\mathcal{O}$.

(a) For such η let $v_\eta : \mathcal{O} \to \mathbb{N}_0 \cup \{\infty\}$ be defined by

$$v_\eta(f) = \sup\{k : f \in (\eta^k)\}.$$

Show that

$$|f|_\eta = p^{-\deg(\eta)v_\eta(f)}$$

defines an absolute value on the ring $\mathcal{O}$.

(b) Let $v_\infty(f) = -\deg(f)$. Show that $|f|_\infty = p^{-v_\infty(f)}$ is an absolute value.

(c) Prove the product formula

$$\prod_{\eta \leq \infty} |f|_\eta = 1.$$

Exercise 4.8 Prove the universal property of Proposition 4.5.6.

Exercise 4.9 Prove Proposition 4.5.11.

Exercise 4.10 Show that the direct limit $\varinjlim A_i$ of abelian groups is abelian. Show that the converse does not hold in general.

Exercise 4.11 Show that the universal property determines the direct limit up to isomorphy. More precisely, let (A_i, φ_i^j) be a direct system of groups and let B be a group and for each i let $\beta_i : A_i \to B$ be a group homomorphism such that (B, β_i) has the universal property of the direct limit $A = \varinjlim A_i$. Show that A and B are isomorphic.

Chapter 5
Adeles and Ideles

In order to understand the field $\mathbb{Q}$ of rational numbers it seems necessary to consider all its completions $\mathbb{Q}_p$ at the same time. This is best done through a single object which contains all the fields $\mathbb{Q}_p$. The first candidate would be the direct product $\prod_p \mathbb{Q}_p$. Unfortunately, the product topology is no longer locally compact, which means there is no Haar integration for addition and multiplication. This difficulty is remedied by using the restricted product which is introduced in the first section of this chapter. It yields a locally compact ring that still contains all $\mathbb{Q}_p$.

5.1 Restricted Products

Tychonov's theorem says that the product of compact spaces is compact (see [DE09], Appendix A). If one replaces the word 'compact' by 'locally compact', the analogous assertion becomes false in general, as the following lemma shows.

Lemma 5.1.1 *Given an index set I and a locally compact Hausdorff space X_i for every $i \in I$. Then the product space $X = \prod_{i \in I} X_i$ is locally compact if and only if almost all of the X_i are compact.*

Proof First note the following observation: if a product $X = \prod_i X_i$ is compact, then so is each factor X_i, since it is the image of the projection $p_i : X \to X_i$, which is a continuous map.

Let $E \subset I$ be a finite subset and for each $i \in E$ let $U_i \subset X_i$ be a given subset. These data determine a *rectangle*

$$R = R\big((U_i)_{i \in E}\big) = \prod_{i \in E} U_i \times \prod_{i \in I \smallsetminus E} X_i.$$

A rectangle is open, respectively closed, if and only if all the U_i are open, respectively closed. By definition of the product topology, every open set in X is a union of open rectangles. Note that the intersection of two open rectangles is again an open

A. Deitmar, *Automorphic Forms*, Universitext,
DOI 10.1007/978-1-4471-4435-9_5, © Springer-Verlag London 2013

rectangle. If X is locally compact, there must exist a non-empty open rectangle with compact closure. The closure of the rectangle $R((U_i)_i)$ is the rectangle $R((\overline{U}_i)_i)$, so that by our remark above almost all X_i are compact.

The converse direction is a direct consequence of Tychonov's theorem and the simple observation that finite products of locally compact spaces are locally compact. □

Let $(X_i)_{i \in I}$ be a family of locally compact spaces and let, for each $i \in I$, a compact open subset $K_i \subset X_i$ be given. Define the *restricted product* as

$$X = \widehat{\prod_{i \in I}}^{K_i} X_i \overset{\text{def}}{=} \left\{ x \in \prod_{i \in I} X_i : x_i \in K_i \text{ for almost all } i \in I \right\}$$

$$= \bigcup_{\substack{E \subset I \\ \text{finite}}} \prod_{i \in E} X_i \times \prod_{i \notin E} K_i.$$

If it is clear which sets K_i have been chosen, we leave out the K_i from the notation, so we simply write $X = \widehat{\prod}_{i \in I} X_i$ then.

We now define the *restricted product topology* as follows. A *restricted open rectangle* is a subset of the restricted product of the form

$$\prod_{i \in E} U_i \times \prod_{i \notin E} K_i,$$

where $E \subset I$ is a finite subset and $U_i \subset X_i$ is an arbitrary open set for $i \in E$. A subset $A \subset \widehat{\prod}_{i \in I} X_i$ is called *open* if it can be written as a union of restricted open rectangles. Note that the intersection of two restricted open rectangles is again a restricted open rectangle. This is the place where we use the fact that the K_i are open sets.

Lemma 5.1.2

(a) *If I is finite, then*

$$\prod_i X_i = \widehat{\prod_i} X_i$$

and the restricted product topology coincides with the product topology.

(b) *For every disjoint decomposition $I = A \cup B$ one has*

$$\widehat{\prod_{i \in I}} X_i \cong \left(\widehat{\prod_{i \in A}} X_i \right) \times \left(\widehat{\prod_{i \in B}} X_i \right).$$

(c) *The inclusion map $\widehat{\prod}_i X_i \hookrightarrow \prod_i X_i$ is continuous, but the restricted product topology only coincides with the subspace topology, if $X_i = K_i$ holds for almost all $i \in I$.*

(d) *If all X_i are locally compact, then so is $X = \widehat{\prod}_i X_i$.*

Proof (a) is clear. For (b) note that both sides describe the same subset of $\prod_i X_i$. The definition of the restricted product topology implies that the left-hand side indeed has the product topology of the two factors on the right.

(c) To prove continuity, we have to show that the preimage of a set of the form $\prod_{i\in E} U_i \times \prod_{i\notin E} X_i$ is open in $\widehat{\prod}_i X_i$, where $E \subset I$ is a finite set and every $U_i \subset X_i$ is open. This follows from (a) and (b). The second assertion is clear.

We now show (d). Let $x \in X$. Then there exists a finite set $E \subset I$ such that $x_i \in K_i$, if $i \notin E$. For every $i \in E$ we choose a compact neighborhood U_i of x_i. Then $\prod_{i\in E} U_i \times \prod_{i\notin E} K_i$ is a compact neighborhood of x, i.e. X is locally compact. □

5.2 Adeles

By a *place* of $\mathbb{Q}$ we mean either a prime number or ∞; the latter is called the *infinite place*. We write $p < \infty$, if p is a prime number and $p \leq \infty$, if p is an arbitrary place.

The wording stems from algebraic geometry, as these 'places' in many ways behave like the points on a curve. A *set of places* is a subset

$$S \subset \{p : \text{prime number}\} \cup \{\infty\}.$$

By $\mathbb{Q}_p$ we denote the completion of $\mathbb{Q}$ at the place p, so in particular,

$$\mathbb{Q}_\infty = \mathbb{R}.$$

We define the set of *finite adeles* as the restricted product

$$\mathbb{A}_{\text{fin}} = \widehat{\prod_{p<\infty}}^{\mathbb{Z}_p} \mathbb{Q}_p,$$

and the set of *adeles* as

$$\mathbb{A} = \mathbb{A}_{\text{fin}} \times \mathbb{R}.$$

We also write

$$\mathbb{A} = \widehat{\prod_{p\leq\infty}} \mathbb{Q}_p,$$

although this is not a restricted product as at the place ∞ there is no restriction given.

For an arbitrary set of places S we write

$$\mathbb{A}_S = \widehat{\prod_{p\in S}} \mathbb{Q}_p \quad \text{and} \quad \mathbb{A}^S = \widehat{\prod_{p\notin S}} \mathbb{Q}_p.$$

Note that

$$\mathbb{A} = \mathbb{A}_S \times \mathbb{A}^S.$$

Theorem 5.2.1

(a) *For every set of places S the ring $\mathbb{A}_S$ is a locally compact topological ring.*
(b) *By the diagonal embedding, $\mathbb{Q}$ is a discrete subset of $\mathbb{A}$ and $\mathbb{A}/\mathbb{Q}$ is compact.*
(c) *$\mathbb{Q}$ is dense in $\mathbb{A}_{\mathrm{fin}}$.*

Proof The space $\mathbb{A}_S$ is locally compact by Lemma 5.1.2. For (a) we have to show that addition and multiplication are continuous maps $\mathbb{A}_S \times \mathbb{A}_S \to \mathbb{A}_S$. We do this for the addition only, as the multiplication is similar. Let $a, b \in \mathbb{A}_S$ and let U be an open neighborhood of $a + b$. We have to show that there are open neighborhoods V, W of a and b, respectively, such that $V + W \subset U$, where $V + W$ is the set of all $v + w$ with $v \in V$ and $w \in W$. By shrinking U we can assume $U = \prod_{p \in E} U_p \times \prod_{p \in S \setminus E} \mathbb{Z}_p$ for some finite set $E \subset S$. For $p \in E$ the addition is continuous on $\mathbb{Q}_p$, so there are open neighborhoods $V_p, W_p \subset \mathbb{Q}_p$ of a_p and b_p, respectively, such that $V_p + W_p \subset U_p$. Setting $V = \prod_{p \in E} V_p \times \prod_{p \in S \setminus E} \mathbb{Z}_p$ and $W = \prod_{p \in E} W_p \times \prod_{p \in S \setminus E} \mathbb{Z}_p$ we conclude that V and W are open neighborhoods of a and b, respectively, and that $V + W \subset U$ as claimed.

For part (b) let

$$U = \left(-\frac{1}{2}, \frac{1}{2}\right) \times \prod_{p < \infty} \mathbb{Z}_p.$$

The set U is an open neighborhood of zero in $\mathbb{A}$. For $r \in \mathbb{Q} \cap U$, one has $|r|_p \le 1$ for every $p < \infty$ and so $r \in \mathbb{Z}$. Further, one has $|r|_\infty < \frac{1}{2}$ and so finally $r = 0$. We have found an open zero neighborhood U with $U \cap \mathbb{Q} = \{0\}$. As $\mathbb{Q}$ is a subgroup of the additive group of $\mathbb{A}$ it follows that $\mathbb{Q}$ is discrete in $\mathbb{A}$. For the compactness it suffices to show that the compact set $K = [0, 1] \times \prod_{p < \infty} \mathbb{Z}_p$ contains a set of representatives of $\mathbb{A}/\mathbb{Q}$, because that means the projection $P : K \to \mathbb{A}/\mathbb{Q}$ is surjective, so $\mathbb{A}/\mathbb{Q}$ is the image of a compact set under a continuous map, hence compact.

So let $x \in \mathbb{A}$. Then there is a finite set of places E such that $p \notin E \Rightarrow x_p \in \mathbb{Z}_p$. For $p \in E$, $p < \infty$ we write

$$x_p = \sum_{j=-N}^{\infty} a_j p^j.$$

Then

$$x_p - \underbrace{\sum_{j=-N}^{-1} a_j p^j}_{=r \in \mathbb{Q}} \in \mathbb{Z}_p.$$

For a second prime number $q \neq p$ we have

$$|r|_q = \left| \sum_{j=-N}^{-1} a_j p^j \right|_q \leq \max\{|a_j p^j|_q\} \leq 1.$$

We replace x by $x - r$, thus reducing E to $E \smallsetminus \{p\}$. Iterating this argument, we arrive at $E = \{\infty\}$, so $x_p \in \mathbb{Z}_p$ for all primes p. This means that $x \in \mathbb{R} \times \prod_{p<\infty} \mathbb{Z}_p$. Modulo $\mathbb{Z}$ one can then move x to $[0, 1] \times \prod_{p<\infty} \mathbb{Z}_p = K$.

For (c) it suffices to show that $\mathbb{Z}$ is dense in $\widehat{\mathbb{Z}} = \prod_{p<\infty} \mathbb{Z}_p$, since $\mathbb{A}_{\mathrm{fin}} = \mathbb{Q}\widehat{\mathbb{Z}}$. So we have to show that $\mathbb{Z}$ meets every open subset of $\widehat{\mathbb{Z}}$. Every such open set is a union of sets of the form

$$U = \prod_{p \in E} B_p \times \prod_{p \notin E} \mathbb{Z}_p$$

for a finite set of primes E, where each B_p is an open ball in $\mathbb{Z}_p$. This means that B_p can be written as $B_p = n_p + p^{k_p} \mathbb{Z}_p$ for some $n_p \in \mathbb{Z}$ and a $k_p \in \mathbb{N}_0$. We have to show that there exists $l \in \mathbb{Z}$, such that for every $p \in E$ one has $l \in n_p + p^{k_p} \mathbb{Z}_p$, or $l \equiv n_p \bmod p^{k_p}$. The existence of such an l follows from the Chinese Remainder Theorem. $\qquad \square$

Since $\mathbb{A}$ is a locally compact ring, so in particular a locally compact group under addition, there is an additive Haar measure dx on $\mathbb{A}$. A *simple function* f on $\mathbb{A}$ is a function of the form $f = \prod_{p \leq \infty} f_p$ with $f_p = \mathbf{1}_{\mathbb{Z}_p}$ for almost all p.

Theorem 5.2.2 *The Haar measure dx on $(\mathbb{A}, +)$ can be normalized such that for every integrable simple function $f = \prod_p f_p$ one has the product formula*

$$\int_{\mathbb{A}} f(x) \, dx = \prod_p \int_{\mathbb{Q}_p} f_p(x_p) \, dx_p.$$

The Haar measure on $\mathbb{Q}_p$ is normalized such that $\mathrm{vol}(\mathbb{Z}_p) = 1$ holds for $p < \infty$ and that dx_∞ is the Lebesgue measure. The product is finite, i.e. almost all factors are equal to one.

This theorem also holds for $\mathbb{A}_S$, where S is any set of places. In the sequel we shall always use the normalization of Haar measures indicated by this theorem.

Proof A *simple set* is a subset A of $\mathbb{A}$ of the form $A = \prod_p A_p$, where every A_p is open in $\mathbb{Q}_p$ and $A_p = \mathbb{Z}_p$ holds for almost all p. For a simple set we define

$$\mu(A) \overset{\text{def}}{=} \prod_p \int_{\mathbb{Q}_p} \mathbf{1}_{A_p} \, dx_p.$$

The simple sets form a generating set of the Borel σ-algebra of $\mathbb{A}$ which is stable under intersections, so this prescription defines a measure, which is easily seen to be a Haar measure.   □

5.3 Ideles

The group of units of the adele ring can be described as follows,

$$\mathbb{A}^\times = \left\{ a \in \mathbb{A} : \begin{matrix} a_p \neq 0 \ \forall p \leq \infty \\ |a_p|_p = 1 \text{ for almost all } p \end{matrix} \right\}.$$

If we equip $\mathbb{A}^\times$ with the subspace topology of $\mathbb{A}$, then the multiplication is continuous, but not the inverse map $x \mapsto x^{-1}$. In order to make $\mathbb{A}^\times$ a topological group, we need more open sets. We need to insist that with every open set U the set $U^{-1} = \{u^{-1} : u \in U\}$ is open as well. The topology of $\mathbb{A}$ is generated by sets of the form

$$\prod_{p \in E} U_p \times \prod_{p \notin E} \mathbb{Z}_p,$$

where E is a finite set of places. The subspace topology of $\mathbb{A}^\times$ is thus generated by all sets of the form

$$U = \left\{ a \in \mathbb{A}^\times : \begin{matrix} a_p \in U_p, \ p \in E \\ |a_p| \leq 1, \ p \notin E \end{matrix} \right\},$$

where we can insist that every U_p lies in $\mathbb{Q}_p^\times$. So we need the condition that the sets of the form

$$U^{-1} = \left\{ a \in \mathbb{A}^\times : \begin{matrix} a_p^{-1} \in U_p, \ p \in E \\ |a_p| \geq 1, \ p \notin E \end{matrix} \right\}$$

are open as well. This implies that the intersection of two sets, one of the form U and one of the form U^{-1} (different U), is open. Such sets are of the form

$$W = \left\{ a \in \mathbb{A}^\times : \begin{matrix} a_p \in W_p, \ p \in E \\ |a_p| = 1, \ p \notin E \end{matrix} \right\},$$

where W_p is any open subset of $\mathbb{Q}_p^\times$. One notes that sets of the form U and of the form U^{-1} can be written as unions of sets of the form W. We have shown the following lemma.

Lemma 5.3.1 *The coarsest topology on $\mathbb{A}^\times$, which contains the subspace topology of $\mathbb{A}$ and which makes $\mathbb{A}^\times$ a topological group, is the topology generated by all sets of the type W above. It is the restricted product topology*

$$\mathbb{A}^\times = \widehat{\prod_{p < \infty}}^{\mathbb{Z}_p^\times} \mathbb{Q}_p^\times \times \mathbb{R}^\times.$$

The locally compact group $\mathbb{A}^\times$ with this topology is called the idele-group of $\mathbb{Q}$. Its elements are called ideles.

Definition 5.3.2 We define the *absolute value of an idele* $a \in \mathbb{A}^\times$ as

$$|a| = \prod_p |a_p|_p.$$

This product is well defined, as only finitely many of the factors are different from one. We extend the definition to all of $\mathbb{A}$ by setting $|a| = 0$ if $a \in \mathbb{A} \smallsetminus \mathbb{A}^\times$. Note that the identity $|a| = \prod_p |a_p|_p$ holds in this case as well, if one interprets the product as the limit $\lim_{N \to \infty} \prod_{p \le N} |a_p|_p$. Let

$$\mathbb{A}^1 = \{a \in \mathbb{A}^\times : |a| = 1\}.$$

According to the product formula one has $\mathbb{Q}^\times \subset \mathbb{A}^1$. We write

$$\widehat{\mathbb{Z}} = \prod_{p < \infty} \mathbb{Z}_p.$$

Then $\widehat{\mathbb{Z}}$ is a compact subring of $\mathbb{A}_{\text{fin}}$. The unit group is

$$\widehat{\mathbb{Z}}^\times = \prod_{p < \infty} \mathbb{Z}_p^\times.$$

Theorem 5.3.3 *The subgroup $\mathbb{Q}^\times$ is discrete in $\mathbb{A}^\times$ and the quotient $\mathbb{A}^1/\mathbb{Q}^\times$ is compact. More precisely, there is a canonical isomorphism*

$$\mathbb{A}^1/\mathbb{Q}^\times \cong \widehat{\mathbb{Z}}^\times.$$

In particular, $F^1 = \widehat{\mathbb{Z}}^\times \times \{1\}$ is a set of representatives of $\mathbb{A}^1/\mathbb{Q}^\times$ and $F = \widehat{\mathbb{Z}}^\times \times (0, \infty)$ is a set of representatives of $\mathbb{A}^\times/\mathbb{Q}^\times$. The absolute value induces an isomorphism of topological groups: $\mathbb{A}^\times \cong \mathbb{A}^1 \times (0, \infty)$. Further, one has $\mathbb{A}^1 \cong \mathbb{A}_{\text{fin}}^\times \times \{\pm 1\}$.

Proof Choose $0 < \varepsilon < 1$ and set

$$U = (1 - \varepsilon, 1 + \varepsilon) \times \prod_{p < \infty} \mathbb{Z}_p^\times.$$

Then U is an open neighborhood of the unit in $\mathbb{A}^\times$. With $r \in \mathbb{Q} \cap U$ we get $|r|_p = 1$ for every prime p, so $r \in \mathbb{Z}$ and $r^{-1} \in \mathbb{Z}$. We have $r \in (1 - \varepsilon, 1 + \varepsilon)$, so $r = 1$.

Consider the map $\eta : \prod_p \mathbb{Z}_p^\times \to \mathbb{A}^1/\mathbb{Q}^\times$ given by $x \mapsto (x, 1)\mathbb{Q}^\times$. We claim that η is an isomorphism of topological groups. The map η is a group homomorphism and since the map $\prod_p \mathbb{Z}_p^\times \hookrightarrow \mathbb{A}^\times$ is continuous, the map η is continuous. The inverse map is given by $x = (x_{\mathrm{fin}}, x_\infty) \mapsto \frac{1}{x_\infty} x_{\mathrm{fin}}$, where we note that for $x \in \mathbb{A}^1$ we have $x_\infty \in \mathbb{Q}^\times$.

The isomorphism $\mathbb{A}^\times$ to $\mathbb{A}^1 \times (0, \infty)$ is given by the map $x \mapsto (\tilde{x}, |x|)$, where $\tilde{x} \in \mathbb{A}$ is defined by

$$(\tilde{x})_p = \begin{cases} x_p & \text{if } p < \infty, \\ \frac{x_\infty}{|x|} & \text{if } p = \infty. \end{cases}$$

Finally, the map $\phi : \mathbb{A}_{\mathrm{fin}}^\times \times \{\pm 1\} \to \mathbb{A}^1$:

$$\phi(a_{\mathrm{fin}}, \varepsilon) = \left(a_{\mathrm{fin}}, \varepsilon |a_{\mathrm{fin}}|^{-1}\right)$$

is an isomorphism.   □

Proposition 5.3.4

(a) *The set* $\mathbb{A}_{\mathrm{fin}}^\times$ *is the disjoint union*

$$\mathbb{A}_{\mathrm{fin}}^\times = \coprod_{q \in \mathbb{Q}_{>0}^\times} q\widehat{\mathbb{Z}}^\times.$$

The set $\widehat{\mathbb{Z}} \cap \mathbb{A}_{\mathrm{fin}}^\times$ *is the disjoint union*

$$\widehat{\mathbb{Z}} \cap \mathbb{A}_{\mathrm{fin}}^\times = \coprod_{k \in \mathbb{N}} k\widehat{\mathbb{Z}}^\times.$$

(b) *For every* $s \in \mathbb{C}$ *with* $\mathrm{Re}(s) > 1$ *the integral*

$$\int_{\widehat{\mathbb{Z}}} |x|^s \, d^\times x$$

converges absolutely and equals the Riemann zeta function $\zeta(s)$. *Here* $d^\times x$ *is the uniquely determined Haar measure on* $\mathbb{A}_{\mathrm{fin}}^\times$, *such that the compact open subgroup* $\widehat{\mathbb{Z}}^\times$ *has volume 1. We also consider this measure as a measure on* $\mathbb{A}_{\mathrm{fin}}$, *which is zero outside* $\mathbb{A}_{\mathrm{fin}}^\times$.

Proof Given $x \in \mathbb{A}_{\mathrm{fin}}^\times$, the absolute value $|x|$ lies in $\mathbb{Q}_{>0}^\times$. Consider the element $|x|x \in \mathbb{A}_{\mathrm{fin}}^\times$. Let p be a prime number. Then one has $x_p = p^k u$ for some $k \in \mathbb{Z}$ and some $u \in \mathbb{Z}_p^\times$. Hence $|x| = p^{-k} r$, where $r \in \mathbb{Q}$ is coprime to p. We conclude that $||x|x_p|_p = 1$, so $|x|x \in \widehat{\mathbb{Z}}^\times$. Setting $q = |x|^{-1}$, we get $x \in q\widehat{\mathbb{Z}}^\times$ and therefore the first claim of (a) is proven. The second follows as well.

To show (b), we use (a) as follows,

$$\int_{\widehat{\mathbb{Z}}} |x|^s \, d^\times x = \sum_{k \in \mathbb{N}} \int_{k\widehat{\mathbb{Z}}^\times} |x|^s \, d^\times x = \sum_{k \in \mathbb{N}} \int_{\widehat{\mathbb{Z}}^\times} |kx|^s \, d^\times x$$

$$= \sum_{k \in \mathbb{N}} k^{-s} \underbrace{\int_{\widehat{\mathbb{Z}}^{\times}} |x|^s \, d^{\times} x}_{=1}.$$

The convergence follows from the convergence of the Dirichlet series $\zeta(s)$. $\square$

5.4 Fourier Analysis on $\mathbb{A}$

In this section we introduce Fourier transformation and Fourier series on the adeles. We finish the section with the most important Poisson Summation Formula.

A *character* of a topological group G is a continuous group homomorphism $\chi : G \to \mathbb{T}$, where

$$\mathbb{T} = \{z \in \mathbb{C} : |z| = 1\}$$

is the *circle group*. The set of characters is a group under point-wise multiplication,

$$\chi \eta(x) = \chi(x) \eta(x).$$

This group is called the *dual group* of G. We denote the dual group of G by $\widehat{G}$.

Examples 5.4.1

- Consider the group $G = \mathbb{Z}$ with the discrete topology. For every $z \in \mathbb{T}$ the map

$$\chi_z : G \to \mathbb{T}$$
$$k \mapsto z^k$$

  is a character of G. The map $z \mapsto \chi_z$ is an isomorphism $\mathbb{T} \cong \widehat{G}$.
- Dually, for $k \in \mathbb{Z}$ we have a character $\varepsilon_k : \mathbb{T} \to \mathbb{T}$ of the group $\mathbb{T}$ given by

$$\varepsilon_k(z) = z^k.$$

  The map $k \mapsto \varepsilon_k$ is an isomorphism $\mathbb{Z} \cong \widehat{\mathbb{T}}$.

We now define a character e of the additive group of $\mathbb{A}$. First, for every prime number p we define $e_p : \mathbb{Q}_p \to \mathbb{T}$ by

$$e_p(x) = e\left(\sum_{j=-N}^{\infty} a_j p^j\right) = \exp\left(-2\pi i \sum_{j=-N}^{-1} a_j p^j\right) = \prod_{j=-N}^{-1} e^{-2\pi i a_j p^j}.$$

The map e_p is a character of the additive group $(\mathbb{Q}_p, +)$, as was shown in Exercise 4.5. Note that $e_p(\mathbb{Z}_p) = 1$.

For $p = \infty$ finally define $e_{\infty} : \mathbb{R} \to \mathbb{T}$ by

$$e_{\infty}(x) = e^{2\pi i x}.$$

Lemma 5.4.2 *For every $x \in \mathbb{A}$ the product*

$$e(x) = \prod_{p \leq \infty} e_p(x_p)$$

has only finitely many factors $\neq 1$, i.e. the product is finite. The ensuing map e : $\mathbb{A} \to \mathbb{T}$ is a character.

Proof For given $x \in \mathbb{A}$ one has $x_p \in \mathbb{Z}_p$ for almost all p, so the product is indeed finite and the map well defined. As the single e_p are group homomorphisms, so is e. It remains to show continuity. As e is a group homomorphism and $\mathbb{A}$ and $\mathbb{T}$ are topological groups, it suffices to show that for every unit neighborhood U in $\mathbb{T}$ there exists a zero neighborhood V in $\mathbb{A}$ such that $e(V) \subset U$. It suffices to assume that U is the set of all $e^{2\pi i t} \in \mathbb{T}$ with $t \in (-\varepsilon, \varepsilon)$ for some $\varepsilon > 0$. Then indeed the set $V = \prod_{p < \infty} \mathbb{Z}_p \times (-\varepsilon, \varepsilon)$ is an open zero neighborhood in $\mathbb{A}$ with $e(V) \subset U$. $\square$

Theorem 5.4.3

(a) *For a given place $p \leq \infty$ and a given character $\chi : \mathbb{Q}_p \to \mathbb{T}$ there exists a uniquely determined $a \in \mathbb{Q}_p$ such that $\chi(x) = e_p(ax)$. The map $a \mapsto e_p(a\cdot)$ is an isomorphism of groups $\mathbb{Q}_p \to \widehat{\mathbb{Q}_p}$.*

(b) *For every character $\chi : \mathbb{A} \to \mathbb{T}$ there exists a uniquely determined $a \in \mathbb{A}$ with $\chi(x) = e(ax)$. The map $\chi \mapsto a$ is an isomorphism of groups $\widehat{\mathbb{A}} \to \mathbb{A}$.*

(c) *The characters of the compact group $\mathbb{A}/\mathbb{Q}$ are the characters χ of $\mathbb{A}$ with $\chi(\mathbb{Q}) = 1$. These are given by $x \mapsto \chi(qx)$ with $q \in \mathbb{Q}$.*

(d) *For given $x \in \mathbb{A}$ one has*

$$e(xy) = 1 \; \forall y \in \mathbb{Q} \quad \Leftrightarrow \quad x \in \mathbb{Q}.$$

Proof Part (a) for $p < \infty$ is Exercise 4.6. We now prove it for $p = \infty$. Let $\chi : \mathbb{R} \to \mathbb{T}$ be a character. By continuity of χ there exists an $\varepsilon > 0$ such that $\chi([-\varepsilon, \varepsilon]) \subset \{\mathrm{Re}(z) > 0\}$. Let a be the uniquely determined element of $[\frac{-1}{4\varepsilon}, \frac{1}{4\varepsilon}]$ such that $\chi(\varepsilon) = e^{2\pi i a\varepsilon}$. We claim that

$$\chi\left(\frac{\varepsilon}{2}\right) = e^{2\pi i a\frac{\varepsilon}{2}}.$$

To show this, we note $\chi(\frac{\varepsilon}{2})^2 = \chi(\varepsilon) = e^{2\pi i a\varepsilon}$, so $\chi(\frac{\varepsilon}{2}) = \pm e^{2\pi i a\frac{\varepsilon}{2}}$ and $-e^{2\pi i a\frac{\varepsilon}{2}}$ has real part < 0. We repeat this argument and we see that $\chi(\frac{\varepsilon}{2^n}) = e^{2\pi i a\frac{\varepsilon}{2^n}}$. For $k \in \mathbb{Z}$ it follows that

$$\chi\left(\frac{k}{2^n}\right) = \chi\left(\frac{1}{2^n}\right)^k = e^{2\pi i a\frac{k}{2^n}}.$$

The set of all $\frac{k}{2^n}$ with $k \in \mathbb{Z}$, $n \in \mathbb{N}$, is dense in $\mathbb{R}$, so that by continuity we can conclude $\chi(x) = e^{2\pi i ax}$ for every $x \in \mathbb{R}$. This gives existence. The uniqueness of a

can be derived, for example, from the fact that the derivative of $x \mapsto e^{2\pi i a x}$ at $x = 0$ is $2\pi i a$.

For part (b) let $\chi_p : \mathbb{Q}_p \hookrightarrow \mathbb{A} \xrightarrow{\chi} \mathbb{T}$, where the embedding of $\mathbb{Q}_p$ is given by $x \mapsto (\dots, 0, x, 0, \dots)$. By (a) there exists a unique $a_p \in \mathbb{Q}_p$ with $\chi_p(x) = e_p(a_p x)$ for every $x \in \mathbb{Q}_p$. We now show that

$$\chi_p(\mathbb{Z}_p) = 1 \quad \text{for almost all } p.$$

Let $U \subset \mathbb{T}$ be the set of all $z \in \mathbb{T}$ with $\mathrm{Re}(z) > 0$. Then U is an open neighborhood of the unit. So $\chi^{-1}(U)$ must contain an open neighborhood of zero in $\mathbb{A}$. There therefore exists a finite set of places E with

$$X = \left(\prod_{p \notin E} \mathbb{Z}_p \right) \times \left(\prod_{p \in E} \{0\} \right) \subset \chi^{-1}(U).$$

The left-hand side X is a subgroup of $\mathbb{A}$, so $\chi(X) \subset U$ is a subgroup of $\mathbb{T}$ contained in U. The set U, however, contains only the trivial subgroup, so that $\chi(X) = 1$, which means $\chi_p(\mathbb{Z}_p) = 1$ for every $p \notin E$. This implies that the map $\tilde{\chi} : \mathbb{A} \to \mathbb{T}$,

$$\tilde{\chi}(x) = \prod_p \chi_p(x_p)$$

is well defined. Indeed, it is a continuous group homomorphism and it coincides with χ on all $\mathbb{Q}_p$. The $\mathbb{Q}_p$ generate a dense subgroup of $\mathbb{A}$ so $\chi = \tilde{\chi}$. Hence for $x \in \mathbb{A}$ we get

$$\chi(x) = \prod_p \chi_p(x_p) = \prod_p e_p(a_p x_p) = e(ax).$$

The uniqueness of a follows from the uniqueness of its coordinates a_p.

The map $\chi \mapsto a$ is a group homomorphism as $\chi(x) = e(ax)$ and $\eta(x) = e(bx)$, so $\chi\eta = e((a+b)x)$. The inverse map is $a \mapsto e(a \cdot)$, so the map is bijective.

Now for (c). The first assertion is clear. Let χ be a character of $\mathbb{A}$ with $\chi(\mathbb{Q}) = 1$. By (b) there exists an $a \in \mathbb{A}$ with $\chi(x) = e(ax)$ and we have to show that $a \in \mathbb{Q}$. Replacing a by $p^{-k}a$, if $|a|_p > 1$ we successively get $|a|_p \le 1$ for every p, so $a_{\mathrm{fin}} \in \hat{\mathbb{Z}}$. We want to show that this implies $a \in \mathbb{Z}$. By $e(a_{\mathrm{fin}}) = 1$ we get $1 = e_\infty(a_\infty) = e^{2\pi i a_\infty}$, and so $a_\infty \in \mathbb{Z}$. We multiply by -1 if necessary, so that we can assume $a_\infty \ge 0$. Let p be a prime. Write $a_p = \sum_{j=0}^\infty c_j p^j$ with $0 \le c_j < p$ and $a_\infty = \sum_{j=0}^N b_j p^j$ with $0 \le b_j < p$. We claim $c_j = b_j$ for all j and all p. For $k \in \mathbb{N}$ one has $e(p^{-k}a) = 1$, so $e_p(p^{-k}a_p)e_\infty(p^{-k}a_\infty) = 1$, which implies that

$$\exp\left(2\pi i \sum_{j=0}^k c_j p^{j-k} \right) = \exp\left(2\pi i \sum_{j=0}^N b_j p^{j-k} \right).$$

As this holds for all k, we get $c_j = b_j$. Part (d) is merely a reformulation of (c). $\qquad \square$

5.4.1 Local Fourier Analysis

Recall the space $\mathcal{S}(\mathbb{R})$ of all infinite differentiable functions $f : \mathbb{R} \to \mathbb{C}$ such that for any two $m, n \in \mathbb{Z}$ with $m, n \geq 0$ the function $x^m f^{(n)}(x)$ is bounded.

For $p < \infty$ a *Schwartz–Bruhat function* is a complex valued function on $\mathbb{Q}_p$, which is locally constant and has compact support. Let $\mathcal{S}(\mathbb{Q}_p)$ denote the vector space of all Schwartz–Bruhat functions on $\mathbb{Q}_p$.

Lemma 5.4.4 *Every $f \in \mathcal{S}(\mathbb{Q}_p)$ is a finite linear combination of functions of the form $\mathbf{1}_{a+p^k\mathbb{Z}_p}$, where $a \in \mathbb{Q}_p$ and $k \in \mathbb{Z}$.*

Proof Since any given $f \in \mathcal{S}(\mathbb{Q}_p)$ is locally constant, for every $z \in \mathbb{C}$ the preimage $f^{-1}(z)$ is an open set in $\mathbb{Q}_p$. Therefore $f^{-1}(0)$ is open and $\mathbb{Q}_p \smallsetminus f^{-1}(0)$ is closed, so that $\mathbb{Q}_p \smallsetminus f^{-1}(0)$ equals the support of f; hence this set is compact. It is covered by the open sets $f^{-1}(z)$ with $z \neq 0$, and by compactness we can choose finitely many, so that the function f has finite image. Every open set in $\mathbb{Q}_p$ is a disjoint union of open balls. Hence so for the open set $f^{-1}(z)$ for $z \neq 0$. Open balls in $\mathbb{Q}_p$ are of the form $a + p^k\mathbb{Z}_p$ as above. The lemma is proven. $\qquad\square$

We now define the Fourier transformation on $\mathbb{Q}_p$ for arbitrary $p \leq \infty$. For a function $f \in \mathcal{S}(\mathbb{Q}_p)$ let

$$\hat{f}(x) = \int_{\mathbb{Q}_p} f(y)e_p(-xy)\,dy$$

be its *Fourier transform.*

Lemma 5.4.5 *Let $f \in \mathcal{S}(\mathbb{Q}_p)$.*

(a) *If $g(x) = f(x)e_p(ax)$ with $a \in \mathbb{Q}_p$, then $\hat{g}(x) = \hat{f}(x-a)$.*
(b) *If $g(x) = f(x-a)$ with $a \in \mathbb{Q}_p$, then $\hat{g}(x) = \hat{f}(x)e_p(-ax)$.*
(c) *If $g(x) = f(\lambda x)$ with $\lambda \in \mathbb{Q}_p^\times$, then $\hat{g}(x) = \frac{1}{|\lambda|}\hat{f}(\frac{x}{\lambda})$.*

Proof These are simple consequences of the definition. $\qquad\square$

Lemma 5.4.6 *Let K be a compact group with Haar measure dx. For a given character $\chi : K \to \mathbb{T}$ we have*

$$\int_K \chi(x)\,dx = \begin{cases} \mathrm{vol}(K) & \text{if } \chi = 1, \\ 0 & \text{if } \chi \neq 1. \end{cases}$$

Proof The case $\chi = 1$ is clear. In the other case there is $k_0 \in K$ with $\chi(k_0) \neq 1$. Then, by left-invariance of the Haar measure,

$$\chi(k_0)\int_K \chi(x)\,dx = \int_K \chi(k_0x)\,dx = \int_K \chi(x)\,dx.$$

Since $\chi(k_0) \neq 1$, we get $\int_K \chi(x)\,dx = 0$. The lemma is proven. $\qquad\square$

Theorem 5.4.7 *For $p \leq \infty$ and $f \in \mathcal{S}(\mathbb{Q}_p)$, the Fourier transform $\hat{f}$ lies in $\mathcal{S}(\mathbb{Q}_p)$ and we have the inversion formula for the Fourier transformation,*

$$\hat{\hat{f}}(x) = f(-x).$$

Proof In the case $p = \infty$, we have already made use of the inversion formula in Theorem 2.4.8. In this case the inversion formula is proven in any of the books [Dei05, Rud87, SW71]. We consider the case $p < \infty$. The function $h = 1_{\mathbb{Z}_p}$ is its own Fourier transform, i.e. $\hat{h} = h$, as the following computation shows,

$$\hat{h}(x) = \int_{\mathbb{Q}_p} h(y) e_p(-xy) \, dy = \int_{\mathbb{Z}_p} e_p(-xy) \, dy.$$

Now $\chi(y) = e_p(-xy)$ is a character of the compact group $\mathbb{Z}_p$ and this character is trivial if and only if $x \in \mathbb{Z}_p$ so we get $\hat{h} = h$ by Lemma 5.4.6.

We introduce the following operators on $\mathcal{S}(\mathbb{Q}_p)$,

$$\Omega_a f(x) = f(x) e_p(ax), \qquad L_a f(x) = f(x-a), \qquad M_\lambda f(x) = f(\lambda x),$$

where $a \in \mathbb{Q}_p$ and $\lambda \in \mathbb{Q}_p^\times$. The assertions of Lemma 5.4.5 can be rephrased as follows:

$$\widehat{\Omega_a f} = L_a \hat{f}, \qquad \widehat{L_a f} = \Omega_{-a} \hat{f}, \qquad \widehat{M_\lambda f} = \frac{1}{|\lambda|} M_{1/\lambda} \hat{f}.$$

We show that the Fourier transformation maps the space $\mathcal{S}(\mathbb{Q}_p)$ to itself. By Lemma 5.4.4 every $f \in \mathcal{S}(\mathbb{Q}_p)$ is a linear combination of functions of the form $f = 1_{a+p^k \mathbb{Z}_p} = L_a M_{p^{-k}} h$. Therefore,

$$\hat{f} = \widehat{L_a M_{p^{-k}} h} = \Omega_{-a} p^{-k} M_{p^k} \hat{h} = \Omega_{-a} p^{-k} M_{p^k} h.$$

So we have $\hat{f}(x) = e_p(-ax) 1_{p^{-k} \mathbb{Z}_p}(x)$. Since the character e_p is locally constant, this function lies in $\mathcal{S}(\mathbb{Q}_p)$.

Consider the vector space A of all $f \in \mathcal{S}(\mathbb{Q}_p)$ with $\hat{\hat{f}}(x) = f(-x)$. As $h(-x) = h(x)$, we have $h \in A$. We now show that the function $f = L_a M_{p^{-k}} h$ lies in A,

$$\hat{\hat{f}} = \widehat{\widehat{L_a M_{p^{-k}} h}} = L_{-a}(\widehat{\widehat{M_{p^{-k}} h}}) = L_{-a} M_{p^{-k}} \hat{\hat{h}} = L_{-a} M_{p^{-k}} h = 1_{-a+p^k \mathbb{Z}_p}.$$

This means $\hat{\hat{f}}(x) = f(-x)$. By linearity and Lemma 5.4.4 the inversion formula follows in the case $p < \infty$. □

5.4.2 Global Fourier Analysis

Let $\mathbb{A}_{\text{fin}} = \widehat{\prod}_{p<\infty} \mathbb{Q}_p$ and let $\mathcal{S}(\mathbb{A}_{\text{fin}})$ be the space of all functions $f : \mathbb{A}_{\text{fin}} \to \mathbb{C}$ which are locally constant of compact support. An elements of $\mathcal{S}(\mathbb{A}_{\text{fin}})$ is called a *Schwartz–Bruhat function* on $\mathbb{A}_{\text{fin}}$.

Lemma 5.4.8 *Every Schwartz–Bruhat function f on $\mathbb{A}_{\text{fin}}$ is a finite linear combination of functions of the form*

$$f = \mathbf{1}_{a+N\widehat{\mathbb{Z}}},$$

where $a \in \mathbb{A}$ and $N \in \mathbb{N}$. In turn, the function $\mathbf{1}_{a+N\widehat{\mathbb{Z}}}$ can be written as a product

$$f = \prod_{p<\infty} \mathbf{1}_{a_p+N\mathbb{Z}_p},$$

where almost all factors are equal to $\mathbf{1}_{\mathbb{Z}_p}$. For every given $x \in \mathbb{A}$ the product

$$f(x) = \prod_{p<\infty} \mathbf{1}_{a_p+N\mathbb{Z}_p}(x_p)$$

is finite, i.e. almost all factors are equal to 1.

Proof The proof of the first assertion is completely analogous to the proof of Lemma 5.4.4. The second comes about by the fact that

$$a + N\widehat{\mathbb{Z}} = a + N \prod_{p<\infty} \mathbb{Z}_p$$

$$= \prod_{p<\infty} (a_p + N\mathbb{Z}_p). \qquad \square$$

Let $\mathcal{S}(\mathbb{A})$ be the space of all functions of the form

$$f(x) = \sum_{j=1}^{n} h_j(x_{\text{fin}}) g_j(x_\infty),$$

where $h_j \in \mathcal{S}(\mathbb{A}_{\text{fin}})$ and $g_j \in \mathcal{S}(\mathbb{R})$. An element of $\mathcal{S}(\mathbb{A})$ is called a *Schwartz–Bruhat function* on $\mathbb{A}$. For $f \in \mathcal{S}(\mathbb{A})$ define the *Fourier transform* by

$$\hat{f}(x) = \int_{\mathbb{A}} f(y) e(-xy)\, dy.$$

We have shown above, that every $f \in \mathcal{S}(\mathbb{A})$ can be written as a finite sum of functions of the form

$$f = \prod_{p\leq\infty} f_p,$$

with $f_p \in \mathcal{S}(\mathbb{Q}_p)$ and $f_p = \mathbf{1}_{\mathbb{Z}_p}$ for almost all p.

Theorem 5.4.9 *For every function f on $\mathbb{A}$ of the form $f = \prod_{p \leq \infty} f_p$ with $f_p \in \mathcal{S}(\mathbb{Q}_p)$ one has*

$$\hat{f} = \prod_p \hat{f}_p.$$

For $f \in \mathcal{S}(\mathbb{A})$ one has $\hat{f} \in \mathcal{S}(\mathbb{A})$ and the inversion formula for the Fourier transformation,

$$\hat{\hat{f}}(x) = f(-x)$$

holds.

Proof Let f be a product as in the theorem. By Theorem 5.2.2 one has

$$\hat{f}(x) = \int_{\mathbb{A}} f(y) e(-xy) \, dy = \prod_p \int_{\mathbb{Q}_p} f_p(y) e_p(-x_p y) \, dy = \prod_p \hat{f}_p(x_p).$$

The global inversion formula therefore follows from the local inversion formulae. □

Theorem 5.4.10 (Poisson Summation Formula, adelic version) *For every $f \in \mathcal{S}(\mathbb{A})$ one has*

$$\sum_{q \in \mathbb{Q}} f(q) = \sum_{q \in \mathbb{Q}} \hat{f}(q),$$

where the series are both absolutely convergent.

Proof We show convergence first. It suffices to show convergence for $f(x) = \mathbf{1}_{a+N\hat{\mathbb{Z}}}(x_{\text{fin}}) f_\infty(x_\infty)$, where $a \in \mathbb{A}_{\text{fin}}$, $N \in \mathbb{N}$, and $f_\infty \in \mathcal{S}(\mathbb{R})$. As $\mathbb{Q}$ is dense in $\mathbb{A}_{\text{fin}}$, we find a rational number r in the set $a + N\hat{\mathbb{Z}}$, so that $a + N\hat{\mathbb{Z}} = r + N\hat{\mathbb{Z}}$. Therefore,

$$\sum_{q \in \mathbb{Q}} f(q) = \sum_{q \in \mathbb{Q} \cap r + N\hat{\mathbb{Z}}} f_\infty(q) = \sum_{q \in \mathbb{Q} \cap N\hat{\mathbb{Z}}} f_\infty(q - r)$$

$$= \sum_{q \in N\mathbb{Z}} f_\infty(q - r) = \sum_{q \in \mathbb{Z}} f_\infty(qN - r).$$

The last sum is clearly absolutely convergent.

Next we show the claimed identity by reducing to the case $f = \mathbf{1}_{\hat{\mathbb{Z}}} f_\infty$. Let $h = \mathbf{1}_{\hat{\mathbb{Z}}}$. Then $\mathbf{1}_{r+N\hat{\mathbb{Z}}} = L_r M_{1/N} h$, so it suffices to show that the claim is stable under the operators L_r and M_r with $r \in \mathbb{Q}$. So assume $\sum_{q \in \mathbb{Q}} f(q) = \sum_{q \in \mathbb{Q}} \hat{f}(q)$

and let $r \in \mathbb{Q}$. Then

$$\sum_{q \in \mathbb{Q}} L_r f(q) = \sum_{q \in \mathbb{Q}} f(q - r) = \sum_{q \in \mathbb{Q}} f(q).$$

As $e(\mathbb{Q}) = 1$, this equals

$$\sum_{q \in \mathbb{Q}} \hat{f}(q) = \sum_{q \in \mathbb{Q}} e(-rq) \hat{f}(q) = \sum_{q \in \mathbb{Q}} \Omega_{-r} \hat{f}(q) = \sum_{q \in \mathbb{Q}} \widehat{L_r f}(q).$$

Using the product formula, we similarly show

$$\sum_{q \in \mathbb{Q}} M_r f(q) = \sum_{q \in \mathbb{Q}} \widehat{M_r f}(q).$$

Assume $f = \mathbf{1}_{\widehat{\mathbb{Z}}} f_\infty$. Then $\hat{f} = \widehat{\mathbf{1}_{\widehat{\mathbb{Z}}}} \hat{f}_\infty = \mathbf{1}_{\widehat{\mathbb{Z}}} \hat{f}_\infty$ and

$$\sum_{q \in \mathbb{Q}} f(q) = \sum_{k \in \mathbb{Z}} f_\infty(k).$$

Therefore the adelic Poisson Summation Formula is reduced to the following proposition. We write $C^\infty(\mathbb{R}/\mathbb{Z})$ for the space of all smooth functions g on $\mathbb{R}$ with $g(x + 1) = g(x)$.

Proposition 5.4.11 (Poisson Summation Formula, classical) *For $f \in \mathcal{S}(\mathbb{R})$ one has*

$$\sum_{k \in \mathbb{Z}} f(k) = \sum_{k \in \mathbb{Z}} \hat{f}(k).$$

Proof For given $f \in \mathcal{S}(\mathbb{R})$, the function $g(x) = \sum_k f(x + k)$ is a smooth function on $\mathbb{R}/\mathbb{Z}$ (see Exercise 5.1) and by Proposition 2.2.7 we infer that

$$\sum_k f(k) = g(0) = \sum_k c_k(g) = \sum_k \int_0^1 \sum_l f(x + l) e^{-2\pi i k x} \, dx$$

$$= \sum_k \int_{-\infty}^\infty f(x) e^{-2\pi i k x} \, dx = \sum_k \hat{f}(k).$$

The proposition and the theorem are proven.   □

There is also a higher-dimensional version of the Poisson Summation Formula, which we give now. Let $n \in \mathbb{N}$ and let $\mathcal{S}(\mathbb{A}_{\mathrm{fin}}^n)$ be the set of all functions $f : \mathbb{A}_{\mathrm{fin}}^n \to \mathbb{C}$ which are locally constant and of compact support. Next, $\mathcal{S}(\mathbb{R}^n)$ is defined to be the set of all infinitely differentiable functions $f : \mathbb{R}^n \to \mathbb{C}$, such that every derivative is rapidly decreasing, so for every $\alpha(\alpha_1, \ldots, \alpha_n) \in \mathbb{N}_0^n$ and every polynomial $p(x) \in \mathbb{C}[x_1, \ldots, x_n]$ the function

$$p(x)\frac{\partial^{\alpha^1}}{\partial x_1^{\alpha_1}} \cdots \frac{\partial^{\alpha^n}}{\partial x_n^{\alpha_n}} f(x)$$

is bounded. Then $\mathcal{S}(\mathbb{A})$ is the space of functions $\mathbb{A} \to \mathbb{C}$ which are finite linear combinations of functions of the form $x \mapsto f_\infty(x_\infty)f_{\mathrm{fin}}(x_{\mathrm{fin}})$ with $f_\infty \in \mathcal{S}(\mathbb{R}^n)$ and $f_{\mathrm{fin}} \in \mathcal{S}(\mathbb{A}^n)$. We establish an additive character $e^n : \mathbb{A}^n \to \mathbb{C}^\times$ given by

$$e^n(x_1, \ldots, x_n) = e(x_1) \cdots e(x_n),$$

and we define the Fourier transform as

$$\hat{f}(x) = \int_{\mathbb{A}^n} f(y)e^n(-xy)\,dy$$

for $f \in \mathcal{S}(\mathbb{A}^n)$. As in the one-dimensional case, we get that the Fourier transformation maps $\mathcal{S}(\mathbb{A}^n)$ to itself and that the *inversion formula for the Fourier transformation*,

$$\hat{\hat{f}}(x) = f(-x)$$

holds.

Theorem 5.4.12 (Poisson Summation Formula in n dimensions) *For every $f \in \mathcal{S}(\mathbb{A}^n)$ one has*

$$\sum_{q \in \mathbb{Q}^n} f(q) = \sum_{q \in \mathbb{Q}^n} \hat{f}(q),$$

where the series are both absolutely convergent.

The proof is essentially the same as in the one-dimensional case.

5.5 Exercises

Exercise 5.1 Let $f \in \mathcal{S}(\mathbb{R})$ and set $g(x) = \sum_{k \in \mathbb{Z}} f(x+k)$. Show that g is a smooth function on $\mathbb{R}$.
(Hint: the estimate $|f(x)| \leq C/(1+x^2)$ for a constant C shows point-wise convergence. The same holds for the nth derivative $f^{(n)}$ instead of f. Now integrate n times.)

Exercise 5.2

(a) Show that the family $(N\widehat{\mathbb{Z}})_{N\in\mathbb{N}}$ is a neighborhood basis of zero in $\mathbb{A}_{\text{fin}}$. That is, show that every $N\widehat{\mathbb{Z}}$ is a neighborhood of zero and that every zero neighborhood contains a set of the form $N\widehat{\mathbb{Z}}$ for some N.
(b) Show that the sets of the form $(1+N\widehat{\mathbb{Z}})\cap\widehat{\mathbb{Z}}^{\times}$, $N\in\mathbb{N}$ are a neighborhood basis of the unit 1 in $\mathbb{A}_{\text{fin}}^{\times}$.

Exercise 5.3 Let U be a compact open subgroup of $\mathbb{A}_{\text{fin}}$ and let $K\subset\mathbb{A}_{\text{fin}}$ be a compact set. Let $\mathcal{S}(U,K)$ be the set of all functions $f:\mathbb{A}\to\mathbb{C}$ such that

- supp $f\subset K\times\mathbb{R}$,
- $f(x+u)=f(x)$ $\forall x\in\mathbb{A}$, $u\in U$,
- for all $m,n\in\mathbb{N}_0$ one has $\sigma_{m,n}(f)=\sup_{x\in\mathbb{A}}|f^{(m)}(x)|\,|x_\infty|_\infty^n<\infty$.

Show

$$\bigcup_{U,K}\mathcal{S}(U,K)=\mathcal{S}(\mathbb{A}).$$

Exercise 5.4 By a *tempered distribution* or simply a *distribution* on $\mathbb{A}$ we mean a linear map $T:\mathcal{S}(\mathbb{A})\to\mathbb{C}$ such that for every sequence f_j in $\mathcal{S}(U,K)$ with $\sigma_{m,n}(f_j)\to 0$ for every pair (m,n) one has $T(f_j)\to 0$.
    Show

- $f\mapsto\delta(f)=f(0)$ is a distribution,
- $f\mapsto I(f)=\int_{\mathbb{A}}f(x)\,dx$ is a distribution,
- $f\mapsto S(f)=\sum_{q\in\mathbb{Q}}f(q)$ is a distribution.

Exercise 5.5 Let p be a prime number, $n\in\mathbb{N}$ and let dx be the additive Haar measure on $M_n(\mathbb{Q}_p)$, so

$$\int_{M_n(\mathbb{Q}_p)}f(x)\,dx=\int_{\mathbb{Q}_p}\cdots\int_{\mathbb{Q}_p}f(x_{i,j})\,dx_{1,1}\cdots dx_{n,n}.$$

(a) Show that $\frac{dx}{|\det x|^n}$ is a left- and right-invariant Haar measure on the group $GL_n(\mathbb{Q}_p)$. Conclude that the group $GL_n(\mathbb{Q}_p)$ is unimodular.
(b) Show that the group $GL_n(\mathbb{A})$ is unimodular.

Exercise 5.6 Let $n,N\in\mathbb{N}$ and let K_N be the set of all invertible $n\times n$ matrices g with entries in $\widehat{\mathbb{Z}}$ such that $g\equiv I\bmod N$.
    Show

- K_N is a compact open subgroup of $GL_n(\widehat{\mathbb{Z}})$,
- $K_N\subset K_d$ if $d|N$,
- the K_N form a neighborhood basis of the unit in $GL_n(\widehat{\mathbb{Z}})$.

Exercise 5.7 Let U be a compact open subgroup of the locally compact group G. Show that for every $g \in G$ the set

$$UgU/U$$

is finite.

Exercise 5.8 Let G be a totally disconnected locally compact group. For a compact open subgroup U and a compact set K let $L(U, K)$ be the set of all functions $f : G \to \mathbb{C}$ with

- supp $f \subset K$ and
- $f(ux) = f(x)$ for every $x \in G$ and every $u \in U$.

Further let $R(U, K)$ be the set of all functions $f : G \to \mathbb{C}$ with

- supp $f \subset K$ and
- $f(xu) = f(x)$ for every $x \in G$ and every $u \in U$.

Show that in general one has $L(U, K) \neq R(U, K)$, but

$$\bigcup_{U,K} L(U, K) = \bigcup_{U,K} R(U, K).$$

Chapter 6
Tate's Thesis

This chapter is devoted to the PhD thesis of John Tate, which has reached cult status in number theory. In Tate's thesis, harmonic analysis of adeles and ideles is used to investigate L-functions. In later chapters we shall do the same with 2×2 matrices and thus obtain analytic continuation of L-functions attached to automorphic forms.

6.1 Poisson Summation Formula and the Riemann Zeta Function

We apply the Poisson Summation Formula on $\mathbb{R}$ to the Gauss function $e^{-\pi x^2}$ and we obtain the theta transformation formula, which can be used to prove the functional equation of the Riemann zeta function.

Lemma 6.1.1

(a) *The function* $f(x) = e^{-\pi x^2}$ *lies in* $\mathcal{S}(\mathbb{R})$. *It is its own Fourier transform, i.e. it satisfies* $\hat{f} = f$.

(b) *For* $f \in \mathcal{S}(\mathbb{R})$ *let* $M_a f(x) = f(ax)$ *for* $a > 0$. *Then* $M_a f$ *is in* $\mathcal{S}(\mathbb{R})$ *and one has* $\widehat{M_a f} = \frac{1}{a} M_{1/a} \hat{f}$.

Proof (a) This has been shown in the proof of Lemma 2.7.9.

(b) The function $M_a f$ clearly lies in $\mathcal{S}(\mathbb{R})$. The change of variables $v = ay$ gives

$$\widehat{M_a f}(x) = \int_{\mathbb{R}} f(ay)e^{-2\pi ixy}\,dy = \frac{1}{a}\int_{\mathbb{R}} f(v)e^{-2\pi ixv/a}\,dv = \frac{1}{a}\hat{f}(x/a). \qquad \square$$

For $t > 0$ set $f_t(x) = e^{-\pi tx^2}$. Then $\hat{f}_t = \frac{1}{\sqrt{t}} f_{1/t}$.

Lemma 6.1.2 *The theta series*

$$\Theta(t) = \sum_{k \in \mathbb{Z}} e^{-\pi tk^2}$$

A. Deitmar, *Automorphic Forms*, Universitext,
DOI 10.1007/978-1-4471-4435-9_6, © Springer-Verlag London 2013

converges uniformly on every interval of the form $[a, \infty)$ *for* $a > 0$. *The function* $\Theta(t)$ *is rapidly decreasing on each such interval. It satisfies the* theta transformation formula,

$$\Theta(t) = \frac{1}{\sqrt{t}}\Theta(1/t)$$

for every $t > 0$.

Proof For fixed $t > 0$ the function $k \mapsto e^{-\pi t k^2}$ is rapidly decreasing, therefore the series converges point-wise. Further, each summand $e^{-\pi t k^2}$ is monotonically decreasing in t. The assertion on convergence follows. The transformation formula is a direct consequence of the Poisson Summation Formula as in Proposition 5.4.11. $\square$

This equation will now lead to a proof of the functional equation of the Riemann zeta function

$$\zeta(s) = \sum_{n=1}^{\infty} \frac{1}{n^s}, \quad \mathrm{Re}(s) > 1.$$

Theorem 6.1.3 *The function* $\hat{\zeta}(s) = \zeta(s)\Gamma(s/2)\pi^{-s/2}$, *defined for* $\mathrm{Re}(s) > 1$, *can be extended to a meromorphic function on* $\mathbb{C}$ *which has simple poles at* $s = 0, 1$ *and is regular everywhere else. The extension satisfies the functional equation*

$$\hat{\zeta}(1 - s) = \hat{\zeta}(s).$$

Proof We compute

$$\zeta(s)\Gamma(s/2)\pi^{-s/2} = \sum_{n=1}^{\infty} \int_0^{\infty} n^{-s} t^{s/2} \pi^{-s/2} e^{-t} \frac{dt}{t} = \sum_{n=1}^{\infty} \int_0^{\infty} \left(\frac{t}{n^2 \pi}\right)^{s/2} e^{-t} \frac{dt}{t}$$

$$= \sum_{n=1}^{\infty} \int_0^{\infty} t^{s/2} e^{-\pi t n^2} \frac{dt}{t} = \frac{1}{2} \int_0^{\infty} t^{s/2} (\Theta(t) - 1) \frac{dt}{t}.$$

The interchange of summation and integration is justified by absolute convergence. Let

$$A(s) = \frac{1}{2} \int_1^{\infty} t^{s/2} (\Theta(t) - 1) \frac{dt}{t}.$$

As $\Theta - 1$ is rapidly decreasing, the integral converges for all $s \in \mathbb{C}$ and defines an entire function. The remaining integral over the interval $(0, 1)$ is computed by changing variables from t to $1/t$,

$$\frac{1}{2}\int_0^1 t^{s/2}\big(\Theta(t)-1\big)\frac{dt}{t} = \frac{1}{2}\int_0^1 t^{s/2}\Theta(t)\frac{dt}{t} - \frac{1}{s} = \frac{1}{2}\int_1^\infty t^{-s/2}\Theta(1/t)\frac{dt}{t} - \frac{1}{s}$$

$$= \frac{1}{2}\int_1^\infty t^{-s/2}\sqrt{t}\,\Theta(t)\frac{dt}{t} - \frac{1}{s}$$

$$= \frac{1}{2}\int_1^\infty t^{-(s-1)/2}\Theta(t)\frac{dt}{t} - \frac{1}{s}$$

$$= \frac{1}{2}\int_1^\infty t^{-(s-1)/2}\big(\Theta(t)-1\big)\frac{dt}{t} - \frac{1}{s} - \frac{1}{1-s}$$

$$= A(1-s) - \frac{1}{s} - \frac{1}{1-s}.$$

The claim follows.   □

6.2 Zeta Functions in the Adelic Setting

We want to repeat the arguments of the last section in the adeles. The role of the theta series is played by the sum $E(f)$ which we define next.

For $f \in \mathcal{S}(\mathbb{A})$ let

$$E(f)(x) = |x|^{\frac{1}{2}} \sum_{q\in\mathbb{Q}^\times} f(qx), \quad x \in \mathbb{A}^\times.$$

For technical reasons we shall also consider the sum $E(|f|)(x) = |x|^{\frac{1}{2}} \times \sum_{q\in\mathbb{Q}^\times} |f(qx)|$.

Lemma 6.2.1 *The sums $E(f)$ and $E(|f|)$ converge locally uniformly on $\mathbb{A}^\times$, so they define continuous functions on $\mathbb{A}^\times/\mathbb{Q}^\times$. These functions are rapidly decreasing in the sense that for every $N \in \mathbb{N}$ there exists $C_N > 0$ with*

$$\big|E(f)(x)\big| \le E(|f|)(x) \le C_N |x|^{-N} \quad \text{for } |x| \ge 1.$$

One has

$$E(f)(x) = E(\hat{f})\left(\frac{1}{x}\right) + |x|^{-\frac{1}{2}}\hat{f}(0) - |x|^{\frac{1}{2}}f(0).$$

Proof We can assume that the function f is of the form $f(x) = \mathbf{1}_R(x_{\text{fin}})f_\infty(x_\infty)$, where R is an open restricted rectangle in $\mathbb{A}_{\text{fin}}$ and $f_\infty \in \mathcal{S}(\mathbb{R})$. As in the proof of the Poisson Summation Formula in Theorem 5.4.10, we can assume that R is a subset of $\widehat{\mathbb{Z}}$. Since $f_\infty \in \mathcal{S}(\mathbb{R})$, there exists $C > 0$ such that $|f_\infty(x)| \le C(1+|x|)^{-2}$ holds for every $x \in \mathbb{R}$.

We have the estimate

$$\sum_{q \in \mathbb{Q}^\times} |f(qx)| \leq C \sum_{\substack{q \in \mathbb{Q}^\times \\ q x_{\mathrm{fin}} \in \widehat{\mathbb{Z}}}} \left(\frac{1}{1 + |q x_\infty|_\infty} \right)^2 .$$

Let $q_0 \in \mathbb{Q}^\times$. We show uniform convergence in a set of the form $q_0 \widehat{\mathbb{Z}}^\times \times U$ with a compact subset U of $\mathbb{R}^\times$. If x is in this set, then $q x_{\mathrm{fin}} \in \widehat{\mathbb{Z}} \Leftrightarrow q \in q_0^{-1} \mathbb{Z}$. Therefore,

$$\sum_{q \in \mathbb{Q}^\times} |f(qx)| \leq C \sum_{q \in q_0^{-1} \mathbb{Z} \smallsetminus \{0\}} \left(\frac{1}{1 + |q x_\infty|_\infty} \right)^2 .$$

As the right-hand side converges uniformly in x_∞, the claimed locally uniform convergence follows.

We next prove the growth estimate. For $x \in \mathbb{A}^\times$ with $1 \leq |x| = |x_{\mathrm{fin}}||x_\infty|$ and $f(x) \neq 0$ we have $|x_{\mathrm{fin}}| \leq 1$ and so $|x_\infty| \geq 1$. Let $N \in \mathbb{N}$, $N \geq 2$ be given. As $f_\infty \in S(\mathbb{R})$, there is a $C_N > 0$ such that $|f_\infty(x_\infty)| \leq C_N |x_\infty|^{-N}$ holds for $|x_\infty| \geq |q_0|_\infty$, $x_\infty \in \mathbb{R}$. Then

$$\sum_{q \in \mathbb{Q}^\times} |f(qx)| \leq C_N \sum_{k \in \mathbb{Z}} |q_0 k x_\infty|^{-N} = C_N |x_\infty|^{-N} \sum_{k \in \mathbb{Z}} |q_0 k|_\infty^{-N} .$$

Because of

$$|x_\infty|^{-N} = |x_{\mathrm{fin}}|^N |x|^{-N} \leq |x|^{-N} ,$$

the estimate for $E(f)(x)$ follows.

The functional equation will follow from the Poisson Summation Formula. Let $f_x(a) = f(ax)$ with $x \in \mathbb{A}^\times$. Then

$$\widehat{f_x}(y) = \int_{\mathbb{A}} f_x(z) e(-zy) \, dz = |x|^{-1} \hat{f}(y/x) .$$

This implies

$$E(f)(x) = |x|^{\frac{1}{2}} \sum_{q \in \mathbb{Q}} f(qx) - |x|^{\frac{1}{2}} f(0) = |x|^{-\frac{1}{2}} \sum_{q \in \mathbb{Q}} \hat{f}(q/x) - |x|^{\frac{1}{2}} f(0)$$

$$= E(\hat{f})(1/x) + |x|^{-\frac{1}{2}} \hat{f}(0) - |x|^{\frac{1}{2}} f(0) .$$

The lemma is proven.   □

For $f \in S(\mathbb{A})$ define the *zeta integral* of f by

$$\zeta(f, s) = \int_{\mathbb{A}^\times} f(x) |x|^s \, d^\times x .$$

Here for a Haar measure $d^\times x$ on $\mathbb{A}^\times = \mathbb{A}_{\text{fin}}^\times \times \mathbb{R}^\times$ the product $d^\times x_{\text{fin}} \times \frac{dt}{t}$ has been chosen with the normalization $d^\times x_{\text{fin}}(\widehat{\mathbb{Z}}^\times) = 1$.

Theorem 6.2.2 *One has*

$$\zeta(f, s) = \int_{\mathbb{A}^\times / \mathbb{Q}^\times} E(f)(x)|x|^{s-\frac{1}{2}} d^\times x.$$

The integral $\zeta(f, s)$ converges locally uniformly for $\mathrm{Re}(s) > 1$ and defines a holomorphic function there, which extends to a meromorphic function on the complex plane $\mathbb{C}$. This function is holomorphic away from the points $s = 0, 1$, where it has at most simple poles of residue $-f(0)$ and $\hat{f}(0)$, respectively. The zeta integral satisfies the functional equation

$$\zeta(f, s) = \zeta(\hat{f}, 1 - s).$$

Proof We first show convergence for $\mathrm{Re}(s) > 1$. For this we replace f by $|f|$ and estimate this function by a scalar multiple of $\mathbf{1}_{n^{-1}\widehat{\mathbb{Z}}}(x_{\text{fin}})(1 + |x|_\infty^N)^{-1}$, where $n, N \in \mathbb{N}$ with $N > \mathrm{Re}(s)$. The claimed convergence follows as, on the one hand, $\int_0^\infty \frac{t^{s-1}}{1+t^N} dt < \infty$, and on the other, $\int_{n^{-1}\widehat{\mathbb{Z}}} |x|^s d^\times x = n^s \int_{\widehat{\mathbb{Z}}} |x|^s d^\times < \infty$, according to Proposition 5.3.4. These estimates hold locally uniformly in s for $\mathrm{Re}(s) > 1$, which shows the convergence.

The set $F = \widehat{\mathbb{Z}}^\times \times (0, \infty)$ is, by Theorem 5.3.3, a set of representatives for $\mathbb{A}^\times / \mathbb{Q}^\times$. By absolute convergence, we can compute for $\mathrm{Re}(s) > 1$,

$$\zeta(f, s) = \int_{\mathbb{A}^\times} f(x)|x|^s d^\times x = \sum_{q \in \mathbb{Q}^\times} \int_{qF} f(x)|x|^s d^\times x$$

$$= \sum_{q \in \mathbb{Q}^\times} \int_F f(qx)|qx|^s d^\times x = \int_{\mathbb{A}^\times / \mathbb{Q}^\times} E(f)(x)|x|^{s-\frac{1}{2}} d^\times x.$$

The singleton $\{1\}$ is a set of measure zero in $\mathbb{R}_+^\times$, so $\mathbb{A}^1 / \mathbb{Q}^\times$ is a set of measure zero in $\mathbb{A}^\times / \mathbb{Q}^\times$. Therefore, we can decompose the zeta integral as $\int = \int_{|x|>1} + \int_{|x|<1}$. For given $N \in \mathbb{N}$ there is $C_N > 0$ such that $|f_\infty(x_\infty)| \le C_N |x_\infty|^{-N}$ holds for $|x_\infty| \ge |q_0|_\infty$, $x_\infty \in \mathbb{R}$. For arbitrary $N \in \mathbb{N}$ we estimate

$$\int_{|x|>1} E(|f|)(x)\, |x|^{\mathrm{Re}(s)-\frac{1}{2}} d^\times x \le C_N \int_{\substack{x \in \mathbb{A}^\times / \mathbb{Q}^\times \\ |x|>1}} |x|^{\mathrm{Re}(s)-\frac{1}{2}-N} d^\times x$$

$$= C_N \int_1^\infty t^{\mathrm{Re}(s)-\frac{1}{2}-N} \frac{dt}{t}.$$

The right-hand side is finite for $\mathrm{Re}(s) < N - \frac{1}{2}$. Since this holds for every N, the integral $\int_{|x|>1}$ converges for every s and defines an entire function.

To extend the integral $\int_{|x|<1}$, we use the functional equation of $E(f)$ and for $\mathrm{Re}(s) > 1$ we get,

$$\int_{|x|<1} E(f)(x)|x|^{s-\frac{1}{2}}\,d^{\times}x = \int_{|x|<1} |x|^{s-\frac{1}{2}} E(\hat{f})(1/x)\,d^{\times}x$$

$$+ \int_{|x|<1} |x|^{s-1}\,d^{\times}x\,\hat{f}(0) - \int_{|x|<1} |x|^{s}\,d^{\times}x f(0).$$

The first integral equals $\int_{|x|>1} E(\hat{f})(x)|x|^{\frac{1}{2}-s}\,d^{\times}x$ and defines an entire function in s. The second converges for $\mathrm{Re}(s) > 1$ and equals

$$\int_{0}^{1} t^{s-1}\frac{dt}{t} = \frac{t^{s-1}}{s-1}\Big|_{0}^{1} = \frac{1}{s-1}.$$

The third converges for $\mathrm{Re}(s) > 0$ and equals $\int_{0}^{1} t^{s}\frac{dt}{t} = \frac{1}{s}$. Together we get

$$\zeta(f,s) = \int_{|x|>1} \left(E(f)(x)|x|^{s-\frac{1}{2}} + E(\hat{f})(x)|x|^{\frac{1}{2}-s} \right) d^{\times}x - \frac{\hat{f}(0)}{1-s} - \frac{f(0)}{s}.$$

The theorem is proven. $\square$

Example 6.2.3 Let

$$f(x) = 1_{\hat{\mathbb{Z}}}(x_{\mathrm{fin}})e^{-\pi x_{\infty}^{2}} = \prod_{p} f_{p}(x),$$

where $f_{p} = 1_{\mathbb{Z}_{p}}$, if $p < \infty$. We note that every local factor f_{p} is self-dual with respect to the Fourier transformation, i.e. that $\hat{f}_{p} = f_{p}$ holds for every $p \leq \infty$. Therefore,

$$\hat{f}(x) = \prod_{p \leq \infty} \hat{f}_{p}(x) = f(x).$$

By

$$\zeta(f,s) = \int_{\mathbb{A}^{\times}} f(x)|x|^{s}\,d^{\times}x = \underbrace{\left(\prod_{p<\infty} \int_{\mathbb{Z}_{p}\setminus\{0\}} |x|^{s}\,d^{\times}x \right)}_{=\prod_{p}\sum_{n=0}^{\infty} p^{-ns}=\zeta(s)} \underbrace{\int_{\mathbb{R}^{\times}} |x|^{s} e^{-\pi x^{2}}\frac{dx}{|x|}}_{=\Gamma(s/2)\pi^{-s/2}} = \hat{\zeta}(s),$$

we conclude that the last theorem implies the functional equation of the Riemann zeta function. So we have found a new proof of this functional equation. Or was it the same as the old proof?

Definition 6.2.4 A *totally disconnected group* is a Hausdorff topological group G possessing a neighborhood base of the unit consisting of open subgroups (see Exercise 6.5).

This means that G is a totally disconnected group, if it is Hausdorff and there is a family $(U_i)_{i \in I}$ of open subgroups, such that for every neighborhood U of the unit in G there is $i \in I$, such that $U_i \subset U$.

Note that an open subgroup H of a topological group G is also closed. This follows from the decomposition of G in left cosets,

$$G = \bigcup_{g \in G} gH,$$

since one can write the complement of H as

$$G \setminus H = \bigcup_{g \in G \setminus H} gH.$$

The coset gH is open since H is and the map $x \mapsto gx$ is a homeomorphism of G. So $G \setminus H$ is a union of open sets, hence open, so H is closed.

Lemma 6.2.5

(a) *If G is a totally disconnected compact group, then every character of G has finite image. The same holds for a finite-dimensional representation $\pi : G \to \mathrm{GL}(V)$. In particular, π is trivial on a compact open subgroup.*

(b) *Every character of the group $\mathbb{A}^1/\mathbb{Q}^\times$ has finite image.*

(c) *Every character χ of the group $\mathbb{A}^\times/\mathbb{Q}^\times$ is of the form $\chi(x) = |x|^{it}\chi_0(x)$, for a uniquely determined $t \in \mathbb{R}$ and a uniquely determined character χ_0 of finite image.*

Proof The set $U = \mathbb{T} \cap \{\mathrm{Re}(z) > 0\}$ is an open neighborhood of the unit. The only subgroup of $\mathbb{T}$, contained in U, is the trivial group. For any character χ the open set $\chi^{-1}(U)$ must contain an open subgroup H. Then $\chi(H)$ is a subgroup of $\mathbb{T}$ contained in U, hence trivial. This means that χ factors through G/H. As G is compact and H is open, G can be covered by finitely many of the open translates gH, therefore the set G/H is finite. Next for the case of a finite-dimensional representation $\pi : G \to \mathrm{GL}(V)$. Every finite-dimensional Banach space V is isomorphic to $\mathbb{C}^n$ with $n = \dim V$. So the representation π is a continuous group homomorphism $\pi : G \to \mathrm{GL}_n(\mathbb{C}) \subset \mathrm{GL}_{2n}(\mathbb{R})$. By Proposition 3.4.1 there exists an open unit neighborhood U in $\mathrm{GL}_{2n}(\mathbb{R})$, which does not contain any subgroup of $\mathrm{GL}_{2n}(\mathbb{R})$ other than $\{1\}$. As in the case of a character we get an open subgroup H such that $\pi(H) \subset U$, hence $\pi(H) = \{1\}$. Therefore, the kernel C of π contains an open subgroup H, but because of $C = \bigcup_{x \in C} xH$ the subgroup C is itself open and so G/C is finite.

The group $\mathbb{A}^1/\mathbb{Q}^\times$ is isomorphic to $\widehat{\mathbb{Z}}^\times$. Hence it has a unit neighborhood base consisting of open subgroups, i.e. it is a totally disconnected group.

Part (c) follows from (b) and $\mathbb{A}^\times/\mathbb{Q}^\times \cong (\mathbb{A}^1/\mathbb{Q}^\times) \times \mathbb{R}_+^\times$. □

6.3 Dirichlet L-Functions

Recall that a character χ of a topological group G is a continuous group homomorphism of G to the circle group $\mathbb{T}$. A finite group G will be equipped with the discrete topology and in this way become a topological group. Then every map from G into any topological space is continuous, so that characters are exactly the group homomorphisms $G \to \mathbb{T}$. A finite abelian group G has exactly as many characters as elements, as is shown in Exercise 6.2.

Let $N \in \mathbb{N}$ and consider the unit group $(\mathbb{Z}/N\mathbb{Z})^\times$ of the finite ring $\mathbb{Z}/N\mathbb{Z}$. A character

$$\chi : (\mathbb{Z}/N\mathbb{Z})^\times \to \mathbb{T}$$

is called a *Dirichlet character* modulo N.

Given a divisor d of N and a Dirichlet character η modulo d, the pre-composition with the projection $(\mathbb{Z}/N\mathbb{Z})^\times \to (\mathbb{Z}/d\mathbb{Z})^\times$ turns η into a Dirichlet character

$$(\mathbb{Z}/N\mathbb{Z})^\times \to (\mathbb{Z}/d\mathbb{Z})^\times \xrightarrow{\eta} \mathbb{T}$$

modulo N.

A given Dirichlet character χ modulo N is called *primitive* if χ is not induced from any character modulo d with $d < N$. In this case the number N is called the *conductor* of χ.

Example 6.3.1 One has $|(\mathbb{Z}/3\mathbb{Z})^\times| = 2$ and $|(\mathbb{Z}/5\mathbb{Z})^\times| = 4$, so $|(\mathbb{Z}/15\mathbb{Z})^\times| = 8$ and there are eight different characters modulo 15. Of these, two are induced from $\mathbb{Z}/3\mathbb{Z}$ and four are induced from $\mathbb{Z}/5\mathbb{Z}$, where exactly one, the trivial character, is induced from both. It follows that there are exactly three primitive characters modulo 15.

Let χ be a primitive Dirichlet character modulo N. One sets $\chi(n) = 0$ if $\gcd(n, N) > 1$ and by composing it with the projection $\mathbb{Z} \to \mathbb{Z}/N\mathbb{Z}$ one considers χ as a function on $\mathbb{Z}$. As such, it is *multiplicative*, i.e. one has $\chi(mn) = \chi(m)\chi(n)$ for all $m, n \in \mathbb{Z}$. The corresponding *Dirichlet series* is defined to be

$$L(\chi, s) = \sum_{n=1}^{\infty} \frac{\chi(n)}{n^s}.$$

As $|\chi(n)| \leq 1$ for all $n \in \mathbb{N}$, this series converges absolutely for $\mathrm{Re}(s) > 1$.

Similar to Exercise 1.5, one can use the multiplicativity of χ to show that $L(\chi, s)$ can be written as an Euler product:

$$L(\chi, s) = \prod_p \frac{1}{1 - \chi(p)p^{-s}}, \qquad \mathrm{Re}(s) > 1.$$

Lemma 6.3.2 *The isomorphism*

$$\mathbb{A}^1/\mathbb{Q}^\times \cong \prod_p \mathbb{Z}_p^\times \cong \varprojlim_{N \in \mathbb{N}} (\mathbb{Z}/N\mathbb{Z})^\times$$

induces a bijection between the set of all characters of the group $\mathbb{A}^1/\mathbb{Q}^\times$ *and the set of all primitive Dirichlet characters as follows*: *A given primitive Dirichlet character* χ *modulo* N_0 *gives a character of* $\mathbb{A}^1/\mathbb{Q}^\times$ *by*

$$\mathbb{A}^1/\mathbb{Q}^\times \cong \varprojlim_{N \in \mathbb{N}} (\mathbb{Z}/N\mathbb{Z})^\times \xrightarrow{\text{Proj}} (\mathbb{Z}/N_0\mathbb{Z})^\times \xrightarrow{\chi} \mathbb{T}.$$

Proof Let ϕ denote the map in question, sending a primitive Dirichlet character χ modulo N_0 to the character

$$\mathbb{A}^1/\mathbb{Q}^\times \to (Z/N_0\mathbb{Z})^\times \xrightarrow{\chi} \mathbb{T}$$

of the group $\mathbb{A}^1/\mathbb{Q}^\times$. We claim that ϕ is a bijection. For injectivity assume $\phi(\chi) = \phi(\chi')$. Let N_1 be the smallest natural number such that $\phi(\chi)$ factors through $(\mathbb{Z}/N_1\mathbb{Z})^\times$. As χ is primitive, we get $N_1 = N_0$ and therefore N_0 is uniquely determined by $\phi(\chi) = \phi(\chi')$, so χ' also is a Dirichlet character modulo N_0 and it follows that $\chi = \chi'$.

For surjectivity let η be a character of $\mathbb{A}^1/\mathbb{Q}^\times$. For a given $N \in \mathbb{N}$ let U_N be the set of all $x \in \prod_p \mathbb{Z}_p^\times$ with $x_p \equiv 1 \bmod N$ for every p. The family of the U_N forms a neighborhood base of the unit in $\prod_p \mathbb{Z}_p^\times$, so there is an N such that U_N lies in the kernel of η. Let N be minimal with this property. Then η factors through $(\mathbb{Z}/N\mathbb{Z})^\times$ and by minimality of N we get that η factors through a primitive Dirichlet character. $\qquad\square$

Using the isomorphism

$$\mathbb{A}^\times/\mathbb{Q}^\times \cong \mathbb{A}^1/\mathbb{Q}^\times \times (0, \infty)$$

one can identify the set of all characters of $\mathbb{A}^1/\mathbb{Q}^\times$ with the set of all characters of $\mathbb{A}^\times/\mathbb{Q}^\times$, which have finite image.

Let χ be a character of $\mathbb{A}^\times/\mathbb{Q}^\times$. We view χ as a character of $\mathbb{A}^\times = \mathbb{A}_{\text{fin}}^\times \times \mathbb{R}^\times$ and we thus write $\chi = \chi_{\text{fin}}\chi_\infty$.

Definition 6.3.3 Let χ be a character of $\mathbb{A}^\times/\mathbb{Q}^\times$ with finite image and let $f \in S(\mathbb{A})$. Define

$$\zeta(f, \chi, s) = \int_{\mathbb{A}^\times} f(x)\chi(x)|x|^s \, d^\times x.$$

Analogous to Theorem 6.2.2 one shows

$$\zeta(f, \chi, s) = \int_{\mathbb{A}^\times/\mathbb{Q}^\times} E(f)(x)\chi(x)|x|^{s-\frac{1}{2}} \, d^\times x,$$

if the integral $\zeta(f, \chi, s)$ converges. For the trivial character $\chi = 1$ one has $\zeta(f, 1, s) = \zeta(f, s)$, so this case has been treated already.

Theorem 6.3.4 *Let $\chi \neq 1$ be a character of $\mathbb{A}^{\times}/\mathbb{Q}^{\times}$ with finite image and conductor $N \in \mathbb{N}$. The integral $\zeta(f,\chi,s)$ converges locally uniformly for $\mathrm{Re}(s) > 1$ and so it defines a holomorphic function in that range, which extends to an entire function on $\mathbb{C}$. One has*

$$\zeta(f,\chi,s) = \zeta(\hat{f},\overline{\chi},1-s).$$

There exists an f with

$$\zeta(f,\chi,s) = L_{\infty}(\chi,s)L(\chi,s),$$

where

$$L_{\infty}(\chi,s) = \Gamma\left(\frac{s+\nu}{2}\right)\pi^{-\frac{s+\nu}{2}}$$

and $\nu \in \{0,1\}$ is defined by $\chi_{\mathrm{fin}}(-1) = \chi_{\infty}(-1) = (-1)^{\nu}$.
Setting $\tilde{L}(\chi,s) = L_{\infty}(\chi,s)L(\chi,s)$ one gets the functional equation,

$$\tilde{L}(\chi,s) = (-i)^{\nu}N^{-s}\overline{\tau(\chi,e)}\tilde{L}(1-s,\overline{\chi}),$$

where $\tau(\chi,e) = \phi(N)\int_{\frac{1}{N}\widehat{\mathbb{Z}}^{\times}}\chi(x)e(x)\,d^{\times}x$ and $\phi(N)$ is the Euler ϕ-function. It follows that $|\tau(\chi,e)| = \sqrt{N}$.

Proof Once we have shown the functional equation, a second application of the functional equation yields $|\tau(\chi,e)\tau(\overline{\chi},e)| = N$. Now

$$\overline{\tau(\chi,e)} = \phi(N)\int_{\frac{1}{N}\widehat{\mathbb{Z}}^{\times}}\overline{\chi}(x)e(-x)\,d^{\times}x = \phi(N)\int_{\frac{1}{N}\widehat{\mathbb{Z}}^{\times}}\overline{\chi}\big((-1)x\big)e(x)\,d^{\times}x$$

$$= \overline{\chi(-1)}\phi(N)\int_{\frac{1}{N}\widehat{\mathbb{Z}}^{\times}}\overline{\chi}(x)e(x)\,d^{\times}x = \overline{\chi(-1)}\tau(\overline{\chi},e).$$

Therefore $\tau(\chi,e)$ and $\tau(\overline{\chi},e)$ have the same absolute value and the equality $|\tau(\chi,e)| = \sqrt{N}$ follows.

The other assertions follow similarly to the proof of the case $\chi = 1$, i.e. Theorem 6.2.2. All works well up to the identity

$$\int_{|x|<1} E(f)(x)\chi(x)|x|^{s-\frac{1}{2}}\,d^{\times}x$$

$$= \int_{|x|<1}|x|^{s-\frac{1}{2}}E(\hat{f})(1/x)\chi(x)\,d^{\times}x$$

$$+ \int_{|x|<1}\chi(x)|x|^{s-1}\,d^{\times}x\,\hat{f}(0) - \int_{|x|<1}\chi(x)|x|^{s}\,d^{\times}x f(0).$$

The first summand on the right-hand side is an entire function in s. We now show that the last two summands are zero. We have $\{|x| < 1\} \cong \mathbb{A}^1/\mathbb{Q}^\times \times (0,1)$. Using this isomorphism, we write the first integral as

$$\int_{(0,1)} t^{s-1} \int_{\mathbb{A}^1/\mathbb{Q}^\times} \chi(y)\, d^\times y \frac{dt}{t}.$$

By Lemma 5.4.6 the inner integral equals zero. The second integral is treated similarly. The theorem follows up to the last point, the existence of an $f \in \mathcal{S}(\mathbb{A})$ with

$$\zeta(f, \chi, s) = L_\infty(\chi, s) L(\chi, s).$$

We now view χ as a character of $\mathbb{A}^\times$. For $p \leq \infty$ let χ_p be the character of $\mathbb{Q}_p^\times$ given by

$$\mathbb{Q}_p^\times \to \mathbb{A}^\times \xrightarrow{\ \chi\ } \mathbb{T},$$

where $\mathbb{Q}_p^\times$ is embedded via $x \mapsto (\ldots, 1, x, 1, \ldots)$. Let $p < \infty$.

A *quasi-character* of the group $\mathbb{Q}_p^\times$ is a continuous group homomorphism $\lambda : \mathbb{Q}_p^\times \to \mathbb{C}^\times$. A quasi-character λ is called *unramified* if $\lambda(\mathbb{Z}_p^\times) = 1$.

Lemma 6.3.5 *A quasi-character λ of $\mathbb{Q}_p^\times$ is unramified if and only if there exists $s \in \mathbb{C}$ with*

$$\lambda(a) = |a|^s.$$

Proof Every quasi-character of the announced form is clearly unramified. So let λ be an unramified quasi-character. We have an exact sequence of abelian groups,

$$1 \to \mathbb{Z}_p^\times \to \mathbb{Q}_p^\times \xrightarrow{\ |\cdot|\ } p^{\mathbb{Z}} \to 1.$$

This means that a quasi-character is unramified if and only if it factors through $|\cdot|$. There exists $s \in \mathbb{C}$ such that $\lambda(p) = p^{-s}$. For an arbitrary $a = p^k u \in \mathbb{Q}_p^\times$ with $u \in \mathbb{Z}_p^\times$ and $k \in \mathbb{Z}$ one has

$$\lambda(a) = \lambda(p^k) = \lambda(p)^k = (p^{-k})^s = |a|^s. \qquad \square$$

Lemma 6.3.6 *For almost all p the quasi-character χ_p is unramified. More precisely, if χ comes from a primitive Dirichlet character modulo N, then χ_p is unramified if and only if p is not a divisor of N. If p is a divisor of $N = p^k n$ with n coprime to p, then k is the smallest natural number with $\chi_p(1 + p^k \mathbb{Z}_p) = 1$.*

Proof If $N = p_1^{k_1} \cdots p_l^{k_l}$, then χ factors through

$$\prod_p \mathbb{Z}_p^\times \to \prod_{j=1}^{l} (\mathbb{Z}/p_j^{k_j}\mathbb{Z})^\times.$$

The second assertion follows from the fact that χ is primitive modulo N. $\qquad \square$

Let S be the set of prime numbers that divide N. We define

- $f_p = \mathbf{1}_{\mathbb{Z}_p}$, if χ_p is unramified,
- $f_p = p^k(1 - \frac{1}{p})\mathbf{1}_{1+p^k\mathbb{Z}_p}$ if $N = p^k N'$ with N' coprime to p,
- $f_\infty(t) = t^\nu e^{-\pi t^2}$,

and finally $f = \prod_{p \le \infty} f_p \in \mathcal{S}(\mathbb{A})$.

With this choice, the zeta integral is a product,

$$\zeta(f, \chi, s) = \int_{\mathbb{A}^\times} f(x)\chi(x)|x|^s \, d^\times x = \prod_p \underbrace{\int_{\mathbb{Q}_p^\times} f_p(x)\chi_p(x)|x|_p^s \, d^\times x}_{=\zeta_p(f_p, \chi_p, s)}$$

$$= \prod_{p \in S} \int_{\mathbb{Q}_p^\times} f_p(x)\chi_p(x)|x|_p^s \, d^\times x \times \prod_{\substack{p \notin S \\ p < \infty}} \int_{\mathbb{Q}_p^\times} f_p(x)\chi_p(x)|x|_p^s \, d^\times x$$

$$\times \int_{\mathbb{R}^\times} f_\infty(x)\chi_\infty(x)|x|_\infty^s \, d^\times x.$$

We compute the factors. For a prime number $p \notin S$ we have

$$\int_{\mathbb{Q}_p^\times} f_p(x)\chi_p(x)|x|_p^s \, d^\times x = \int_{\mathbb{Z}_p \smallsetminus \{0\}} \chi_p(x)|x|_p^s \, d^\times x$$

$$= \sum_{j=0}^\infty \chi_p(p)^j p^{-js} = \frac{1}{1 - \chi_p(p)p^{-s}}.$$

Let $p \in S$ and write $N = p^k N'$ with N' coprime to p. Then

$$\int_{\mathbb{Q}_p^\times} f_p(x)\chi_p(x)|x|_p^s \, d^\times x = p^k\left(1 - \frac{1}{p}\right) \int_{1+p^k\mathbb{Z}_p} \chi_p(x)|x|_p^s \, d^\times x$$

$$= p^k\left(1 - \frac{1}{p}\right) \int_{1+p^k\mathbb{Z}_p} d^\times x = 1.$$

For $p = \infty$ and $\nu = 0$ we get the contribution

$$\int_{\mathbb{R}^\times} e^{-\pi x^2}|x|^s \frac{dx}{|x|} = \Gamma\left(\frac{s}{2}\right)\pi^{-\frac{s}{2}}.$$

In the case $\nu = 1$ we get

$$\int_{\mathbb{R}^\times} e^{-\pi x^2} x \underbrace{\chi_\infty(x)}_{=\mathrm{sgn}(x)} |x|^s \frac{dx}{|x|} = \int_{\mathbb{R}^\times} e^{-\pi x^2}|x|^{s+1} \frac{dx}{|x|} = \Gamma\left(\frac{s+1}{2}\right)\pi^{-\frac{s+1}{2}}.$$

This shows that for the chosen f the equation

$$\zeta(f,\chi,s) = L_\infty(\chi,s)L(\chi,s)$$

holds.

To finish the proof of the theorem, we have to determine the Fourier transform of f. Consider $p < \infty$. If p is unramified, then $\hat{f} = f$. If p is ramified, however, write $N = p^k N'$ with N' coprime to p. Then we have

$$\hat{f}_p(x) = \int_{\mathbb{Q}_p} f_p(y) e_p(-xy)\, dy = p^k\left(1 - \frac{1}{p}\right)\int_{1+p^k\mathbb{Z}_p} e_p(-xy)\, dy$$

$$= p^k\left(1 - \frac{1}{p}\right)p^{-k}e_p(-x)\int_{\mathbb{Z}_p} e_p(-p^k xy)\, dy$$

$$= \left(1 - \frac{1}{p}\right)\overline{e_p(x)}\mathbf{1}_{p^{-k}\mathbb{Z}_p}(x).$$

Finally, if $p = \infty$ and $v = 0$, then $\hat{f}_\infty = f_\infty$. In the case $v = 1$, one has

$$\hat{f}_\infty(x) = \int_{\mathbb{R}} y e^{-\pi y^2 - 2\pi i xy}\, dy = \frac{1}{-2\pi i}\frac{\partial}{\partial x}\int_{\mathbb{R}} e^{-\pi y^2 - 2\pi i xy}\, dy$$

$$= \frac{1}{-2\pi i}\frac{\partial}{\partial x}e^{-\pi x^2} = \frac{x}{i}e^{-\pi x^2} = -i f_\infty(x).$$

We see that

$$\tilde{L}(\chi,s) = \zeta(\hat{f},\overline{\chi},1-s) = \prod_p \zeta_p(\hat{f}_p,\overline{\chi}_p,1-s).$$

We compute the local contributions for $\mathrm{Re}(1-s) > 1$. If p is a prime number such that χ_p is unramified, then

$$\zeta_p(\hat{f},\overline{\chi},1-s) = \frac{1}{1 - \overline{\chi(p)}p^{1-s}}.$$

In the case $p = \infty$ one gets

$$\zeta_p(\hat{f},\overline{\chi},1-s) = (-i)^v L_\infty(\overline{\chi},1-s).$$

There remains the case of a ramified prime p. Let $N = p^k N'$ with N' coprime to p. Then

$$\zeta_p(\hat{f},\overline{\chi},1-s) = \left(1 - \frac{1}{p}\right)\int_{p^{-k}\mathbb{Z}_p\setminus\{0\}} \overline{e_p(x)\chi_p(x)}|x|^{1-s}\, d^\times x$$

$$= \left(1 - \frac{1}{p}\right)\sum_{j=-k}^{\infty} p^{js-j}\int_{p^j\mathbb{Z}_p^\times} \overline{e_p(x)\chi_p(x)}\, d^\times x.$$

We show that $\int_{p^j \mathbb{Z}_p^\times} \overline{e_p(x)\chi_p(x)} \, d^\times x = 0$ for $j > -k$. If $j \geq 0$, then e_p equals the constant 1 on the domain of integration. So one has $\int_{p^j \mathbb{Z}_p^\times} \overline{\chi_p(x)} \, d^\times x = 0$, as $\chi_p(\mathbb{Z}_p^\times) \neq 1$. If $-k < j < 0$, then χ_p is non-trivial on the multiplicative group $1 + p^{k-1}\mathbb{Z}_p$. On the other hand, $p^j(1 + p^{k-1}\mathbb{Z}_p) = p^j + p^{k-1+j}\mathbb{Z}_p$ and e_p is trivial on $p^{k-1+j}\mathbb{Z}_p \subset \mathbb{Z}_p$. Therefore the integral vanishes by the same argument. We infer that

$$\zeta_p(\hat{f}, \overline{\chi}, 1-s) = p^k \left(1 - \frac{1}{p}\right)(p^k)^{-s} \int_{p^{-k}\mathbb{Z}_p^\times} \overline{e_p(x)\chi_p(x)} \, d^\times x.$$

The product over all ramified p is just $N^{-s}\overline{\tau(\chi, e)}$ and the theorem is proven. □

The subject of Tate's thesis is the quotient $\mathbb{A}^\times / \mathbb{Q}^\times = \mathrm{GL}_1(\mathbb{A}) / \mathrm{GL}_1(\mathbb{Q})$. In the theory of automorphic forms one considers more generally the quotient

$$\mathrm{GL}_n(\mathbb{A}) / \mathrm{GL}_n(\mathbb{Q}).$$

In this book we shall restrict to the case $n = 2$.

6.4 Galois Representations and L-Functions

This section is a survey without proper proofs.

The Langlands conjectures assert, among other things, that L-functions attached to Galois representations should be automorphic. In this section we shall define these L-functions, consider the number-theoretic background and the motivation given by the one-dimensional case.

Consider a finite field extension $L/\mathbb{Q}$ and write $\mathcal{O}_L$ for the *ring of integers* of L, i.e. $\mathcal{O}_L$ is the set of all $\alpha \in L$ such that $f(\alpha) = 0$ for some polynomial $f \in \mathbb{Z}[x]$ with leading coefficient 1. One has

- $\mathcal{O}_L$ is a subring of L.
- There is a basis $v_1, \ldots, v_n$ of the vector space L over $\mathbb{Q}$ such that $\mathcal{O}_L = \mathbb{Z}v_1 \oplus \mathbb{Z}v_2 \oplus \cdots \oplus \mathbb{Z}v_n$.
- $\mathcal{O}_L \cap \mathbb{Q} = \mathbb{Z}$.
- $\mathcal{O}$ is a *Dedekind ring*, i.e. every ideal is a product of prime ideals. Further, for every prime ideal $P \neq 0$, the quotient $\mathcal{O}/P$ is a finite field.
- If $P \neq 0$ is a prime ideal of $\mathcal{O}_L$, then $P \cap \mathbb{Z}$ is a prime ideal $\neq 0$ of $\mathbb{Z}$, so there exists exactly one prime number p with $P \cap \mathbb{Z} = (p)$. In this case one says that P lies over p, or P divides p.

We now assume that $L/\mathbb{Q}$ is a Galois extension and write $\mathrm{Gal}(L/\mathbb{Q})$ for its Galois group. The group $\mathrm{Gal}(L/\mathbb{Q})$ acts on $\mathcal{O}_L$ and on the set Prim_L of all prime ideals of $\mathcal{O}_L$. We write Gal_K for the Galois group $\mathrm{Gal}(\overline{K}/K)$, where K is any field and $\overline{K}$ an algebraic closure of K. So Gal_K is the group of all field automorphisms of $\overline{K}$, which leave K point-wise fixed. This group does not depend on the choice of the algebraic closure $\overline{K}$ up to isomorphy.

For a given prime number p let $\mathrm{Prim}_{L,p}$ be the set of all prime ideals of $\mathcal{O}_L$ lying above p. Then $\mathrm{Prim}_{L,p}$ is a non-empty finite set. The Galois group $\mathrm{Gal}(L/\mathbb{Q})$ acts on $\mathrm{Prim}_{L,p}$.

For a given $P \in \mathrm{Prim}_{L,p}$ let

$$\mathrm{Gal}(L/\mathbb{Q})_P = \left\{ g \in \mathrm{Gal}(L/\mathbb{Q}) : gP = P \right\}$$

be the *decomposition group* of P. Then $\mathrm{Gal}(L/\mathbb{Q})_P$ acts on the finite field $\mathbb{F}_P = \mathcal{O}_L/P$, so we have a group homomorphism

$$\phi_P : \mathrm{Gal}(L/\mathbb{Q})_P \to \mathrm{Gal}(\mathbb{F}_P/\mathbb{F}_p).$$

The kernel I_P of this homomorphism is called the *inertia group* in P. Since $\mathbb{F}_P$ is a finite field, the Galois group is a finite cyclic group generated by the *Frobenius homomorphism*

$$\mathrm{Frob}_p : x \mapsto x^p.$$

One can show that the homomorphism ϕ_P is surjective. So one can view Frob_p as an element of $\mathrm{Gal}(L/\mathbb{Q})_P/I_P$.

By a *Galois representation* over $\mathbb{C}$ we mean a group homomorphism

$$\rho : \mathrm{Gal}(L/\mathbb{Q}) \to \mathrm{GL}(V),$$

where V is a finite-dimensional $\mathbb{C}$-vector space. The group $\mathrm{Gal}(L/\mathbb{Q})_P/I_P$ acts on the space

$$V^{I_P} = \{v \in V : gv = v \ \forall g \in I_P\}.$$

The Galois group $\mathrm{Gal}(L/\mathbb{Q})$ acts transitively on the set $\mathrm{Prim}_{L,p}$, so the characteristic polynomial

$$\det\left(1 - x\,\mathrm{Frob}_p\,|V^{I_P}\right) = \det\left(1 - x\rho(\mathrm{Frob}_p)|_{V^{I_P}}\right)$$

depends only on p. Define the *Artin L-function* of the Galois representation ρ as

$$L(\rho, s) = \prod_p \frac{1}{\det(1 - p^{-s}\,\mathrm{Frob}_p\,|V^{I_P})}.$$

One can show that the product converges absolutely for $\mathrm{Re}(s) > \dim V$ and that it can be extended to a meromorphic function on the complex plane $\mathbb{C}$.

A Galois representation ρ is called *irreducible* if it does not admit any subrepresentation, i.e. for every subspace $U \subset V$ one has: if $\rho(g)U \subset U$ for every $g \in \mathrm{Gal}(L/\mathbb{Q})$, then $U = 0$ or $U = V$.

Artin Conjecture If ρ is irreducible and $\neq 1$, then $L(\rho, s)$ is entire.

We now prove the Artin conjecture for one-dimensional representations. For this we cite the classical theorem of Kronecker and Weber.

Theorem 6.4.1 (Kronecker–Weber) *Let $K/\mathbb{Q}$ be a finite Galois extension with abelian Galois group. Then K is a subfield of $\mathbb{Q}(\xi)$, where ξ is a root of unity. In other words, every abelian number field is cyclotomic.*

Note the following: if $K/\mathbb{Q}$ is a Galois extension, then Gal_K is a normal subgroup of $\mathrm{Gal}_\mathbb{Q}$ and one has $\mathrm{Gal}(K/\mathbb{Q}) = \mathrm{Gal}_\mathbb{Q} / \mathrm{Gal}_K$.

For an arbitrary group G let $[G, G]$ denote the *commutator subgroup*. This is the subgroup generated by all *commutators*, i.e. all elements of the form

$$[a, b] = aba^{-1}b^{-1}.$$

It is easy to see that $[G, G]$ is a normal subgroup of G. The quotient $G^{\mathrm{ab}} = G/[G, G]$ is called the *abelianization* of G. It is the largest abelian quotient of G, in the sense that if N is a normal subgroup such that G/N is abelian, then $N \supset [G, G]$, or, equivalently, the map $G \to G/N$ factors through $G/[G, G]$.

For given $n \in \mathbb{N}$, let μ_n be the set of all nth roots of unity in $\mathbb{C}$; this is the set of all complex numbers ξ satisfying $\xi^n = 1$. Further let μ_∞ be the union of all μ_n as n runs through $\mathbb{N}$. Then μ_∞ can also be described as the set of complex numbers of the form $e^{2\pi i \alpha}$ with $\alpha \in \mathbb{Q}/\mathbb{Z}$. Let $K_0 = \mathbb{Q}(\mu_\infty)$ be the field generated by all $\xi \in \mu_\infty$. The field extension $K_0/\mathbb{Q}$ is an infinite Galois extension with Galois group $\mathrm{Gal}_\mathbb{Q}^{\mathrm{ab}}$, as the Kronecker–Weber theorem implies. One concludes

$$\mathrm{Gal}_\mathbb{Q}^{\mathrm{ab}} = \mathrm{Aut}(\mu_\infty) = \mathrm{Aut}(\mathbb{Q}/\mathbb{Z})$$

$$= \mathrm{Aut}\left(\bigoplus_p \varinjlim_{k \in \mathbb{N}} p^{-k}\mathbb{Z}/\mathbb{Z} \right)$$

$$= \prod_p \varprojlim_{k \in \mathbb{N}} \underbrace{\mathrm{Aut}(\mathbb{Z}/p^k\mathbb{Z})}_{=(\mathbb{Z}/p^k\mathbb{Z})^\times}$$

$$= \prod_p \mathbb{Z}_p^\times \cong \mathbb{A}^1/\mathbb{Q}^\times.$$

Let's turn to the Artin conjecture. Given a finite Galois extension L of $\mathbb{Q}$ with Galois group $\mathrm{Gal}(L/\mathbb{Q})$ and a character $\rho : \mathrm{Gal}(L/\mathbb{Q}) \to \mathbb{T}$. By Galois theory, $\mathrm{Gal}(L/\mathbb{Q})$ is the quotient $\mathrm{Gal}_\mathbb{Q} / \mathrm{Gal}_L$, so ρ can be extended to a character $\rho : \mathrm{Gal}_\mathbb{Q} \to \mathbb{T}$. As $\mathbb{T}$ is abelian, ρ factors uniquely through the abelian quotient $\mathrm{Gal}_\mathbb{Q}^{\mathrm{ab}}$. By the Kronecker–Weber isomorphism given above, ρ yields a character of $\mathbb{A}^1/\mathbb{Q}$, i.e. a Dirichlet character χ_ρ.

Lemma 6.4.2 *The Artin L-function of ρ coincides with the Dirichlet L-function of χ_ρ, so*

$$L(\rho, s) = L(\chi_\rho, s).$$

Sketch of Proof In the definition of $L(\rho, s)$ one can replace the field L by the fix field of the group $\ker(\rho)$. This field has an abelian Galois group over $\mathbb{Q}$, and hence is a subfield of $\mathbb{Q}(\mu_N)$ for some $N \in \mathbb{N}$. Enlarging the field, one assumes $L = \mathbb{Q}(\mu_N)$ with minimal N. Then $\mathcal{O}_L = \mathbb{Z}[\mu_N]$ and for a prime ideal P above p one gets $\mathcal{O}_L/P \cong \mathbb{F}_p[\mu_{N'}]$, where N' is the part of N which is coprime to p, i.e. $N = p^k N'$. If $p|N$, then $\mathrm{Gal}(L/\mathbb{Q})_P = I_P$ and $\rho(\mathrm{Gal}(L/\mathbb{Q})_P) \neq 1$, so the Euler factor of $L(\rho, s)$ equals 1. If $p \nmid N$, the Frobenius operator is the unique element of $\mathrm{Gal}(L/\mathbb{Q})$, given by $\xi \mapsto \xi^p$ for every $\xi \in \mu_N$. Then $\rho(\mathrm{Frob}_p) = \chi_\rho(p)$ and the lemma follows. $\square$

If $\rho \neq 1$, then we conclude that $L(\rho, s)$ is entire and we have shown the Artin conjecture for one-dimensional representations!

The absolute Galois group $G = \mathrm{Gal}_\mathbb{Q}$ of $\mathbb{Q}$ and its representations contain some of the deepest and farthest reaching mysteries of all of mathematics. Since, as we have seen, the abelian quotient G^{ab} of the Galois group is isomorphic to $\mathbb{A}^1/\mathbb{Q}^\times$, one can identify the one-dimensional unitary Galois representations with the characters of $\mathbb{A}^1/\mathbb{Q}^\times = \mathrm{GL}_1(\mathbb{A})^1/\mathrm{GL}_1(\mathbb{Q})$. In order to understand higher-dimensional Galois representations, one analogously should consider $\mathrm{GL}_n(\mathbb{A})^1/\mathrm{GL}_n(\mathbb{Q})$, where

$$\mathrm{GL}_n(\mathbb{A})^1 = \left\{ g \in \mathrm{GL}_n(\mathbb{A}) : \left|\det(g)\right| = 1 \right\}.$$

Since, however, $\mathrm{GL}_n(\mathbb{Q})$ is not normal in $\mathrm{GL}_n(\mathbb{A})^1$, the quotient is not a group, and hence has no representations. So what should be the generalization of the characters of $\mathbb{A}^1/\mathbb{Q}^\times$?

The answer to this question is in the Fourier analysis of the compact group $\mathbb{A}^1/\mathbb{Q}^\times$. Every character $\chi : \mathbb{A}^1/\mathbb{Q}^\times \to \mathbb{T} \subset \mathbb{C}$ is an element of $L^2(\mathbb{A}^1/\mathbb{Q}^\times)$ and the characters form an orthonormal basis of this Hilbert space, i.e. one has

$$L^2(\mathbb{A}^1/\mathbb{Q}^\times) = \bigoplus_\chi \mathbb{C}\chi.$$

By analogy, the irreducible subrepresentations π of the $\mathrm{GL}_n(\mathbb{A})^1$ representation on the space $L^2(\mathrm{GL}_n(\mathbb{A})^1/\mathrm{GL}_n(\mathbb{Q}))$ are the right replacement of the characters χ. The global Langlands conjecture asserts that for every n-dimensional Galois representation ρ there exists such a representation π, such that $L(\rho, s) = L(\pi, s)$. Note that $L(\pi, s)$ has not been defined as yet, but we shall do that later. We also show that the *automorphic L-function* $L(\pi, s)$ extends to an entire function; thus the Artin conjecture follows from the Langlands conjecture.

6.5 Exercises

Exercise 6.1 Fix a Dirichlet character $\chi : (\mathbb{Z}/N\mathbb{Z})^\times \to \mathbb{T}$ modulo N. The *conductor* $f_\chi \in \mathbb{N}$ of χ is the smallest divisor of N, such that χ factors through the projection $\mathbb{Z}/N\mathbb{Z} \to \mathbb{Z}/f_\chi\mathbb{Z}$.

Show that for Dirichlet characters χ, η modulo N with $(f_\chi, f_\eta) = 1$, on has $f_{\chi\eta} = f_\chi f_\eta$.

Exercise 6.2 Let A be a finite abelian group. We equip the vector space $C(A)$ of all functions $f : A \to \mathbb{C}$ with the inner product $\langle f, g \rangle = \frac{1}{|A|} \sum_{a \in A} f(a)\overline{g(a)}$. Let $\widehat{A}$ be the set of all group homomorphisms $\chi : A \to \mathbb{T}$.

(a) Show that $|\widehat{A}| = |A|$. (Use the fact that every finite abelian group is a product of cyclic groups.)
(b) Show that the $\chi \in \widehat{A}$ form an orthonormal basis of $C(A)$.
(c) Let $N \in \mathbb{N}$ and let $a \in A = (\mathbb{Z}/N\mathbb{Z})^\times$. Show that there are coefficients $c(\chi) \in \mathbb{C}$, such that

$$\sum_{\chi \in \widehat{A}} c(\chi) L(\chi, s) = \sum_{\substack{n \in \mathbb{N} \\ n \equiv a \bmod N}} \frac{1}{n^s}.$$

Exercise 6.3 Let $K \subset G$ be a compact subgroup of the locally compact group G and let $L^1(K\backslash G/K)$ be the set of all $f \in L^1(G)$, such that $f(k_1 x k_2) = f(x)$ holds for all $k_1, k_2 \in K$. Show that $L^1(K\backslash G/K)$ is a convolution subalgebra of $L^1(G)$ (see Proposition 3.1.11).

Exercise 6.4 Let $\tau : K \to GL(V)$ be a continuous representation of the compact group K on the Banach space V. Show that the map $P : V \to V$, given by

$$P(v) = \int_K \tau(k) v \, dk$$

is a projection with image $V_\pi(\tau)$.

Exercise 6.5 A group G is called *pro-finite* if it can be written as a projective limit of finite groups. In this case one equips the finite groups with the discrete topology and G with the projective limit topology, which is the topology induced from the product topology.

A topological group G is called *totally disconnected* if it has a neighborhood base of the unit consisting of open subgroups (see Definition 6.2.4). Show:

   G is pro-finite $\Leftrightarrow$ G is compact, Hausdorff and totally disconnected.

(Hint: let G be compact, Hausdorff and totally disconnected and let H be an open subgroup. Show that G/H is a finite set and conclude that H contains a normal open subgroup N of finite index. Order the set I of all normal open subgroups by inverse inclusion and show that G is isomorphic with $\lim_N G/N$.)

Exercise 6.6 Let H be a subgroup of the locally compact group G. We extend any given $f \in C(H)$ by zero to a function on G and view $C(H)$ as a linear subspace of the space of all maps from G to $\mathbb{C}$. Show:

$$C(H) \subset C(G) \quad \Leftrightarrow \quad H \text{ open in } G.$$

Show that if H is not open in G, then

$$C(H) \cap C(G) = 0.$$

Exercise 6.7 Show that the complex algebra $M_n(\mathbb{C})$ of $n \times n$ matrices has no two-sided ideals other than 0 and $M_n(\mathbb{C})$.
(Hint: if J is a two-sided ideal and $A \in J$, then the diagonalizable matrix AA^* lies in J. If $J \neq 0$, then J must contain invertible elements.)

Exercise 6.8 Let p be a prime number and let M_p be the set of all integral 2×2 matrices of determinant p. Show that $M_p = \Gamma \left(\begin{smallmatrix} p & \\ & 1 \end{smallmatrix} \right) \Gamma$, where $\Gamma = \mathrm{SL}_2(\mathbb{Z})$.

Exercise 6.9 Let G be a group and H a subgroup of finite index, so $|G/H| < \infty$. Show that H contains a subgroup N, of finite index and normal in G.
(Hint: write $G = \bigcup_{j=1}^n g_j H$ and consider $N = \bigcap_j g_j H g_j^{-1}$.)

Chapter 7
Automorphic Representations of $GL_2(\mathbb{A})$

We introduce a useful notation. Given a ring R, commutative with unit, we write G_R for the group $GL_2(R)$ of invertible 2×2 matrices with entries from R. These are exactly those matrices, whose determinant is a unit in the ring R. In particular, the group $G_\mathbb{A} = GL_2(\mathbb{A})$ is the restricted product of the groups $G_{\mathbb{Q}_p}$, see Exercise 7.1. We also write G_p for the group $G_{\mathbb{Q}_p} = GL_2(\mathbb{Q}_p)$, so in particular $G_\infty = G_\mathbb{R} = GL_2(\mathbb{R})$. More generally, for an arbitrary set S of places, we write $G_S = G_{\mathbb{A}_S} = GL_2(\mathbb{A}_S)$. This group equals the restricted product of the G_p with $p \in S$, see Exercise 7.1. Analogously we write G^S for $GL_2(\mathbb{A}^S)$, so that we have $G_\mathbb{A} = G_S \times G^S$. If S is the set of all prime numbers, we write $G_{\mathrm{fin}} = G_S = G_{\mathbb{A}_{\mathrm{fin}}}$.

For any ring R we write Z_R for the set of all diagonal matrices $\begin{pmatrix} r & \\ & r \end{pmatrix}$ with $r \in R^\times$. For $p \le \infty$ we also write Z_p for $Z_{\mathbb{Q}_p}$. Finally, we write $G_\mathbb{A}^1 = GL_2(\mathbb{A})^1$ for the set of all $x \in GL_2(\mathbb{A})$ with $|\det(x)| = 1$. The injection $G_\mathbb{A}^1 \hookrightarrow G_\mathbb{A}$ yields an identification of $G_\mathbb{Q} \backslash G_\mathbb{A}^1$ with $(G_\mathbb{Q} Z_\mathbb{R}) \backslash G_\mathbb{A}$.

7.1 Principal Series Representations

In this section, we describe an important class of representations of the group $G_p = GL_2(\mathbb{Q}_p)$ for a prime number p, the *principal series representations*.

The group G_p gets a locally compact topology from the inclusion $G_p = GL_2(\mathbb{Q}_p) \subset M_2(\mathbb{Q}_p) \cong \mathbb{Q}_p^4$. With this topology it is a locally compact group.

Let A_p be the group of diagonal matrices in G_p and N_p the group of all matrices of the form $\begin{pmatrix} 1 & x \\ & 1 \end{pmatrix}$ with $x \in \mathbb{Q}_p$. Finally, let K_p be the compact open subgroup $GL_2(\mathbb{Z}_p)$. We give G_p the unique Haar measure, which assigns the volume 1 to the compact open subgroup K_p.

Let $\lambda : A_p \to \mathbb{C}^\times$ be a quasi-character. We write a^λ for its value $\lambda(a)$ and we consider the group of quasi-characters as an additive group. So for quasi-characters λ and λ', the quasi-character $\lambda + \lambda'$ is given by

$$a^{\lambda+\lambda'} = a^\lambda a^{\lambda'}.$$

A. Deitmar, *Automorphic Forms*, Universitext,
DOI 10.1007/978-1-4471-4435-9_7, © Springer-Verlag London 2013

Example 7.1.1 Given complex numbers λ_1, λ_2, the map $a \mapsto a^\lambda$ with

$$\begin{pmatrix} a_1 & \\ & a_2 \end{pmatrix}^\lambda \stackrel{\text{def}}{=} |a_1|^{\lambda_1} |a_2|^{\lambda_2}$$

is a quasi-character of the group A_p. In this way the additive group $\mathbb{C}^2$ is a sub-group of the quasi-character group of A_p. Of particular importance will be the quasi-character $a^\delta = |a_1|/|a_2|$. Note that the modular function of the group $B_p = A_p N_p$ is given by $\Delta_{B_p}(an) = a^\delta$ (Exercise 7.3).

Proposition 7.1.2 *The group G_p is unimodular. One has the* Iwasawa *decomposition*

$$G_p = A_p N_p K_p.$$

The Haar measures can be normalized in such a way that the Iwasawa *integral formula*

$$\int_{G_p} h(x)\, dx = \int_{A_p} \int_{N_p} \int_{K_p} h(ank)\, dk\, dn\, da$$

holds for every $h \in L^1(G_p)$. There are continuous functions

$$\underline{a} : G_p \to A_p, \quad \underline{n} : G_p \to N_p, \quad \underline{k} : G_p \to K_p,$$

such that for every $g \in G_p$ the identity $g = \underline{a}(g)\underline{n}(g)\underline{k}(g)$ holds. For every choice of such functions, every $f \in L^1(K_p)$ and every $y \in G$ one has the integral formula

$$\int_{K_p} f(k)\, dk = \int_{K_p} \underline{a}(ky)^\delta f\big(\underline{k}(ky)\big)\, dk.$$

For simplicity we also write $\underline{an}(x) = \underline{a}(x)\underline{n}(x)$.

Note that, unlike the real case, the Iwasawa decomposition is not a direct product decomposition, as the group K_p has non-trivial intersection with both other factors A_p and N_p.

Proof Take an element $g = \begin{pmatrix} a & b \\ c & d \end{pmatrix}$ of G_p. In the case $|c| > |d|$ we multiply g from the right with the matrix $\begin{pmatrix} & 1 \\ 1 & \end{pmatrix} \in K_p$, so we can assume $|c| \leq |d|$. As g is invertible, one has $d \neq 0$ and

$$\begin{pmatrix} a & b \\ c & d \end{pmatrix} = \underbrace{\begin{pmatrix} a - \frac{bc}{d} & b \\ & d \end{pmatrix}}_{\in A_p N_p} \underbrace{\begin{pmatrix} 1 & \\ c/d & 1 \end{pmatrix}}_{\in K_p}.$$

This shows that $G_p = A_p N_p K_p$. The group $B = A_p N_p$ coincides with the subgroup of G_p consisting of upper triangular matrices, and hence is a closed subgroup. Further $da\, dn$ is a Haar measure on B, see Exercise 8.2. Therefore the Iwasawa integral formula follows from Theorem 3.1.15. The maps $\underline{a}, \underline{n}, \underline{k}$ are defined by the above formula. They are continuous on the open sets $\{|c| > |d|\}$ and $\{|c| \leq |d|\}$ and therefore are continuous everywhere.

A quasi-character χ of the group A_p is called *unramified* if it is trivial on $A_p \cap K_p$. This holds if and only if there are $\lambda_1, \lambda_2 \in \mathbb{C}$ with $\chi \left({}^{a_1} \ {}_{a_2} \right) = |a_1|^{\lambda_1} |a_2|^{\lambda_2}$, see Exercise 7.4.

Next we show unimodularity. Let $\Delta : G_p \to \mathbb{R}_+^\times$ be the modular function of G_p. We have to show that Δ is trivial. First note that $\Delta(K_p) = 1$, since K_p is compact. So the restriction of Δ to A_p is an unramified character. Therefore there are $\lambda_1, \lambda_2 \in \mathbb{C}$ with $\Delta \left({}^{a_1} \ {}_{a_2} \right) = |a_1|^{\lambda_1} |a_2|^{\lambda_2}$. Let $w = \left({}_1 \ {}^1 \right) \in G_p$. Conjugation by w interchanges the entries a_1 and a_2. By $\Delta(waw^{-1}) = \Delta(a)$ it follows that $\lambda_1 = \lambda_2$. Write $\lambda \in \mathbb{C}$ for this number. By definition the function Δ is trivial on the center of G_p, so

$$1 = \Delta \left({}^{a} \ {}_{a} \right) = |a|^{2\lambda}.$$

We infer that $p^{2\lambda} = 1$. As $\Delta \geq 0$, we get $p^\lambda = 1$ and therefore $\Delta(A_p) = 1$. For a given $n = \left({}^1 \ {}_1^x \right) \in N_p$ there exists a $k \in \mathbb{N}$ such that $n^{p^k} = \left({}^1 \ {}_1^{p^k x} \right) \in K_p$. Hence $1 = \Delta(n^{p^k}) = \Delta(n)^{p^k}$ and as $\Delta \geq 0$, we get $\Delta(n) = 1$. By Iwasawa decomposition one therefore has $\Delta(G_p) = \Delta(A_p)\Delta(N_p)\Delta(K_p) = 1$, so the group G_p is unimodular.

The space of continuous functions $C(K_p)$ is dense in $L^1(K_p)$. So it suffices to show the last formula for $f \in C(K_p)$. Choose a function $\eta \in C_c(A_p N_p)$ such that $\eta \geq 0$ and $\int_{A_p N_p} \eta(an)\, da\, dn = 1$. Set $g(x) = \eta(\underline{an}(x)) f(\underline{k}(x))$. Then $g \in C_c(G)$ and

$$\int_{G_p} g(x)\, dx = \int_{A_p N_p} \eta(an)\, dan \int_{K_p} f(k)\, dk = \int_{K_p} f(k)\, dk.$$

On the other hand, as G_p is unimodular, the integral also equals

$$\int_{G_p} g(xy)\, dx = \int_{G_p} \eta(\underline{an}(xy)) f(\underline{k}(xy))\, dx$$

$$= \int_{A_p N_p} \int_{K_p} \eta(\underline{an}(anky)) f(\underline{k}(ky))\, dk\, da\, dn$$

$$= \int_{K_p} \left(\int_{A_p N_p} \eta(\underline{an}(anky))\, dan \right) f(\underline{k}(ky))\, dk$$

$$= \int_{K_p} \left(\int_{A_p N_p} \eta(an\underline{an}(ky))\, dan \right) f(\underline{k}(ky))\, dk$$

$$= \int_{K_p} \underline{a}(ky)^\delta \left(\int_{A_p N_p} \eta(an)\, dan \right) f(\underline{k}(ky))\, dk$$

$$= \int_{K_p} \underline{a}(ky)^\delta f(\underline{k}(ky))\, dk.$$

The proposition is proven.   □

By V_λ we denote the space of all measurable functions $\varphi : G_p \to \mathbb{C}$ with

- $\varphi(anx) = a^{\lambda+\delta/2}\varphi(x)$, $a \in A_p$, $n \in N_p$, $x \in G_p$ and
- $\int_{K_p} |\varphi(k)|^2 \, dk < \infty$.

We identify two functions which differ only on a set of measure zero. Thus we get a Hilbert space with inner product

$$\langle \varphi, \psi \rangle \overset{\text{def}}{=} \int_{K_p} \varphi(k)\overline{\psi(k)} \, dk.$$

The representation π_λ of G_p on the space V_λ is defined by

$$\big(\pi_\lambda(y)\varphi\big)(x) \overset{\text{def}}{=} \varphi(xy),$$

so it is given by right translation.

Theorem 7.1.3 *The map π_λ is a representation of the group G_p on the Hilbert space V_λ. It is unitary if and only if λ is a character, i.e. if $a^\lambda \in \mathbb{T}$ holds for every $a \in A_p$. The restriction of π_λ to the compact group K_p is independent of λ and is always unitary.*

These representations are called the principal series representations *of G_p.*

Note that a^λ lies in $\mathbb{T}$ if and only if $a^{\lambda+\overline{\lambda}} = 1$.

Proof First we have to show that $\pi_\lambda(y)\varphi$ indeed lies in the space V_λ, if $\varphi \in V_\lambda$ does. The first defining property of V_λ is clearly satisfied by $\pi_\lambda(y)\varphi$, since

$$\pi_\lambda(y)\varphi(anx) = \varphi(anxy) = a^{\lambda+\delta/2}\varphi(xy) = a^{\lambda+\delta/2}\pi_\lambda(y)\varphi(x).$$

The second property is more subtle. We have to show

$$\big\|\pi_\lambda(y)\varphi\big\|^2 = \int_{K_p} |\varphi(ky)|^2 \, dk < \infty.$$

The following lemma contains a much sharper assertion.

Lemma 7.1.4 *For a given compact subset $U \subset G$ there exists a $C > 0$ such that for every $y \in U$ and every $\varphi \in V_\lambda$ one has*

$$\big\|\pi_\lambda(y)\varphi\big\| \leq C\|\varphi\|.$$

Proof The property $\varphi(ky) = \underline{a}(ky)^{\lambda+\delta/2}\varphi(\underline{k}(ky))$ implies

$$\int_{K_p} |\varphi(ky)|^2 \, dk = \int_{K_p} \underline{a}(ky)^{\lambda+\overline{\lambda}+\delta} |\varphi(\underline{k}(ky))|^2 \, dk.$$

The continuous function $(y, k) \mapsto \underline{a}(ky)^{\lambda+\bar{\lambda}}$ is bounded on the compact set $U \times K_p$, so there is $C > 0$ with $|\underline{a}(ky)^{\lambda+\bar{\lambda}}| \leq C^2$ for all $y \in U$ and all $k \in K_p$. By Proposition 7.1.2 this implies

$$\|\pi_\lambda(y)\varphi\|^2 \leq C^2 \int_{K_p} \underline{a}(ky)^\delta |\varphi(\underline{k}(ky))|^2 \, dk = C^2 \int_{K_p} |\varphi(k)|^2 \, dk = C^2 \|\varphi\|^2.$$

The lemma is proven.   □

We next have to show that the map $G_p \times V_\lambda \to V_\lambda$, given by $(g, \varphi) \mapsto \pi_\lambda(g)\varphi$, is continuous. So let $y_n \to y$ be a convergent sequence in G_p and $\varphi_n \to \varphi$ a convergent sequence in V_λ. We claim that the sequence $\pi_\lambda(y_n)\varphi_n$ in V_λ converges to $\pi_\lambda(y)\varphi$. For the compact set $U = \{y_n : n \in \mathbb{N}\} \cup \{y\} \subset G_p$ there exists a constant $C > 0$ as in the lemma. One has

$$\|\pi_\lambda(y_n)\varphi_n - \pi_\lambda(y)\varphi b\| \leq \|\pi_\lambda(y_n)\varphi_n - \pi_\lambda(y_n)\varphi\| + \|\pi_\lambda(y_n)\varphi - \pi_\lambda(y)\varphi\|$$
$$\leq C \underbrace{\|\varphi_n - \varphi\|}_{\to 0} + \|\pi_\lambda(y_n)\varphi - \pi_\lambda(y)\varphi\|.$$

The first summand tends to zero as $n \to \infty$. We consider the square of the second summand:

$$\|\pi_\lambda(y_n)\varphi - \pi_\lambda(y)\varphi\|^2 = \int_{K_p} |\varphi(ky_n) - \varphi(ky)|^2 \, dk.$$

Assume first that the function φ is bounded on K_p. Then it is bounded on $K_p U$, say by a constant $D > 0$. Then $|\varphi(ky_n) - \varphi(ky)|^2 \leq 4D^2$ and the constant function $4D^2$ is integrable on the compact set K_p, so, by the dominated convergence theorem, one concludes that the integral $\int_{K_p} |\varphi(ky_n) - \varphi(ky)|^2 \, dk$ converges to zero for $n \to \infty$. If the function ϕ is not bounded on K_p, it is still square integrable and so for given $\varepsilon > 0$ there exists a function $\psi \in V_\lambda$, bounded on K_p, such that $\|\varphi - \psi\| < \varepsilon/4C$. There exists n_0, such that for all $n \geq n_0$ one has $\|\pi_\lambda(y_n)\psi - \pi_\lambda(y)\psi\| < \varepsilon/2$. Thus for every $n \geq n_0$,

$$\|\pi_\lambda(y_n)\varphi - \pi_\lambda(y)\varphi\| \leq \|\pi_\lambda(y_n)\varphi - \pi_\lambda(y_n)\psi\|$$
$$+ \|\pi_\lambda(y_n)\psi - \pi_\lambda(y)\psi\| + \|\pi_\lambda(y)\psi - \pi_\lambda(y)\varphi\|$$
$$\leq 2C\|\varphi - \psi\| + \|\pi_\lambda(y_n)\psi - \pi_\lambda(y)\psi\| < \frac{\varepsilon}{2} + \frac{\varepsilon}{2} = \varepsilon.$$

We have shown that π_λ is indeed a representation.

For unitarity note that, on the one hand,

$$\int_{K_p} |\varphi(ky)|^2 \, dk = \int_{K_p} \underline{a}(ky)^{\lambda+\bar{\lambda}+\delta} |\varphi(\underline{k}(ky))|^2 \, dk$$

for every $\varphi \in V_\lambda$. Then Proposition 7.1.2 asserts that, on the other hand,

$$\int_{K_p} |\varphi(k)|^2 \, dk = \int_{K_p} \underline{a}(ky)^\delta |\varphi(\underline{k}(ky))|^2 \, dk.$$

So unitarity of π_λ is equivalent to the identity

$$\int_{K_p} \underline{a}(ky)^\delta \left| \varphi\big(\underline{k}(ky)\big) \right|^2 dk = \int_{K_p} \underline{a}(ky)^{\lambda+\bar\lambda} \underline{a}(ky)^\delta \left| \varphi\big(\underline{k}(ky)\big) \right|^2 dk$$

for every $\varphi \in V_\lambda$. This in turn is equivalent to $a^{\lambda+\bar\lambda} = 1$ for every $a \in A_p$.

The restriction to the subgroup K_p is a subrepresentation of the right regular representation on $L^2(K_p)$. It is therefore unitary and the theorem is proven. $\square$

7.2 From Real to Adelic

The group $G_\mathbb{A} = GL_2(\mathbb{A})$ is a direct product

$$G_\mathbb{A} = G_{\mathrm{fin}} \times G_\mathbb{R}.$$

The group G_{fin} is the restricted product

$$G_{\mathrm{fin}} = \widehat{\prod_{p<\infty}}^{K_p} G_p$$

(see Exercise 7.1). We equip G_{fin} with the restricted product topology. The group

$$G_{\widehat{\mathbb{Z}}} = \prod_{p<\infty} K_p$$

is a maximal compact (and open) subgroup of G_{fin}. On G_{fin} we choose the unique Haar measure giving $G_{\widehat{\mathbb{Z}}}$ the volume 1. Instead of GL_2, we can do the same with SL_2, so $SL_2(\mathbb{A}) = SL_2(\mathbb{A}_{\mathrm{fin}}) \times SL_2(\mathbb{R})$ and $SL_2(\mathbb{A}_{\mathrm{fin}})$ is the restricted product of the groups $SL_2(\mathbb{Q}_p)$ with respect to the compact open subgroups $SL_2(\mathbb{Z}_p)$.

Theorem 7.2.1 *The group $SL_2(\mathbb{Q})$ is dense in $SL_2(\mathbb{A}_{\mathrm{fin}})$. For GL_2 one has*

$$G_{\mathrm{fin}} = G_\mathbb{Q} G_{\widehat{\mathbb{Z}}}.$$

The group $G_\mathbb{Q}$ is discrete in $G_\mathbb{A}$. It is contained in $G_\mathbb{A}^1 = \{g \in G_\mathbb{A} : |\det g| = 1\}$. The quotient $G_\mathbb{Q} \backslash G_\mathbb{A}^1$ is not compact, but has finite Haar measure. The same holds for $SL_2(\mathbb{Q}) \backslash SL_2(\mathbb{A})$.

Proof Let X be the closure of $SL_2(\mathbb{Q})$ in $SL_2(\mathbb{A}_{\mathrm{fin}})$. Since $\mathbb{Q}$ is dense in $\mathbb{A}_{\mathrm{fin}}$, all matrices of the form $\begin{pmatrix} 1 & x \\ & 1 \end{pmatrix}$ or $\begin{pmatrix} 1 & \\ x & 1 \end{pmatrix}$ lie in the group X. Hence all matrices of the form

$$\begin{pmatrix} 1 & x \\ & 1 \end{pmatrix} \begin{pmatrix} 1 & \\ y & 1 \end{pmatrix} \begin{pmatrix} 1 & z \\ & 1 \end{pmatrix} = \begin{pmatrix} 1+xy & z+xyz+x \\ y & 1+zy \end{pmatrix}$$

lie in X. Let $\begin{pmatrix} a & b \\ c & d \end{pmatrix}$ in $SL_2(\mathbb{A}_{\mathrm{fin}})$, where c is a unit in $\mathbb{A}_{\mathrm{fin}}$. Set $y = c$ and $x = (a-1)/y$, as well as $z = (d-1)/y$. Then one gets $b = z+xyz+x$, as both matrices have determinant 1. So we see that all matrices of the form $\begin{pmatrix} a & b \\ c & d \end{pmatrix}$, where $c \in \mathbb{A}_{\mathrm{fin}}^\times$

lie in X. For arbitrary $\begin{pmatrix} a & b \\ c & d \end{pmatrix} \in \mathrm{SL}_2(\mathbb{A}_{\mathrm{fin}})$ one considers $\begin{pmatrix} 1 & \\ x & 1 \end{pmatrix}\begin{pmatrix} a & b \\ c & d \end{pmatrix} = \begin{pmatrix} a & b \\ c+ax & b+dx \end{pmatrix}$. Considering one prime at a time, one sees that there exists $x \in \mathbb{A}_{\mathrm{fin}}$, such that $c + ax$ is a unit. So $\mathrm{SL}_2(\mathbb{Q})$ is dense in $\mathrm{SL}_2(\mathbb{A}_{\mathrm{fin}})$. As $\mathbb{Q}\widehat{\mathbb{Z}}^{\times} = \mathbb{A}_{\mathrm{fin}}$, we also get $G_{\mathrm{fin}} = G_{\mathbb{Q}}G_{\widehat{\mathbb{Z}}}$.

We now show that $G_{\mathbb{Q}}$ is discrete in $G_{\mathbb{A}}$. For this let $U \subset G_{\mathbb{R}}$ be an open unit neighborhood such that $U \cap \mathrm{GL}_2(\mathbb{Z}) = \{1\}$. Then $V = G_{\widehat{\mathbb{Z}}} \times U$ is an open unit neighborhood in $G_{\mathbb{A}}$. Since $G_{\mathbb{Q}} \cap G_{\widehat{\mathbb{Z}}} = \mathrm{GL}_2(\mathbb{Z})$, it follows that $G_{\mathbb{Q}} \cap V = \{1\}$, so $G_{\mathbb{Q}}$ is discrete. By the product formula we have $|\det(g)| = 1$ for $g \in G_{\mathbb{Q}}$.

Let $G_{\mathbb{R}}^1 = \mathrm{GL}_2(\mathbb{R})^1$ be the set of all $g \in \mathrm{GL}_2(\mathbb{R})$ with $|\det(g)| = 1$. We know that $\mathrm{SL}_2(\mathbb{Z})\backslash\mathrm{SL}_2(\mathbb{R})$ and $\mathrm{GL}_2(\mathbb{Z})\backslash G_{\mathbb{R}}^1$ are not compact, but have finite Haar measure. Therefore the next claim of the theorem follows from the following proposition.

Proposition 7.2.2 *The map* $\phi : G_{\mathbb{Z}}x \mapsto G_{\mathbb{Q}}(1,x)G_{\widehat{\mathbb{Z}}}$ *is a* $G_{\mathbb{R}}$*-equivariant homeomorphism*

$$G_{\mathbb{Z}}\backslash G_{\mathbb{R}} \xrightarrow{\cong} G_{\mathbb{Q}}\backslash G_{\mathbb{A}}/G_{\widehat{\mathbb{Z}}}.$$

The map $\mathrm{SL}_2(\mathbb{Z})x \mapsto \mathrm{SL}_2(\mathbb{Q})(1,x)\mathrm{SL}_2(\widehat{\mathbb{Z}})$ *is an* $\mathrm{SL}_2(\mathbb{R})$*-equivariant homeomorphism*

$$\mathrm{SL}_2(\mathbb{Z})\backslash\mathrm{SL}_2(\mathbb{R}) \xrightarrow{\cong} \mathrm{SL}_2(\mathbb{Q})\backslash\mathrm{SL}_2(\mathbb{A})/\mathrm{SL}_2(\widehat{\mathbb{Z}}).$$

Proof We show the first assertion, as the second is proved analogously. We first have to show well-definedness of the map ϕ. For this let $x, y \in G_{\mathbb{R}}$ with

$$\mathrm{GL}_2(\mathbb{Z})x = \mathrm{GL}_2(\mathbb{Z})y.$$

This means $x = \gamma y$ for some $\gamma \in \mathrm{GL}_2(\mathbb{Z})$. Note $\mathrm{GL}_2(\mathbb{Z}) = G_{\mathbb{Q}} \cap G_{\widehat{\mathbb{Z}}}$. Using square brackets to denote double cosets, we see that in $G_{\mathbb{Q}}\backslash G_{\mathbb{A}}^1/G_{\widehat{\mathbb{Z}}}$ the following identity holds:

$$[1,x] = [1,\gamma y] = [\gamma\gamma^{-1}, \gamma y] = [\gamma^{-1}, y] = [1,y].$$

It follows that the map ϕ is well defined.

To show injectivity assume $\phi(x) = \phi(y)$. Then

$$G_{\mathbb{Q}}(1,x)G_{\widehat{\mathbb{Z}}} = G_{\mathbb{Q}}(1,y)G_{\widehat{\mathbb{Z}}}.$$

This means that there is a $\gamma \in G_{\mathbb{Q}}$ and a $k \in G_{\widehat{\mathbb{Z}}}$ with

$$(1,x) = \gamma(1,y)k = (\gamma k, \gamma y).$$

Comparing the finite coordinates we get $\gamma k = 1$, so $\gamma = k^{-1}$ lies in $G_{\mathbb{Q}} \cap G_{\widehat{\mathbb{Z}}} = G_{\mathbb{Q}} \cap \prod_{p<\infty} K_p = \mathrm{GL}_2(\mathbb{Z})$. Because of $x = \gamma y$ it follows that $\mathrm{GL}_2(\mathbb{Z})x = \mathrm{GL}_2(\mathbb{Z})y$, so ϕ is injective.

For surjectivity we have to show that $G_{\mathbb{A}} = G_{\mathbb{Q}}G_{\mathbb{R}}G_{\widehat{\mathbb{Z}}}$. For this let $x = (x_{\mathrm{fin}}, x_{\infty}) \in G_{\mathbb{A}}$. As $G_{\mathbb{Q}}$ is dense in G_{fin}, there exists $\gamma \in G_{\mathbb{Q}}$ such that $k = \gamma^{-1}x_{\mathrm{fin}} \in G_{\widehat{\mathbb{Z}}}$. So we have $x = \gamma(1, \gamma^{-1}x_{\infty})k$ and the claim follows. It is easy to see that ϕ is continuous and open. $\qquad\square$

To continue the proof of the theorem, we conclude the finiteness of the Haar measure of $G_\mathbb{Q}\backslash G_\mathbb{A}^1$ as follows. The Haar measure induces a $G_\mathbb{R}^1$-invariant measure on

$$G_\mathbb{Q}\backslash G_\mathbb{A}^1 / G_{\widehat{\mathbb{Z}}} \cong GL_2(\mathbb{Z})\backslash G_\mathbb{R}^1.$$

By uniqueness of Haar measures this must be a scalar multiple of the Haar measure on $G_\mathbb{R}^1$. We already know that the quotient

$$GL_2(\mathbb{Z})\backslash G_\mathbb{R}^1 \cong SL_2(\mathbb{Z})\backslash SL_2(\mathbb{R})$$

has finite Haar measure. Hence $G_\mathbb{Q}\backslash G_\mathbb{A}^1 / G_{\widehat{\mathbb{Z}}}$ has finite Haar measure. As the group $G_{\widehat{\mathbb{Z}}}$ is compact, it also has finite Haar measure, so $G_\mathbb{Q}\backslash G_\mathbb{A}^1$ has finite Haar measure. □

For $N \in \mathbb{N}$ let

$$\Gamma(N) = \left\{ g \in GL_2(\mathbb{Z}) : g \equiv \begin{pmatrix} 1 & \\ & 1 \end{pmatrix} \bmod N \right\}.$$

This is the kernel of the group homomorphism

$$\phi : GL_2(\mathbb{Z}) \to GL_2(\mathbb{Z}/N\mathbb{Z}).$$

Therefore $\Gamma(N)$ is a normal subgroup of finite index. Note that for $N \geq 3$ the group $\Gamma(N)$ is contained in $SL_2(\mathbb{Z})$. A subgroup $\Gamma \subset GL_2(\mathbb{Z})$ is called a *congruence subgroup* if it contains $\Gamma(N)$ for some $N \in \mathbb{N}$.

Lemma 7.2.3 *Let Γ be a congruence subgroup. Then the closure K_Γ of Γ in $G_{\widehat{\mathbb{Z}}}$ is an open subgroup of the compact group $G_{\widehat{\mathbb{Z}}}$. One has*

$$\Gamma = K_\Gamma \cap G_\mathbb{Q}.$$

Conversely, for an open subgroup K of $G_{\widehat{\mathbb{Z}}}$ the group $\Gamma = K \cap G_\mathbb{Q}$ is a congruence subgroup such that K is the closure of Γ.

Proof For $N \in \mathbb{N}$ define

$$K_N = \left\{ g \in G_{\widehat{\mathbb{Z}}} : g \equiv \begin{pmatrix} 1 & \\ & 1 \end{pmatrix} \bmod N \right\}.$$

The group K_N is open in $G_{\widehat{\mathbb{Z}}}$ and satisfies $K_N \cap G_\mathbb{Q} = \Gamma(N)$. As $G_\mathbb{Q}$ is dense in G_{fin}, the group $\Gamma(N)$ is dense in K_N, so K_N is the closure of $\Gamma(N)$.

Now for general congruence subgroups. On the one hand, each congruence subgroup Γ contains a group of the form $\Gamma(N)$ with finite index. On the other hand, each compact open subgroup K of $G_{\widehat{\mathbb{Z}}}$ contains a group of the form K_N with finite index, as the latter form a neighborhood base of unity by Exercise 5.6. If now $\Gamma = \bigcup_{j=1}^n \gamma_j \Gamma(N)$, then its closure satisfies

$$K_\Gamma = \overline{\Gamma} = \bigcup_{j=1}^n \gamma_j \overline{\Gamma(N)} = \bigcup_{j=1}^n \gamma_j K_N.$$

Therefore, $K_\Gamma \cap G_\mathbb{Q} = \bigcup_{j=1}^n \gamma_j K_N \cap G_\mathbb{Q} = \bigcup_{j=1}^n \gamma_j \Gamma(N) = \Gamma$. Conversely, if $K = \bigcup_{j=1}^n k_j K_N$, then we may choose $k_j \in G_\mathbb{Q}$ as the latter is dense. It follows that

$$\Gamma_K = K \cap G_\mathbb{Q} = \bigcup_{j=1}^n k_j K_N \cap G_\mathbb{Q} = \bigcup_{j=1}^n k_j \Gamma(N).$$

Hence Γ_K is a congruence subgroup with closure $\overline{\Gamma} = \bigcup_{j=1}^N k_j \overline{\Gamma(N)} = \bigcup_{j=1}^N k_j K_n = K$. $\qquad\square$

Proposition 7.2.4 *Let Γ be a congruence subgroup and let K_Γ be the closure of Γ in $G_{\widehat{\mathbb{Z}}}$. The map $\phi : \Gamma x \mapsto G_\mathbb{Q}(1, x)K_\Gamma$ is a $G_\mathbb{R}^1$-equivariant homeomorphism*

$$\Gamma \backslash G_\mathbb{R}^1 \xrightarrow{\cong} G_\mathbb{Q} \backslash G_\mathbb{A}^1 / K_\Gamma.$$

If $\Gamma \subset \mathrm{SL}_2(\mathbb{Q})$, then K_Γ is a subgroup of $\mathrm{SL}_2(\widehat{\mathbb{Z}})$ and the map

$$\Gamma x \mapsto \mathrm{SL}_2(\mathbb{Q})(1, x)K_\Gamma$$

is a $\mathrm{SL}_2(\mathbb{R})$-equivariant homeomorphism

$$\Gamma \backslash \mathrm{SL}_2(\mathbb{R}) \xrightarrow{\cong} \mathrm{SL}_2(\mathbb{Q}) \backslash \mathrm{SL}_2(\mathbb{A}) / K_\Gamma.$$

Proof Analogous to the proof of Proposition 7.2.2. $\qquad\square$

Theorem 7.2.5 *Let Γ be a congruence subgroup and let K_Γ be the closure of Γ in $G_{\widehat{\mathbb{Z}}}$. The restriction induces a unitary $G_\mathbb{R}^1$-isomorphism*

$$L^2(G_\mathbb{Q} Z_\mathbb{R} \backslash G_\mathbb{A})^{K_\Gamma} \xrightarrow{\cong} L^2(\Gamma \backslash G_\mathbb{R}^1).$$

Proof Because of $L^2(G_\mathbb{Q} Z_\mathbb{R} \backslash G_\mathbb{A})^{K_\Gamma} \cong L^2(G_\mathbb{Q} Z_\mathbb{R} \backslash G_\mathbb{A} / K_\Gamma)$ the claim follows from the last proposition. $\qquad\square$

Theorem 7.2.6 *The classical holomorphic modular forms, as well as the Maaß wave forms, can be interpreted canonically as elements of $L^2(G_\mathbb{Q} Z_\mathbb{R} \backslash G_\mathbb{A})$.*

Further, this correspondence is compatible with L-functions in the following sense: To every holomorphic or Maaß cusp form f we have attached an L-function $L(f, s)$. The spaces of cusp forms have bases consisting of such functions f, the images of which in $L^2(G_\mathbb{Q} Z_\mathbb{R} \backslash G_\mathbb{A} / K_\Gamma)$ generate irreducible subrepresentations π_f of the group $G_\mathbb{A}$. In the next chapter we shall attach L-functions $L(\pi, s)$ to those representations π, and we shall show that $L(\pi, s) = L(f, s)$, if the representation π is generated by the cusp form f. We shall not prove this fact in full generality, but do the computations only for cusp forms of the full modular group SL$_2(\mathbb{Z})$.

Proof of the Theorem Both the wave forms as well as the modular forms lie in the Hilbert space $L^2(\Gamma\backslash\mathbb{H})$, where Γ is a congruence subgroup. The canonical map $g \mapsto gi$ gives an SL$_2(\mathbb{R})$-equivariant bijection SL$_2(\mathbb{R})/$SO(2) $\cong \mathbb{H}$. The same works for $G^1_{\mathbb{R}}$, if we extend the action of SL$_2(\mathbb{R})$ on $\mathbb{H}$ to a $G^1_{\mathbb{R}}$ action as follows: If $g = \begin{pmatrix} a & b \\ c & d \end{pmatrix} \in G^1_{\mathbb{R}}$ already has determinant 1, we have defined the action by $z \mapsto gz = \frac{az+b}{cz+d}$. If $\det(g) = -1$, the complex number $\frac{az+b}{cz+d}$ lies in the lower half plane $\overline{\mathbb{H}}$. In this case we define for $z \in \mathbb{H}$,

$$gz = \overline{\left(\frac{az+b}{cz+d}\right)} = \frac{a\bar{z}+b}{c\bar{z}+d}.$$

This defines an action of the group $G^1_{\mathbb{R}}$ on $\mathbb{H}$ and the map $g \mapsto gi$ is a bijection $G^1_{\mathbb{R}}/$O(2) $\cong \mathbb{H}$.

So we get a canonical identification

$$L^2(\Gamma\backslash\mathbb{H}) \cong L^2\big(\Gamma\backslash G^1_{\mathbb{R}}/\mathrm{O}(2)\big) \cong L^2\big(\Gamma\backslash G^1_{\mathbb{R}}\big)^{\mathrm{O}(2)}.$$

By the last theorem there is a canonical isometric embedding $L^2(\Gamma\backslash G^1_{\mathbb{R}}) \hookrightarrow L^2(G_{\mathbb{Q}}Z_{\mathbb{R}}\backslash G_{\mathbb{A}})$. □

Definition 7.2.7 In this book we define an *automorphic form* to be an element of $L^2(G_{\mathbb{Q}}Z_{\mathbb{R}}\backslash G_{\mathbb{A}})$.

The theorem says that holomorphic modular forms as well as Maaß wave forms are automorphic forms. In the literature one finds other definitions of automorphic forms—some wider and some narrower than ours. For instance, some authors leave out the L^2-integrability and replace it by a growth condition. In this case also non-holomorphic Eisenstein series will be automorphic forms.

7.3 Bochner Integral, Compact Operators and Arzela–Ascoli

In this section we provide some techniques which will be used later.

The *Bochner integral* is also called the *vector-valued integral*. For a function $f : X \to V$ with values in a Banach space V we want to define an integral $\int_X f\,d\mu \in V$, such that for every continuous linear functional α on V one has

$$\alpha\left(\int_X f\,d\mu\right) = \int_X \alpha(f)\,d\mu,$$

where $\alpha(f)$ means $\alpha \circ f$.

We need the notion of operator norm.

Definition 7.3.1 Let V and W be Banach spaces. A linear map $T : V \to W$ is also called a *linear operator* from V to W. The *operator norm* of T is defined as

$$\|T\|_{\mathrm{op}} \overset{\mathrm{def}}{=} \sup_{v \in V \smallsetminus \{0\}} \frac{\|Tv\|}{\|v\|} = \sup_{\|v\|=1} \|Tv\|.$$

The operator T is called a *bounded operator* if it has finite operator norm, i.e. if $\|T\|_{\mathrm{op}} < \infty$.

Lemma 7.3.2 *Let T be a linear operator from the Banach space V to the Banach space W. Then $\|Tv\| \le \|T\|_{\mathrm{op}}\|v\|$, if $v \in V$. The operator T is continuous if and only if it is bounded.*

Proof The first assertion is clear. Let T be continuous. Assume $\|T\|_{\mathrm{op}} = \infty$. Then there exists a sequence $v_j \in V$ with $\|v_j\| = 1$ and $\|Tv_j\| \to \infty$. We can assume $\|Tv_j\| \ne 0$ for every j. Then the sequence $\frac{1}{\|Tv_j\|}v_j$ tends to zero, so the sequence $\|T(\frac{1}{\|Tv_j\|}v_j)\| = \frac{\|Tv_j\|}{\|Tv_j\|}$ converges to zero, which is a contradiction.

For the converse direction assume $C = \|T\|_{\mathrm{op}} < \infty$ and let (v_j) be a sequence in V, converging to $v \in V$. Then $\|Tv_j - Tv\| \le C\|v_j - v\|$, so Tv_j converges to Tv and therefore T is continuous. $\qquad\square$

Let $(V, \|\cdot\|)$ be a Banach space and let $(X, \mathcal{A}, \mu)$ be a measure space. A *simple function* is a map $s : X \to V$, which can be written in the form

$$s = \sum_{j=1}^{n} \mathbf{1}_{A_j} b_j,$$

where $A_1, \dots, A_n$ are pairwise disjoint measurable sets of finite measure, i.e. $\mu(A_j) < \infty$, and $b_j \in V$. We define the integral of a simple function as

$$\int_X s\, d\mu \overset{\mathrm{def}}{=} \sum_{j=1}^{n} \mu(A_j) b_j \in V.$$

Note that $\|\int_X s\, d\mu\| \le \int_X \|s\|\, d\mu$ and that for every linear map $T : V \to W$ for a Banach space W one has $T(\int_X s\, d\mu) = \int_X T(s)\, d\mu$, where $T(s)$ means $T \circ s$.

We equip V with the Borel σ-algebra. A measurable function $f : X \to V$ is called *integrable* if there exists a sequence s_n of simple functions such that

$$\lim_{n \to \infty} \int_X \|f - s_n\|\, d\mu = 0.$$

In this case we call (s_n) an *approximating sequence*.

Theorem 7.3.3

(a) *If f is integrable and (s_n) is an approximating sequence, then the sequence of vectors $\int_X s_n \, d\mu$ converges in V. The limit of this sequence is independent of the choice of the approximating sequence. We define the* integral of *f as this limit:*

$$\int_X f \, d\mu \overset{\text{def}}{=} \lim_{n \to \infty} \int_X s_n \, d\mu.$$

(b) *For every integrable function f one has*

$$\left\| \int_X f \, d\mu \right\| \le \int_X \| f \| \, d\mu < \infty.$$

(c) *For every integrable function f and every continuous linear operator $T : V \to W$ to a Banach space W one has*

$$T\left(\int_X f \, d\mu \right) = \int_X T(f) \, d\mu.$$

(d) *In the case $V = \mathbb{C}$ the Bochner integral coincides with the usual integral.*

Proof It suffices to show that for any approximating sequence (s_n) the sequence of integrals $\int_X s_n \, d\mu$ converges. Then the limit is uniquely determined, as for a second approximating sequence (t_n) also the sequence (r_n) with $r_{2n} = s_n$ and $r_{2n-1} = t_n$ is an approximating sequence. Since $\int_X r_n \, d\mu$ converges, the limits of $\int_X s_n \, d\mu$ and $\int_X t_n \, d\mu$ must coincide.

To show convergence, it suffices to show that $\int_X s_n \, d\mu$ is a Cauchy sequence. For $m, n \in \mathbb{N}$ consider

$$\left\| \int_X s_m \, d\mu - \int_X s_n \, d\mu \right\| = \left\| \int_X s_m - s_n \, d\mu \right\| \le \int_X \| s_m - s_n \| \, d\mu$$

$$\le \int_X \| s_m - f \| \, d\mu + \int_X \| f - s_n \| \, d\mu.$$

The right-hand side tends to zero for $m, n \to \infty$. Therefore $\int_X s_n \, d\mu$ is indeed a Cauchy sequence. This implies part (a).

For (b) note that the inequality $|\, \| f \| - \| s_n \| \,| \le \| f - s_n \|$ implies that the $\mathbb{C}$-valued function $\| f \|$ is integrable and that the sequence $\| s_n \|$ converges to $\| f \|$ in $L^1(X)$. Therefore,

$$\left\| \int_X f \, d\mu \right\| = \lim_n \left\| \int_X s_n \, d\mu \right\| \le \lim_n \int_X \| s_n \| \, d\mu = \int_X \| f \| \, d\mu.$$

Finally for part (c). Continuity and linearity of T implies

$$T\left(\int_X f \, d\mu \right) = \lim_n \int_X T(s_n) \, d\mu.$$

We show that $T(f) = T \circ f$ is integrable and that the right-hand side equals $\int_X T(f) \, d\mu$. Let (s_n) be an approximating sequence for f. As T is continuous, there exists $C > 0$ such that $|T(v)| \leq C\|v\|$ holds for every $v \in V$. We conclude

$$\int_X \|T(f) - T(s_n)\| \, d\mu = \int_X \|T(f - s_n)\| \, d\mu \leq C \int_X \|f - s_n\| \, d\mu.$$

The right-hand side tends to zero, so the claim follows. Part (d) is clear, since an approximating sequence converges to f in the L^1-topology and the integral is a continuous linear functional with respect to that topology. □

Definition 7.3.4 A topological space Y is said to be *separable* if Y contains a countable dense subset. A subset $A \subset Y$ of a topological space Y is called separable if it is separable in the subset topology. This is equivalent to the existence of a countable set $C \subset A$ such that A lies in the closure $\overline{C}$.

A map $f : X \to Y$, where Y is a topological space, is called a *separable map* if the image $f(X)$ is a separable subset of Y.

Examples 7.3.5

- The set $\mathbb{R}$ of real numbers with its usual topology is separable, as it contains the countable dense subset $\mathbb{Q}$ of rational numbers.
- If (V, d) is a metric space, then every compact subset $K \subset V$ is separable. This can be seen by covering K with finitely many open balls of radius $1/n$ and centers in K. Letting $n \in \mathbb{N}$ tend to infinity, the countably many centers of these balls form a dense subset.
- A Hilbert space H is separable if and only if it possesses a countable orthonormal basis $(e_j)_{j \in \mathbb{N}}$. In this case the set of all $\mathbb{Q}(i)$-linear combinations of the basis vectors e_j is a countable dense subset.
- For a given measure space $(X, \mathcal{A}, \mu)$ assume that the σ-algebra $\mathcal{A}$ has a countable set of generators. Then the Hilbert space $L^2(X, \mu)$ is separable.
- Let (V, d) be a separable metric space. Then every map $f : X \to V$ is separable, see Exercise 7.2.
- Let X be a topological space and (V, d) a metric space. Then every continuous function $f : X \to V$ of compact support is separable.

Definition 7.3.6 For a measure space (X, μ) a map $f : X \to V$ to a topological space V is called *essentially separable* if there exists a set $N \subset X$ of measure zero, such that f is separable on $X \setminus N$.

Theorem 7.3.7 *Let X be a measure space and V a Banach space. For a measurable function $f : X \to V$ the following are equivalent:*

- *f is integrable,*
- *f is essentially separable and $\int_X \|f\| \, d\mu < \infty$.*

Proof For integrable f, the function $\|f\|$ is also integrable by Theorem 7.3.3(b). We have to show that f is essentially separable. Let (s_n) be an approximating sequence. Every s_n is a simple function, so the Banach space E, generated by the images $s_n(X)$ for all $n \in \mathbb{N}$, is separable. The set $N = f^{-1}(V \smallsetminus E)$ is a countable union $N = \bigcup_n N_n$, where $N_n = \{x \in X : \|f(x) - e\| \geq \frac{1}{n} \,\forall e \in E\}$. Since $\int_X \|f - s_n\| \, d\mu$ tends to zero, the set N_n is a set of measure zero for each $n \in \mathbb{N}$ and therefore N is a set of measure zero, so f is essentially separable.

For the converse assume that f is essentially separable and that $\int_X \|f\| \, d\mu < \infty$ holds. Let $C = \{c_n : n \in \mathbb{N}\}$ be a countable dense subset of $f(X)$. For $n \in \mathbb{N}$ and $\delta > 0$ let A_n^δ be the set of all $x \in X$ such that $\|f(x)\| \geq \delta$ and $\|f(x) - c_n\| < \delta$. As f is measurable, this is a measurable set. We make this sequence pairwise disjoint by defining

$$D_n^\delta \overset{\text{def}}{=} A_n^\delta \smallsetminus \bigcup_{k<n} A_k^\delta.$$

The set $\bigcup_{n\in\mathbb{N}} A_n^\delta = \bigcup_{n\in\mathbb{N}} D_n^\delta$ equals $f^{-1}(V \smallsetminus B_\delta(0))$, since C is dense. The function $\|f\|$ being integrable, the set $\bigcup_{n\in\mathbb{N}} D_n^\delta$ is of finite measure. Let $s_n = \sum_{j=1}^n \mathbf{1}_{D_j^{1/n}} c_j$. Then s_n is a simple function. We show that the sequence (s_n) converges point-wise to f. Let $x \in X$. If $f(x) = 0$, then $s_n(x) = 0$ for every n. So assume $f(x) \neq 0$. Then $\|f(x)\| \geq \frac{1}{n}$ for some $n \in \mathbb{N}$. For every $m \geq n$ one has $x \in \bigcup_{\nu\in\mathbb{N}} D_\nu^{1/m}$, so that for every $m \geq n$ there exists a unique ν_0 with $x \in D_{\nu_0}^{1/m}$, i.e. $s_m(x) = c_{\nu_0}$ and $\|f(x) - c_{\nu_0}\| < \frac{1}{m}$, which yields $s_n \to f$ as claimed. By construction we get $\|s_n\| \leq 2\|f\|$. It follows that $\|f - s_n\|$ tends to 0 point-wise and $\|f - s_n\| \leq \|f\| + \|s_n\| \leq 3\|f\|$. The dominated convergence theorem gives $\int_X \|f - s_n\| \, d\mu \to 0$. $\qquad\square$

Corollary 7.3.8 *Let μ be a Radon measure on the locally compact space X. Then every continuous function $f : X \to V$ of compact support is integrable.*

Proof The $\mathbb{C}$-valued function $\|f\|$ is continuous as well and has compact support, and is therefore integrable. Further, the image of f is compact, and therefore separable by Example 7.3.5. The claim now follows from Theorem 7.3.7 and the second example of 7.3.5. $\qquad\square$

Let (π, V_π) be a continuous representation of a locally compact group G on a separable Banach space V_π, and let f be a $\mathbb{C}$-valued measurable function on G. Assume that the function

$$x \mapsto f(x)\pi(x)$$

with values in the Banach space $\mathcal{B}(V_\pi)$ of all bounded operators is integrable. Then the operator

$$\pi(f) \overset{\text{def}}{=} \int_G f(x)\pi(x) \, dx$$

is continuous and satisfies

$$\left\| \pi(f) \right\|_{\mathrm{op}} \le \int_G |f(x)| \left\| \pi(x) \right\|_{\mathrm{op}} dx.$$

Note that this holds for example if $f \in C_c(G)$.

According to the remarks following Definition 2.8.10, for every bounded operator T on a Hilbert space V there exists a bounded operator T^* such that

$$\langle Tv, w \rangle = \langle v, T^*w \rangle$$

holds for all $v, w \in V$. The operator T^* is uniquely determined and is called the *adjoint* of T.

Lemma 7.3.9 *Let* (π, V_π) *be a unitary representation of the locally compact group* G. *If* $f \in L^1(G)$, *then the function* $x \mapsto f(x)\pi(x)$ *with values in the Banach space* $\mathcal{B}(V_\pi)$ *of all bounded operators on* V_π *is integrable and the integral satisfies*

$$\left\| \pi(f) \right\|_{\mathrm{op}} \le \|f\|_1.$$

For $f, g \in L^1(G)$ *one has* $\pi(f * g) = \pi(f)\pi(g)$ *and* $\pi(f^*) = \pi(f)^*$.

Proof We first show the integrability of $f(x)\pi(x)$. As π is unitary, one has $\|f(x)\pi(x)\|_{\mathrm{op}} = |f(x)|$ and therefore

$$\int_G \left\| f(x)\pi(x) \right\| dx = \int_G |f(x)| \, dx < \infty.$$

Next we show that $f(x)\pi(x)$ is separable. Since the function f is integrable, there exists an open subgroup H of G, such that f is zero outside H and H is a union $H = \bigcup_{n \in \mathbb{N}} K_n$ of compact sets (see Corollary 1.3.5(d) in [DE09]). On the compact set K_n, the continuous map $f(x)\pi(x)$ is separable, so it is separable in general. The remaining assertions are easily verified. □

An operator T on a Hilbert space H is called a *compact operator* if T maps bounded sets to relatively compact sets. Recall that a set $A \subset H$ is called *relatively compact* if its closure is compact.

One can reformulate the definition as follows: an operator T is compact if and only if for any given bounded sequence $v_j \in H$, the sequence Tv_j has a convergent subsequence. If T compact and S is a bounded operator, then TS and ST are both compact.

An operator T on a Hilbert space is called a *normal operator* if it commutes with its adjoint, i.e. if $TT^* = T^*T$ holds.

Theorem 7.3.10 (Spectral theorem for compact normal operators) *Let T be a compact normal operator on the Hilbert space H. Then there is a sequence of complex numbers $\lambda_n \neq 0$, which is either finite or tends to zero, such that the space H decomposes orthogonally*

$$H = \ker(T) \oplus \overline{\bigoplus_n \mathrm{Eig}(T, \lambda_n)}.$$

Each eigenspace $\mathrm{Eig}(T, \lambda_n) = \{v \in H : Tv = \lambda_n v\}$ is finite-dimensional and the eigenspaces are pairwise orthogonal.

Proof The proof is found in many books on functional analysis, such as [DE09, Rud91, Wei80, Yos95]. □

Let G be a unimodular locally compact group. A *Dirac sequence* in G is a sequence of functions $f_j \in C_c(G)$ such that

- $\mathrm{supp}\, f_{j+1} \subset \mathrm{supp}\, f_j$,
- for every unit neighborhood U there is $j \in \mathbb{N}$ with $\mathrm{supp}\, f_j \subset U$,
- $f_j \geq 0$, $f_j(x^{-1}) = \Delta_G(x) f_j(x)$ and $\int_G f_j(x)\, dx = 1$.

There exists a Dirac sequence if the group has a *countable unit neighborhood base*, i.e. if there is a sequence U_j of unit neighborhoods such that for every unit neighborhood U there exists $j \in \mathbb{N}$ with $U_j \subset U$. This is an immediate consequence of Urysohn's Lemma A.3.2.

Examples 7.3.11

- If the group G is *metrizable*, i.e. if there exists a metric d on G, which generates the topology, then G has a countable unit neighborhood base, for instance, one can take U_j to be the open ball $B_{1/j}(1)$ of radius $1/j$ around the unit element.
- The group $GL_2(\mathbb{R})$ has a countable uni-neighborhood-base as it is metrizable, since the topology of $\mathbb{R}^4$ is given by a metric.
- For every prime number p the group $G_p = GL_2(\mathbb{Q}_p)$ has a countable unit neighborhood base given by $U_j = 1 + p^j \mathrm{M}_2(\mathbb{Z}_p)$.
- If G is the restricted product of countably many groups G_k, each with countable unit neighborhood base, then G has a countable unit neighborhood base. So in particular the group $G_{\mathbb{A}}$ has a countable unit neighborhood base.

Lemma 7.3.12 *Let G be a locally compact group that admits a countable unit neighborhood base. Let (f_j) be a Dirac sequence in G and let (π, V_π) be a unitary representation. Then for every given $v \in V_\pi$, the sequence $\pi(f_j)v$ converges to v in the Banach space V_π.*

Proof Let $\varepsilon > 0$ and let U_ε be the open ε-neighborhood of $v \in V_\pi$. By continuity of the representation, there exists a unit neighborhood U in G with $\pi(U)v \subset U_\varepsilon$.

For the Dirac sequence f_j there exists an index j_0, such that for $j \geq j_0$ one has supp $f_j \subset U$, and so

$$\|\pi(f_j)v - v\| = \left\| \int_U f_j(x)\big(\pi(x)v - v\big)\,dx \right\| \leq \int_U f_j(x)\|\pi(x)v - v\|\,dx < \varepsilon.$$

$\square$

On $C_c(G)$ consider the involution $f^*(x) = \overline{f(x^{-1})}$. An element $f \in C_c(G)$ is called *self-adjoint* if $f^* = f$.

Proposition 7.3.13 *Let (η, V_η) be a unitary representation of G such that there exists a Dirac sequence (f_j) on G with $\eta(f_j)$ being a compact operator for every $j \in \mathbb{N}$. Then η is a direct orthogonal sum of irreducible representations, each occurring with finite multiplicity.*

Proof Zorn's lemma gives us a subspace U, maximal with the property that it is a direct sum of irreducible subrepresentations. The assumption of the current proposition holds for the orthogonal complement $U^\perp$ of U in $V = V_\eta$ as well. This orthogonal complement does not contain an irreducible subspace. We have to show that it is zero. In other words, we have to show that a representation $\eta \neq 0$ that satisfies the assumptions of the proposition contains an irreducible subrepresentation.

Let $j \in \mathbb{N}$. Then $f = f_j$ is self-adjoint and so $\eta(f)$ is self-adjoint and compact. By the spectral theorem the space $V = V_\eta$ decomposes

$$V = V_{f,0} \oplus \bigoplus_{\nu=1}^{\infty} V_{f,\nu},$$

where $V_{f,0}$ is the kernel of $\eta(f)$ and $V_{f,\nu}$ for $\nu > 0$ is an eigenspace of $\eta(f)$ to some eigenvalue $\lambda_\nu \neq 0$. We include the possibility that there are only finitely many eigenspaces, since we allow that $V_{f,\nu}$ be the zero space.

The sequence $(\lambda_{f,\nu})_{\nu \in \mathbb{N}}$ of eigenvalues tends to zero and each of the eigenspaces $V_{f,\nu}$ is finite-dimensional. Fix a closed, G-invariant subspace $0 \neq V' \subset V$. One gets a decomposition

$$V' = V'_{f,0} \oplus \bigoplus_{\nu=1}^{\infty} V'_{f,\nu},$$

where $V'_{f,\nu} = V' \cap V_{f,\nu}$ is the λ_ν-eigenspace of $\eta(f)$ acting on the space V'. We claim that there exists $f = f_j$ such that for some $\nu \geq 1$ the space $V'_{f,\nu}$ is non-trivial. Assume the contrary. Then $V' \subset \ker(\eta(f_j))$ for every j. For $0 \neq v' \in V'$ we have $\eta(f_j)v' = 0$ for all $j \in \mathbb{N}$ which contradicts $\eta(f_j)v' \to v'$ as $j \to \infty$. So the claim follows.

Fix some $f = f_j$ and ν in $\mathbb{N}$, and consider the set of all non-trivial intersections $V_{f,\nu} \cap U$, where U runs over all closed, G-stable linear subspaces of V. Among all these sections choose one, $W = V_{f,\nu} \cap U$, of minimal dimension. Let

$$U^1 = \bigcap_{U : U \cap V_{f,\nu} = W} U,$$

where the intersection extends over all closed, G-stable subspaces U, whose intersection with $V_{f,\nu}$ equals W. Then U^1 is a closed, G-stable linear subspace itself. We claim that it is irreducible. For this assume you are given an orthogonal decomposition $U^1 = E \oplus F$ with closed, G-stable linear subspaces E and F. If $E_{f,\nu} = E \cap V_{f,\nu} = 0 = F \cap V_{f,\nu} = F_{f,\nu}$, then $W = (E \oplus F)_{f,\nu} = E_{f,\nu} \oplus F_{f,\nu} = 0$, a contradiction. By minimality of W one has $W \subset E$ or $W \subset F$, such that one of the two spaces E or F is the zero space. Therefore U^1 is irreducible. We have shown that V indeed contains an irreducible subspace. As shown above, this implies that V is a direct sum of irreducible subspaces.

It remains to show that the multiplicities are finite. For this let I be any index set and let $V_i \subset V_\eta$, $i \in I$, be linearly independent subrepresentations, such that $\sigma_i = \eta_i |_{V_i}$ are pairwise unitarily isomorphic. Let $\lambda \in \mathbb{C} \setminus \{0\}$ be an eigenvalue of one of the $\sigma_i(f_j)$. Then λ is an eigenvalue of every $\sigma_i(f)$ for $i \in I$. Since the eigenvalues of $\eta(f)$ have finite multiplicity, the set I must be finite. $\square$

7.3.1 The Arzela–Ascoli Theorem

Let (X, d) be a separable metric space. Let $C(X)$ be the set of all continuous complex valued functions on X.

Definition 7.3.14 A subset $\mathcal{F} \subset C(X)$ is called *normal* if every sequence in $\mathcal{F}$ admits a locally uniformly convergent subsequence.

Definition 7.3.15 A subset $\mathcal{F}$ of $C(X)$ is called *equicontinuous* at $x \in X$, if for every $\varepsilon > 0$ there exists a $\delta > 0$ such that for every $f \in F$ and every $y \in X$ with $d(x, y) < \delta$ one has $|f(x) - f(y)| < \varepsilon$. We put this in one formula. The set $\mathcal{F}$ is equicontinuous at x if and only if

$$\forall_{\varepsilon > 0} \exists_{\delta > 0} \forall_{f \in \mathcal{F}} : d(x, y) < \delta \quad \Rightarrow \quad |f(x) - f(y)| < \varepsilon.$$

(This is the usual ε, δ criterion for continuity at the point x, except that the value of $\delta > 0$ does not depend on the chosen function $f \in \mathcal{F}$.)

Note that equicontinuity in the case of the manifold $G_\infty = \mathrm{GL}_2(\mathbb{R})$ can be verified through a differential criterion: Suppose that $\mathcal{F}$ lies inside $C^1(G_\infty)$ and suppose that for every $x \in G_\infty$ and every $X \in \mathrm{M}_2(\mathbb{R})$ the set

$$X\mathcal{F}(G_\infty) = \left\{ \tilde{R}_X f(x) : f \in \mathcal{F}, \ x \in G_\infty \right\}$$

is bounded by a constant $C(X, x)$, in such a way that for fixed $X \in M_2(\mathbb{R})$ the map $x \mapsto C(X, x)$ is continuous. Then the set $\mathcal{F}$ is equicontinuous at every $x \in G_\infty$. (This follows from the mean value theorem of differential calculus, as the assumption implies that for every compact set $U \subset G_\infty$ there exists a bound $C(U) > 0$, such that $\|Df(x)\| < C(U)$ holds for every $f \in \mathcal{F}$ and every $x \in U$, where $Df(x)$ is the differential of the map f.)

Theorem 7.3.16 (Arzela–Ascoli) *Let $\mathcal{F}$ be a subset of $C(X)$, such that*

- *for every $x \in X$ the set $\mathcal{F}(x) = \{f(x) : f \in \mathcal{F}\}$ is bounded and*
- *$\mathcal{F}$ is equicontinuous at every point x of X.*

Then $\mathcal{F}$ is normal.

Proof of the Theorem Choose a dense sequence $(x_n)_{n \in \mathbb{N}}$ in X. Since $\mathcal{F}(x_1)$ is bounded, there exists a subsequence (f_j^1) of (f_j) such that the sequence $f_j^1(x_1)$ converges. Next pick a subsequence (f_j^2) of (f_j^1) such that $f_j^2(x_2)$ converges as well. Repeating, for every $n \in \mathbb{N}$ we obtain a subsequence $(f_j^{n+1})_{j \in \mathbb{N}}$ of $(f_j^n)_{j \in \mathbb{N}}$, such that $(f_j^{n+1}(x_{n+1}))_{j \in \mathbb{N}}$ converges. Set

$$g_j = f_j^j.$$

Then $g_j(x_n)$ converges for every $n \in \mathbb{N}$. Call the limit $g(x_n)$. We claim that the sequence g_j converges locally uniformly. To show this, let $x \in X$ and $\varepsilon > 0$. Then there exists $\delta > 0$ such that for $y \in X$ the inequality $d(x, y) < \delta$ implies $|g_j(x) - g_j(y)| < \varepsilon/3$ for every $j \in \mathbb{N}$.

The sequence x_n being dense, there exists $n \in \mathbb{N}$ with $d(x_n, x) < \delta$, and there is j_0 such that for $j \geq j_0$ one has

$$\left|g_j(x_n) - g(x_n)\right| < \frac{\varepsilon}{6}.$$

For $i, j \geq j_0$ this implies $|g_i(x_n) - g_j(x_n)| < \varepsilon/3$, and

$$\left|g_i(x) - g_j(x)\right| \leq \left|g_i(x) - g_i(x_n)\right| + \left|g_i(x_n) - g_j(x_n)\right| + \left|g_j(x_n) - g_j(x)\right|$$
$$< \frac{\varepsilon}{3} + \frac{\varepsilon}{3} + \frac{\varepsilon}{3} = \varepsilon.$$

So $g_j(x)$ is a Cauchy sequence; hence it converges to a complex number which we call $g(x)$. Finally, to show that this convergence is locally uniform, let $y \in X$ with $d(x, y) < \delta/2$. Then one has $d(x_n, y) < \delta$ and the computation above shows that for $j \geq j_0$ one obtains $|g_j(y) - g(y)| \leq \varepsilon$. $\qquad\square$

7.4 Cusp Forms

For any ring R denote by N_R the group of all 2×2 matrices of the form $\left(\begin{smallmatrix} 1 & x \\ & 1 \end{smallmatrix}\right)$
with $x \in R$. Then N_R is a group isomorphic to the additive group of the ring R.
A function $f \in L^2(G_\mathbb{Q}\backslash G_\mathbb{A}^1)$ is called a *cusp form* if

$$\int_{N_\mathbb{Q}\backslash N_\mathbb{A}} f(nx)\, dn = 0$$

holds for almost all $x \in G_\mathbb{A}^1$. The set of cusp forms is a linear subspace, denoted
$L^2_{\text{cusp}} = L^2_{\text{cusp}}(G_\mathbb{Q}\backslash G_\mathbb{A}^1)$ of $L^2(G_\mathbb{Q}\backslash G_\mathbb{A}^1)$. This space is stable under the representation of $G_\mathbb{A}^1$ given by right translation, i.e. with

$$R(y)f(x) = f(xy)$$

one has $R(y)L^2_{\text{cusp}} = L^2_{\text{cusp}}$ for every $y \in G_\mathbb{A}^1$.
    There is an isomorphism

$$G_\mathbb{A}^1 \times \mathbb{R}_+^\times \xrightarrow{\cong} G_\mathbb{A}$$

given by $(g,t) \mapsto \sqrt{t}g$, where $\sqrt{t}$ is $\sqrt{t}\left(\begin{smallmatrix}1 & \\ & 1\end{smallmatrix}\right)$ at the infinite place and $\left(\begin{smallmatrix}1 & \\ & 1\end{smallmatrix}\right)$ at the
finite places.

Definition 7.4.1 Let $C_c^\infty(G_\mathbb{A})$ be the set of all functions on $G_\mathbb{A}$, which are linear
combinations of functions of the form $f = \prod_p f_p$, where for every $p \le \infty$ the func-
tion f_p lies in $C_c^\infty(G_{\mathbb{Q}_p})$ and for almost all p the equality $f_p = \mathbf{1}_{K_p}$ holds. For $f \in$
$C_c^\infty(G_\mathbb{A})$ the operator $R(f)$ is defined by integration $R(f) = \int_{G_\mathbb{A}} f(x)R(x)\, dx$.

    Let $K_\infty = O(2)$ and $K_p = GL_2(\mathbb{Z}_p)$ for $p < \infty$. Further let $K_{\text{fin}} = GL_2(\widehat{\mathbb{Z}}) =$
$\prod_{p<\infty} K_p$ and $K_\mathbb{A} = K_{\text{fin}} \times K_\infty$. For any ring R let P_R be the set of all upper
triangular matrices in $GL_2(R)$. For $c > 0$ let A_c be the set of all matrices of the form
$\left(\begin{smallmatrix} y & \\ & 1 \end{smallmatrix}\right)$, where $y \in \mathbb{R}$ and $y \le c$. A *Siegel domain* is a set of the form $S = S(\Omega, c) =$
$\Omega A_c K_\mathbb{A}$, where $c > 0$ and Ω is a compact subset of $P_\mathbb{R}$.

Lemma 7.4.2 *There exists a Siegel domain S such that*

$$G_\mathbb{A} = G_\mathbb{Q}\, Z_\mathbb{R}\, S.$$

Proof By Proposition 7.2.2 we get

$$G_\mathbb{Q} Z_\mathbb{R}\backslash G_\mathbb{A}/K_\mathbb{A} \cong G_\mathbb{Z} Z_\mathbb{R}\backslash G_\mathbb{R}/K_\infty.$$

The claim is that there exists a compact set $\Omega \subset P_\mathbb{R}$ and $c > 0$ such that the image
of ΩA_c on the left-hand side is the whole set on the left-hand side. It suffices to
show this on the right-hand side. We therefore have to show

$$G_\mathbb{R}/K_\infty = G_\mathbb{Z} Z_\mathbb{R} \Omega A_c K_\infty/K_\infty$$

for some Ω and some c. If Ω contains the set $\{\left(\begin{smallmatrix}1 & x \\ & 1\end{smallmatrix}\right) : |x| \le \frac{1}{2}\}$ and if $c \le \frac{\sqrt{3}}{2}$,
then the set $\Omega A_c K_\infty/K_\infty$ contains a fundamental set for $SL_2(\mathbb{Z})$ in the upper half
plane $\mathbb{H}$. The claim follows. $\square$

A function φ on $Z_{\mathbb{R}}G_{\mathbb{Q}}\backslash G_{\mathbb{A}}$ is called *rapidly decreasing* if there is a compact set $\Omega \subset P_{\mathbb{R}}$ and $c > 0$, such that Lemma 7.4.2 is satisfied and for every $n \in \mathbb{N}$ there exists $C_n > 0$ such that

$$\left| \varphi \left(\omega \begin{pmatrix} y & \\ & 1 \end{pmatrix} k \right) \right| \leq C_n y^{-n}$$

holds for every $\omega \begin{pmatrix} y & \\ & 1 \end{pmatrix} k$ in the Siegel domain $\Omega A_c K_{\mathbb{A}}$.

Proposition 7.4.3 *Let* $f \in C_c^\infty(G_{\mathbb{A}})$.

(a) *There is a constant* $C > 0$ *such that for every* $\varphi \in L^2_{\text{cusp}}(G_{\mathbb{Q}}\backslash G^1_{\mathbb{A}}) \cong L^2_{\text{cusp}}(G_{\mathbb{Q}}Z_{\mathbb{R}}\backslash G_{\mathbb{A}})$ *one has*

$$\sup_{x \in G_{\mathbb{A}}} |R(f)\varphi(x)| \leq C\|\varphi\|_2.$$

The function $R(f)\varphi$ *is rapidly decreasing.*
(b) *The operator* $R(f)$ *is compact on the space* $L^2_{\text{cusp}}(G_{\mathbb{Q}}Z_{\mathbb{R}}\backslash G_{\mathbb{A}})$.

Proof We compute

$$R(f)\varphi(x) = \int_{G_{\mathbb{A}}} f(y)\varphi(xy)\,dy = \int_{G_{\mathbb{A}}} f(x^{-1}y)\varphi(y)\,dy$$

$$= \int_{N_{\mathbb{Q}}\backslash G_{\mathbb{A}}} \underbrace{\sum_{\gamma \in N_{\mathbb{Q}}} f(x^{-1}\gamma y)}_{=K(x,y)} \varphi(y)\,dy.$$

Set $K_0(x,y) = \int_{\mathbb{A}} f(x^{-1}\begin{pmatrix} 1 & a \\ & 1 \end{pmatrix}y)\,da$. We have

$$\int_{N_{\mathbb{Q}}\backslash G_{\mathbb{A}}} K_0(x,y)\varphi(y)\,dy = \int_{N_{\mathbb{Q}}\backslash G_{\mathbb{A}}} \int_{\mathbb{A}} f\left(x^{-1}\begin{pmatrix} 1 & a \\ & 1 \end{pmatrix}y\right)\,da\,\varphi(y)\,dy$$

$$= \int_{N_{\mathbb{Q}}\backslash G_{\mathbb{A}}} \int_{\mathbb{A}/\mathbb{Q}} \sum_{\gamma \in N_{\mathbb{Q}}} f\left(x^{-1}\gamma\begin{pmatrix} 1 & a \\ & 1 \end{pmatrix}y\right)\,da\,\varphi(y)\,dy$$

$$= \int_{\mathbb{A}/\mathbb{Q}} \int_{N_{\mathbb{Q}}\backslash G_{\mathbb{A}}} \sum_{\gamma \in N_{\mathbb{Q}}} f(x^{-1}\gamma y)\varphi\left(\begin{pmatrix} 1 & a \\ & 1 \end{pmatrix}y\right)\,dy\,da$$

$$= \int_{N_{\mathbb{Q}}\backslash G_{\mathbb{A}}} \sum_{\gamma \in N_{\mathbb{Q}}} f(x^{-1}\gamma y) \underbrace{\int_{\mathbb{A}/\mathbb{Q}} \varphi\left(\begin{pmatrix} 1 & a \\ & 1 \end{pmatrix}y\right)\,da}_{=0}\,dy$$

$$= 0.$$

Putting $K'(x,y) = K(x,y) - K_0(x,y)$, we obtain

$$R(f)\varphi(x) = \int_{N_{\mathbb{Q}}\backslash G_{\mathbb{A}}} K'(x,y)\varphi(y)\,dy.$$

Let $F_{x,y}(t) = f(x^{-1}\begin{pmatrix} 1 & t \\ & 1 \end{pmatrix}y)$, $t \in \mathbb{A}$. We get $K_0(x,y) = \widehat{F}_{x,y}(0)$ and the Poisson Summation Formula implies that

$$K'(x,y) = \sum_{q \in \mathbb{Q}^\times} \widehat{F}_{x,y}(q) = \sum_{q \in \mathbb{Q}^\times} \int_{\mathbb{A}} f\left(x^{-1}\begin{pmatrix} 1 & t \\ & 1 \end{pmatrix}y\right) e(-qt)\,dt.$$

Let $S = \Omega A_c K_{\mathbb{A}}$ be a Siegel domain satisfying Lemma 7.4.2. For $x \in S$, $x = p_x\begin{pmatrix} a_x & \\ & 1 \end{pmatrix}k_x$ one has

$$K'(x,y) \neq 0 \quad \Rightarrow \quad y \in N_{\mathbb{Q}} X_P \begin{pmatrix} a_x & \\ & 1 \end{pmatrix} K_{\mathbb{A}} \subset N_{\mathbb{Q}} \begin{pmatrix} a_x & \\ & 1 \end{pmatrix} X_G,$$

where $X_P \subset P(\mathbb{A})$ and $X_G \subset G_{\mathbb{A}}$ are fixed compact sets. We can therefore write

$$K'(x,y) = \sum_{q \in \mathbb{Q}^\times} \int_{\mathbb{A}} f\left(x^{-1}\begin{pmatrix} 1 & t \\ & 1 \end{pmatrix}y\right) e(-qt)\,dt$$

$$= \sum_{q \in \mathbb{Q}^\times} \int_{\mathbb{A}} f\left(\begin{pmatrix} 1 & ta_x^{-1} \\ & 1 \end{pmatrix}\omega_{x,y}\right) e(-qt)\,dt,$$

and here $\omega_{x,y}$ stays in a fixed compact set. Changing variables to $v = ta_x^{-1}$ we get

$$K'(x,y) = |a_x| \sum_{q \in \mathbb{Q}^\times} \int_{\mathbb{A}} f\left(\begin{pmatrix} 1 & v \\ & 1 \end{pmatrix}\omega_{x,y}\right) e(-qa_x v)\,dv.$$

The function $\lambda \mapsto \int_{\mathbb{A}} f(\begin{pmatrix} 1 & v \\ & 1 \end{pmatrix}\omega_{x,y})e(-\lambda v)\,dv$ is the Fourier transform of a Schwartz–Bruhat function and therefore is a Schwartz–Bruhat function itself. So the sum runs over the intersection of $\mathbb{Q}^\times$ with a compact set in $\mathbb{A}_{\text{fin}}$ and for every $\nu \in \mathbb{N}$ there exists a constant $C_\nu > 0$ such that

$$\left| K'(x,y) \right| \leq C_\nu |a_x|(1 + |a_x|)^{-\nu}.$$

The Cauchy–Schwarz inequality implies

$$\left| R(f)\varphi(x) \right| \leq C_\nu |a_x|(1 + |a_x|)^{-\nu} \int_{\begin{pmatrix} a_x & \\ & 1 \end{pmatrix} X_G} |\varphi(y)|\,dy \leq C_\nu |a_x|(1 + |a_x|)^{-\nu} \|\varphi\|_2.$$

This implies part (a). To prove part (b), we use the Arzela–Ascoli theorem. Let $f \in C_c^\infty(G_{\mathbb{A}})$. There exists a compact open subgroup K of G_{fin} such that f is invariant under K in the sense that $f(k_1 x k_2) = f(x)$ for all $k_1, k_2 \in K$. With $\Gamma = K \cap G_{\mathbb{Q}}$ it follows that

$$R(f)L^2(G_{\mathbb{Q}} \backslash G_{\mathbb{A}}^1) \subset C^\infty(G_{\mathbb{Q}} \backslash G_{\mathbb{A}}^1)^K = C^\infty(\Gamma \backslash G_{\mathbb{R}}^1).$$

By part (a) the image of $\{\varphi \in L^2_{\text{cusp}} : \|\varphi\|_2 = 1\}$ is globally bounded and likewise for the image of $\tilde{R}_X R(f) = R(\tilde{L}_X f)$ for every $X \in M_2(\mathbb{R})$, where we use the notation of Sect. 3.4. By the Arzela–Ascoli theorem every sequence in $R(f)L^2_{\text{cusp}}$ has a point-wise convergent subsequence and by part (a) this sequence is dominated by a constant. Therefore the sequence converges in the L^1-norm and as it is bounded, it converges in the L^2-norm. So the operator $R(f)$ on L^2_{cusp} is indeed compact. $\square$

Using Proposition 7.3.13 we infer the following theorem.

> **Theorem 7.4.4** *The space* $L^2_{\text{cusp}}(G_\mathbb{Q} Z_\mathbb{R} \backslash G_\mathbb{A})$ *is a direct sum of irreducible subspaces, each occurring with finite multiplicity,*
>
> $$L^2_{\text{cusp}}(G_\mathbb{Q} Z_\mathbb{R} \backslash G_\mathbb{A}) = \bigoplus_{\pi \in \widehat{G_\mathbb{A}}} N_{\text{cusp}}(\pi)\pi.$$

7.5 The Tensor Product Theorem

One of the most important results in the theory of automorphic forms is the tensor product theorem, which says that every admissible irreducible representation of the group $G_\mathbb{A}$ is an infinite tensor product of irreducible representations of the local groups G_p with $p \leq \infty$. In this section we shall first define infinite tensor products in the subsection 'Synthesis'. Then, in 'Analysis', we shall prove the tensor product theorem.

7.5.1 Synthesis

In this subsection we construct infinite tensor products and show that they are irreducible representations of the global group $G_\mathbb{A}$.

Definition 7.5.1 A **-algebra* is a pair $(A, *)$ consisting of a $\mathbb{C}$-algebra A and a map $A \to A$; $a \mapsto a^*$ having the following properties:

$$(a + b)^* = a^* + b^*, \quad (\lambda a)^* = \overline{\lambda} a^*, \quad (ab)^* = b^* a^*,$$

if $a, b \in A$ and $\lambda \in \mathbb{C}$. Finally, one insists that

$$a^{**} = \left(a^*\right)^* = a$$

holds for every $a \in A$. The map $a \mapsto a^*$ is called the *involution* of the **-algebra A.

Examples 7.5.2

- $\mathbb{C}$ is a *-algebra with the involution $z^* = \overline{z}$.
- The algebra $M_n(\mathbb{C})$ of complex $n \times n$ matrices is a *-algebra with $A^* = \overline{A}^t$ for $A \in M_n(\mathbb{C})$.
- Let V be a Hilbert space. The algebra $\mathcal{B}(V)$ of bounded operators $T : V \to V$ is a *-algebra, when we define T^* to be the adjoint operator of T. This operator is uniquely determined by

$$\langle Tv, w \rangle = \langle v, T^* w \rangle$$

for all $v, w \in V$.

- Let G be a locally compact group and let A be the convolution algebra $C_c(G)$. This is a *-algebra with the involution $f^*(x) = \Delta(x^{-1})\overline{f(x^{-1})}$, where $\Delta(x)$ is the modular function of the group G.

Definition 7.5.3 Let A be a *-algebra. A *-*representation* of A on a Hilbert space V is an algebra homomorphism $\pi : A \to \mathcal{B}(V)$ such that $\pi(a)^* = \pi(a^*)$ holds for every $a \in A$. A *-representation is called *irreducible* if the only A-stable closed subspaces of V are the zero space and V itself.

In other words, π is irreducible, if for every closed linear subspace $U \subset V$ one has

$$\pi(A)U \subset U \quad \Rightarrow \quad (U = 0 \text{ or } U = V).$$

Here we have written $\pi(A)U$ for the linear subspace of V, generated by all vectors of the form $\pi(a)u$ for $a \in A$ and $u \in U$.

Two *-representations π, η of A on Hilbert spaces V_π and V_η are called *unitarily equivalent* if there exists a unitary operator $T : V_\pi \to V_\eta$ such that

$$T\big(\pi(f)v\big) = \eta(f)T(v)$$

holds for every $v \in V_\pi$ and every $f \in A$.

Proposition 7.5.4 *Let G be a locally compact group and let A be the *-algebra $C_c(G)$. Let (π, V_π) be a unitary representation of G on a Hilbert space V_π. Then for every $f \in C_c(G)$ the Bochner integral*

$$\pi(f) = \int_G f(x)\pi(x)\,dx$$

*exists in the Banach space $\mathcal{B}(V_\pi)$ of all bounded operators on V_π. The map $f \mapsto \pi(f)$ is a *-representation of A. Two unitary representations of G are unitarily equivalent if and only if their induced *-representations of A are unitarily equivalent.*

Proof For $f \in C_c(G)$ the function $x \mapsto f(x)\pi(x)$ is continuous and has compact support, therefore is integrable by Corollary 7.3.8. It is a representation of A, as we can compute, formally at first:

$$\pi(f * g) = \int_G f * g(x)\pi(x)\,dx$$

$$= \int_G \int_G f(y)g\big(y^{-1}x\big)\,dy\,dx$$

$$= \int_G f(y)\int_G g\big(y^{-1}x\big)\pi(x)\,dx\,dy$$

$$= \int_G f(y)\int_G g(x)\pi(yx)\,dx\,dy$$

$$= \int_G f(y)\pi(y)\,dy\int_G g(x)\pi(x)\,dx$$

$$= \pi(f)\pi(g).$$

As we did not prove a Fubini-type theorem for Bochner integrals, it remains to justify the interchange of order of integration. We choose $v, w \in V_\pi$ and compute, using the classical Fubini theorem,

$$\left\langle \int_G \int_G f(y)g(y^{-1}x)\pi(x)\,dx\,dy\,v, w \right\rangle$$

$$= \int_G \int_G f(y)g(y^{-1}x)\langle\pi(x)v, w\rangle\,dx\,dy$$

$$= \int_G f(y) \int_G g(y^{-1}x)\langle\pi(x)v, w\rangle\,dx\,dy$$

$$= \left\langle \int_G f(y) \int_G g(y^{-1}x)\pi(x)\,dx\,dy\,v, w \right\rangle.$$

As this holds for all $v, w \in V_\pi$, the interchange is justified. Finally,

$$\pi(f^*) = \int_G f^*(x)\pi(x)\,dx = \int_G \Delta(x^{-1})\overline{f(x^{-1})}\pi(x)\,dx$$

$$= \int_G \overline{f(x)}\pi(x^{-1})\,dx$$

$$= \int_G \overline{f(x)}\pi(x)^*\,dx = \left(\int_G f(x)\pi(x)\,dx\right)^*,$$

so indeed, $f \mapsto \pi(f)$ is a *-representation.

For the final assertion assume that $T : V_\pi \to V_\eta$ is a unitary map with $T\pi(x) = \eta(x)T$ for every $x \in G$. Multiplying with $f(x)$ and integrating over x, this identity yields $T\pi(f) = \eta(f)T$ for every $f \in C_c(G)$. For the converse, assume that we have $T\pi(f) = \eta(f)T$ for every $f \in C_c(G)$ and let $x_0 \in G$. Assume that $\pi(x_0) \neq T^{-1}\eta(x_0)T$, say $\|\pi(x_0) - T^{-1}\eta(x_0)T\| > \varepsilon$ for some $\varepsilon > 0$, where the norm is the operator norm. So there exists $v, w \in V_\pi$ and $a \in \mathbb{R}$ with $\mathrm{Re}(\langle\pi(x_0)v, w\rangle) < a < \mathrm{Re}(\langle T^{-1}\eta(x_0)Tv, w\rangle)$. The set U of all $x \in G$ with $\mathrm{Re}(\langle\pi(x_0)v, w\rangle) < a < \mathrm{Re}(\langle T^{-1}\eta(x_0)Tv, w\rangle)$ is open. So let $0 \leq f \in C_c(G)$ be non-zero. Then

$$\mathrm{Re}\int_U f(x)\langle\pi(x)v, w\rangle\,dx < a\|f\|_1 < \mathrm{Re}\int_U f(x)\langle T^{-1}\eta(x)Tv, w\rangle\,dx.$$

On the other hand, the two outer sides must agree, a contradiction. □

If A is a *-algebra and $\pi : A \to \mathcal{B}(V)$ an irreducible representation, we also say that V is an *irreducible module* of A. This notion is not to be mixed up with the notion of a *simple module*:

Recall that a *module* of a $\mathbb{C}$-algebra A is a complex vector space M together with an algebra homomorphism $A \to \mathrm{End}(M)$, which is usually written as $(a, m) \mapsto am$. A *simple module* is a module M, which has no A-stable linear subspaces other than the trivial ones 0 and M. The difference from the notion of an irreducible module is that in the case of a simple module no topology is considered and in the case of an irreducible module only *closed* stable subspaces are excluded. So an irreducible module might still admit non-trivial stable subspaces.

An A-module M is called a *regular module* if $AM \neq 0$, i.e. a module is regular if the A-operation is not the zero operation.

A key assertion is the following *lemma of Schur*:

Lemma 7.5.5 *Let (π, V) be an irreducible unitary representation of a group G or an irreducible *-representation of a *-algebra A. Then every bounded operator $T : V \to V$, which commutes with all $\pi(x)$, is a scalar multiple of the identity, i.e. of the form $\lambda \operatorname{Id}$ for some $\lambda \in \mathbb{C}$.*

If V is finite-dimensional and irreducible, then the inner product, with respect to which V is a unitary or *-representation, is uniquely determined up to scalar multiples.

Proof We give a complete proof if V is finite-dimensional. For the general case we provide a sketch of a proof and references. Consider first an irreducible unitary representation of a group G. Let $A \subset \mathcal{B}(V)$ be the algebra generated by all operators $\pi(g)$ with $g \in G$. Then A is stable under the involution $T \mapsto T^*$, so the inclusion $A \hookrightarrow \mathcal{B}(V)$ is an irreducible *-representation. Since every $a \in A$ is a linear combination of operators of the form $\pi(g)$, $g \in G$, an operator $T \in \mathcal{B}(V)$ commutes with all $a \in A$ if and only if it commutes with all $\pi(g)$, $g \in G$. Further, A is commutative if G is. It therefore suffices to consider the case of an algebra A only.

If an operator T commutes with all $a \in A$, then, because of $aT^* = (Ta^*)^* = (a^*T)^* = T^*a$, the adjoint operator T^* also commutes with all $a \in A$. Hence the self-adjoint operators $T + T^*$ and $(iT) + (iT)^*$ commute with all $a \in A$. By $T = \frac{1}{2}(T + T^*) + \frac{1}{2i}((iT) + (iT)^*)$ it suffices, to show the claim for a self-adjoint operator T.

We assume that V is finite-dimensional. Then every self-adjoint operator T is diagonalizable. In particular, T has an eigenvalue $\lambda \in \mathbb{C}$. Let E_λ be the corresponding eigenspace. We show that E_λ is stable under the algebra A. So let $v \in E_\lambda$ and $a \in A$. Then $T\pi(a)v = \pi(a)Tv = \pi(a)\lambda v = \lambda \pi(a)v$, which means that $\pi(a)v \in E_\lambda$. So the closed linear subspace E_λ of V is left stable by A and it is non-zero. As V is irreducible, we get $E_\lambda = V$ or $T = \lambda \operatorname{Id}$, as claimed

If V is infinite-dimensional there is not necessarily an eigenvalue. Nevertheless, one can use the spectral theorem to get a similar argument to work in that case as well. One may find this in these books: [Heu06, Rud91, Wei80, Yos95]. In the book [DE09] there is a different proof of the lemma of Schur which uses continuous functional calculus.

We finally prove the addendum on the uniqueness of the inner product. It suffices to consider the case of an algebra. So let $\langle .,. \rangle_1$ and $\langle .,. \rangle_2$ be two inner products on the finite-dimensional space V, such that π is a *-representation with respect to both of them. By the Riesz Representation Theorem for every $w \in V$ there exists a unique vector Tw such that

$$\langle v, w \rangle_2 = \langle v, Tw \rangle_1$$

holds for every $v \in V$. The map $w \mapsto Tw$ is easily seen to be linear. For $a \in A$ one gets

$$\langle v, \pi(a)Tw \rangle_1 = \langle \pi(a^*)v, Tw \rangle_1 = \langle \pi(a^*)v, w \rangle_2 = \langle v, \pi(a)w \rangle_2 = \langle v, T\pi(a)w \rangle_1.$$

As this holds for all v, w, one gets $\pi(a)T = T\pi(a)$ for every $a \in A$, so T is a scalar, as was to be shown.   □

Corollary 7.5.6 *Let A be a *-algebra which is commutative. Then every irreducible *-representation of A is one-dimensional.*

Proof Let (π, V_π) be an irreducible *-representation of A. Let $b \in A$. Then for every $a \in A$ we have

$$\pi(b)\pi(a) = \pi(ba) = \pi(ab) = \pi(a)\pi(b).$$

Therefore the operator $T = \pi(b)$ commutes with every $\pi(a)$, $a \in A$. By the lemma of Schur, T is a scalar. As this applies to all $b \in A$, $\pi(b)$ is always a scalar, so every linear subspace $U \subset V_\pi$ is stable under A. Being irreducible, V_π can have only two closed subspaces, 0 and V. This implies that V is one-dimensional.   □

An irreducible unitary representation (π, V_π) of the group G_p with $p < \infty$ is called *unramified* if the vector space of K_p-invariants,

$$V_\pi^{K_p} = \left\{ v \in V_\pi : \pi(k)v = v \; \forall k \in K_p \right\}$$

is not the zero space, where K_p denotes the compact open subgroup $GL_2(\mathbb{Z}_p)$.

Definition 7.5.7 A function f on G_p is called K_p-*bi-invariant* if

$$f(k_1 x k_2) = f(x) \quad \text{holds for all } x \in G_p, \, k_1, k_2 \in K_p.$$

Lemma 7.5.8 *If π is unramified, then* $\dim V_\pi^{K_p} = 1$.

Proof Write $G = G_p$ and $K = K_p$ and assume that π is unramified. Let $\mathcal{H}_p = \mathcal{H}_G$ be the space of all locally constant functions of compact support. This is an algebra under convolution. Let $\mathcal{H}_p^{K_p}$ be the subalgebra of all K_p-bi-invariant functions and let P be the orthogonal projection $V_\pi \to V_\pi^{K_p}$. In Exercise 7.13 it is shown that $P = \pi(1_{K_p})$. The algebra $\mathcal{H}_p^{K_p}$ is commutative by Exercise 7.12. Further one has

$$\mathcal{H}_p^{K_p} = 1_K * \mathcal{H}_p * 1_K.$$

The algebra $\mathcal{H}_p^{K_p}$ acts via π on the space $V_\pi^{K_p}$. We show that this action is irreducible. For this let $v \in V_\pi^{K_p} \setminus \{0\}$. Then $\pi(\mathcal{H}_p)v$ is a non-zero G-submodule of V_π, and therefore is dense in V_π as the latter is irreducible. Therefore $P\pi(\mathcal{H}_p)v = \pi(1_K * \mathcal{H}_p * 1_K)v = \pi(\mathcal{H}_p^{K_p})v$ is a dense $\mathcal{H}_p^{K_p}$-submodule of $V_\pi^{K_p}$. So for every $v \in V_\pi^{K_p}$, the closed $\mathcal{H}_p^{K_p}$-submodule generated by v equals the entire space $V_\pi^{K_p}$.

Hence the representation of the algebra $\mathcal{H}_p^{K_p}$ on $V_\pi^{K_p}$ is irreducible. By Corollary 7.5.6, the space $V_\pi^{K_p}$ is one-dimensional. $\square$

For every $p \leq \infty$ let there be given an irreducible representation π_p of G_p. Assume that almost all π_p are unramified. We want to combine these representations into an irreducible representation

$$\pi = \widehat{\bigotimes_p} \pi_p$$

of $G_\mathbb{A}$, by extending tensor products of Hilbert spaces. For this we first have to introduce the latter.

Definition 7.5.9 Let V, W be Hilbert spaces. On the algebraic tensor product $V \otimes W$ we define an inner product by Hermitian extension of

$$\langle v \otimes w, v' \otimes w' \rangle \overset{\text{def}}{=} \langle v, v' \rangle \langle w, w' \rangle.$$

(See Exercise 7.5.) Then $V \otimes W$ is a pre-Hilbert space. We denote the completion of this space by $V \hat{\otimes} W$. In particular we have the following: if $(e_i)_{i \in I}$ is an orthonormal base of V and $(f_j)_{j \in J}$ one of W, then $(e_i \otimes f_j)_{(i,j) \in I \times J}$ is an orthonormal base of $V \hat{\otimes} W$.

For finitely many Hilbert spaces $V_1, \ldots, V_n$ one defines the tensor product $V_1 \hat{\otimes} \cdots \hat{\otimes} V_n$ by iteration, where one can show, that the order of tensoring is irrelevant. Indeed, as in the case of the algebraic tensor product, the spaces $(V_1 \hat{\otimes} V_2) \hat{\otimes} V_3$ and $V_1 \hat{\otimes} (V_2 \hat{\otimes} V_3)$ are canonically isomorphic. Further the space $V_1 \hat{\otimes} V_2$ is canonically isomorphic to $V_2 \hat{\otimes} V_1$.

Notation In the sequel, we shall leave out the hat above the tensor product sign, as it is clear that we are only interested in Hilbert spaces. So we simply write $V \otimes W$ instead of $V \hat{\otimes} W$.

Example 7.5.10 Let V be a finite-dimensional Hilbert space. Then the map $\varphi : v \mapsto \langle \cdot, v \rangle$ is an anti-linear bijection from V to the dual space V^*. We define an inner product on V^* by

$$\langle \alpha, \beta \rangle = \langle \varphi^{-1}(\beta), \varphi^{-1}(\alpha) \rangle.$$

The map ψ given by

$$\psi(L \otimes v)(w) = L(w)v$$

is an isomorphism $\psi : V^* \otimes V \to \text{End}(V)$, where $\text{End}(V)$ is the vector space of all linear maps $T : V \to V$. For $\alpha, \beta \in V^* \otimes V$ the ensuing inner product is

$$\langle \alpha, \beta \rangle = \text{tr}\big(\psi(\alpha)\psi(\beta)^*\big),$$

where $\psi(\beta)^*$ is the adjoint of $\psi(\beta)$.

Definition 7.5.11 Let (π, V_π) be a unitary representation of a topological group G, and let (η, V_η) be a unitary representation of a topological group H. We define a unitary representation $\rho = \pi \otimes \eta$ of the group $G \times H$ on the space $V_\pi \otimes V_\eta$ by

$$\rho(g, h)v \otimes w \stackrel{\text{def}}{=} \pi(g)v \otimes \eta(h)w.$$

The continuity of this representation will be shown in Exercise 7.7.

The representation $\pi \otimes \eta$ is known as the *exterior tensor product* of π and η.

Let G be a locally compact group and let (η, V_η) be an irreducible unitary representation. For a Hilbert space W we consider the tensor product representation of $G = G \times \{1\}$ on the space $V_\eta \otimes W$. As G acts on the first tensor factor only, we also call this the *first factor representation*.

Lemma 7.5.12 *If (η, V_η) is an irreducible unitary representation of G, then every closed G-stable subspace of the first factor representation of G on $V_\eta \otimes W$ is of the form $V_\eta \otimes W_1$ for a closed linear subspace W_1 of W.*

In particular, for two locally compact groups G, H and irreducible unitary representations π, τ of G, respectively, H, the representation $\pi \otimes \tau$ of $G \times H$ is irreducible and for another pair (π', τ') of irreducible unitary representations one gets that $\pi \otimes \tau \cong \pi' \otimes \tau'$ implies $\pi \cong \pi'$ and $\tau \cong \tau'$.

Proof Let $U \subset V_\eta \otimes W$ be a closed, G-stable linear subspace. Then its orthogonal space is G-stable as well and so the orthogonal projection P onto U is a bounded operator on $V_\eta \otimes W$, which commutes with all $\eta(g)$, $g \in G$. We show that every such operator T is of the form $1 \otimes S$ for an operator $S \in \mathcal{B}(W)$. So let T be a bounded operator on $V_\eta \otimes W$ with

$$T\big(\eta(g) \otimes 1\big) = \big(\eta(g) \otimes 1\big)T$$

for every $g \in G$. Let $(e_i)_{i \in I}$ be an orthonormal basis of W and for $j \in I$ let P_j be the orthogonal projection onto the subspace $\mathbb{C}e_j$. For $i, j \in I$ consider the operator $T_{i,j}$ given as a composition:

$$V_\eta \to V_\eta \otimes e_i \xrightarrow{T} V_\eta \otimes W \xrightarrow{1 \otimes P_j} V_\eta \otimes e_j \to V_\eta.$$

This operator is composed of operators which commute with the G-actions, so it commutes with the G-action as well. Also, being a composition of continuous operators, it is continuous. By the lemma of Schur there exists a complex number $a_{i,j}$ with $T_{i,j} = a_{i,j}\,\mathrm{Id}$. As $\sum_j P_j = \mathrm{Id}_W$, it follows that

$$T(v \otimes e_i) = \sum_j a_{i,j}v \otimes e_j = v \otimes S(e_i),$$

where $S(e_i) = \sum_j a_{i,j}e_j$ and the sum converges in W. By continuity of T the map S extends to a unique continuous operator $S : W \to W$, such that

$$T(v \otimes w) = v \otimes S(w)$$

holds for all $v \in V_\eta$, $w \in W$. This gives $T = 1 \otimes S$ as claimed.

We apply this to the projection P onto the invariant subspace U, which therefore is of the form $P = 1 \otimes P_1$. Then $P_1^2 = P_1$, so P_1 is a projection. Also, P_1 is self-adjoint, since P is. Together it follows that P_1 is an orthogonal projection. Let W_1 be its image. Then $U = V_\eta \otimes W_1$.

Now for the proof of irreducibility of the exterior tensor product $\pi \otimes \tau$. The first part implies that a closed, $G \times H$-stable linear subspace U has to be of the form $U = V_\pi \otimes U_\tau$ as well as of the form $U = U_\pi \otimes V_\tau$, where $U_\tau \subset V_\tau$ and $\pi \subset V_\pi$ are closed linear subspaces. These subspaces must be invariant themselves, and hence they are each the respective entire space.

Finally assume $\pi \otimes \tau \cong \pi' \otimes \tau'$, so there exists a unitary isomorphism $T : V_\pi \otimes V_\tau \to V_{\pi'} \otimes V_{\tau'}$, commuting with the $G \times H$-actions. Let $w \in V_\tau$ of norm one. Then $V_\pi \otimes w$ is an irreducible G-subrepresentation, so there is $w' \in V_{\tau'}$ of norm one, such that $T(V_\pi \otimes w) = V_{\pi'} \otimes w'$. The resulting map

$$V_\pi \to V_\pi \otimes w \xrightarrow{T} V_{\pi'} \otimes w' \to V_{\pi'}$$

is a unitary G-isomorphism, so $\pi \cong \pi'$ and analogously one gets $\tau \cong \tau'$.   □

For every $p \le \infty$ let there be given an irreducible unitary representation (π_p, V_p) of G_p, such that π_p is unramified for almost all p. Let $S_0 \ni \infty$ be the set of places, where π_p is ramified. For every $p \notin S_0$ fix a vector $v_p^0 \in V_p^{K_p}$ of norm one. For every finite set of places $S \supset S_0$ let $V_S = \bigotimes_{p \in S} V_p$. For two sets of places S and S' with $S_0 \subset S \subset S'$ define an isometric linear map $\varphi_S^{S'} : V_S \to V_{S'}$ by $w \mapsto w \otimes \bigotimes_{p \in S' \smallsetminus S} v_p^0$. Let I be the set of all finite sets $S \supset S_0$ of places. Then I is a directed set with the partial order given by inclusion. The maps $\varphi_S^{S'}$ form a direct system of Hilbert spaces. Consider the direct limit

$$\widetilde{\bigotimes_p} V_p = \varinjlim_S V_S.$$

Since the morphisms of the direct system are isometric, the right-hand side is in a natural way a pre-Hilbert space. We write its completion as $\bigotimes_p V_p$.

The space $\bigotimes_p V_p$ depends on the choice of vectors v_p^0, but only up to canonical isomorphism as follows: for a second choice of vectors v_p^1 for every $p \notin S_0$ there is a unique complex number θ_p of absolute value 1, such that $v_p^1 = \theta_p v_p^0$. For every finite set of places $S \supset S_0$ the multiplication with $\prod_{p \in S \smallsetminus S_0} \theta_p$ induces a unitary isomorphism of $V_S = \bigotimes_{p \in S} V_p$ into itself, commuting with the structural maps $\varphi_S^{S'}$. This induces the claimed isomorphism between the infinite tensor products.

Let $g \in G_{\mathbb{A}}$. There exists a finite set of places $S \ni \infty$ such that $g_p \in K_p$ for all $p \notin S$. The group element g acts by a unitary operator on every $V_{S'}$ such that $S' \supset S$ is a finite set of places. Further, as $g_p \in K_p$ for $p \notin S$, the structural maps $\varphi_S^{S'}$ commute with the respective action of g. Therefore g acts unitarily on $W = \bigotimes_p V_p$. We get a map $\pi : G_{\mathbb{A}} \to \mathrm{GL}(W)$, such that $\pi(g)$ is unitary for every $g \in G_{\mathbb{A}}$. This map is a unitary representation (see Exercise 7.8). We denote this representation by $\bigotimes_p \pi_p$.

Theorem 7.5.13 *The representation $\pi = \bigotimes_p \pi_p$ of $G_{\mathbb{A}}$ on the space $\bigotimes_p V_p$ is irreducible.*

Proof Let $U \neq 0$ be a closed, $G_{\mathbb{A}}$-stable linear subspace. As the convolution algebra $C_c^\infty(G_{\mathbb{A}})$ contains a Dirac sequence, the space $\tilde{U} = \pi(C_c^\infty(G_{\mathbb{A}}))U$ is dense in U. Let $u \in \tilde{U}$. Then there exists an open subgroup H of G_{fin} stabilizing u, i.e. $\pi(h)u = u$ for every $h \in H$, for suppose $u = \pi(f)v$ for some $f \in C_c^\infty(G_{\mathbb{A}})$. There exists an open subgroup H such that $L_h f(x) = f(h^{-1}x) = f(x)$ holds for all $x \in G_{\mathbb{A}}$. But then $\pi(h)u = \pi(h)\pi(f)v = \pi(L_h f)v = \pi(f)v = u$. Here we consider G_{fin} as a subgroup of $G_{\mathbb{A}}$ via the embedding $h \mapsto (h, 1)$.

As H is open, there exists a finite set of places $S \supset S_0$, such that the group K_p stabilizes u if $p \notin S$. It follows that for $p \notin S$, the local contribution of u at p is a multiple of the vector v_p^0, so that $u = u_S \otimes \bigotimes_{p \notin S} v_p^0$ for some $u_S \in V_S$. The representation $\pi|_{G_S}$ of the group G_S is of the form $V_S \otimes V^S$, where $V^S = \bigotimes_{p \notin S} V_p$. By Lemma 7.5.12 the space U is of the form $V_S \otimes U^S$, where U^S is a closed subspace of V^S. We have just remarked that U^S contains the vector $\bigotimes_{p \notin S} v_p$ which generates V^S as G^S-representation space. This implies $U = V$ as claimed. $\qquad\square$

7.5.2 Analysis

In this subsection we show that every irreducible unitary representation of $G_{\mathbb{A}}$, which is admissible in a precise sense to be defined, is isomorphic to an infinite tensor product as in the last subsection. We first define admissibility.

Definition 7.5.14 For a locally compact group G let $\hat{G}$ denote the *unitary dual* of G, i.e. the set[1] of all isomorphy classes of irreducible unitary representations.

A representation (π, V_π) is called *unitarizable* if there exists an inner product making V_π a Hilbert space and π unitary.

---

[1] There is a set-theoretic problem here. What we really mean is a set of representatives for the isomorphy classes of unitary representations. But it is not immediate that this is a set at all. This is a problem that occurs at many places in mathematics. The remedy is this: let α be the cardinality of G. For an irreducible representation (π, V_π) and any vector $v \neq 0$ the set $\pi(G)v$ generates a dense linear subspace. This implies that the cardinality of an orthonormal basis of H (= Hilbert dimension) cannot exceed α. Hence, fixing one Hilbert space V of dimension α, one can find a representative of every isomorphy class realized on a subspace of V. Hence one finds representatives for all isomorphy classes in the set $\text{Map}(G, \mathcal{B}(V))$ of all maps from G to the algebra of bounded operators on V.

Theorem 7.5.15 (Representations of compact groups)

(a) *Every irreducible representation of a compact group K is finite-dimensional and unitarizable.*
(b) *Every representation of a compact group K is isomorphic to a direct sum of irreducible representations.*

Note the following simple consequence. For given unitary representations (τ, V_τ) and (η, V_η) of K, the vector spaces $\mathrm{Hom}_K(V_\tau, V_\eta)$ and $\mathrm{Hom}_K(V_\eta, V_\tau)$ have the same dimension.

Proof The proof of this standard fact of representation theory can be found for example in [DE09, Kat04]. □

Theorem 7.5.16 (Peter–Weyl) *Let K be a compact group. For every irreducible representation (τ, V_τ) of K fix an orthonormal basis $e_1, \ldots, e_{d(\tau)}$ of V_τ and define the matrix coefficients*

$$\tau_{i,j}(k) = \langle \tau(k)e_i, e_j \rangle$$

if $1 \le i, j \le d(\tau)$. Then

$$\left(\sqrt{d(\tau)} \tau_{i,j} \right)_{\tau, i, j}$$

is an orthonormal basis of $L^2(K)$.

Proof This assertion can also be found in the standard literature [DE09, Kat04]. □

Definition 7.5.17 Let L be a closed subgroup of a compact group K. If (γ, V_γ) is a representation of L, we define the *continuously induced representation* $I_L^K(\gamma) = (I, V_I)$ as follows. The representation space V_I is the vector space of all continuous functions $f : K \to V_\gamma$ with the property that $f(lk) = \gamma(l)f(k)$ holds for all $l \in L, k \in K$. The norm $\|f\|_K = \sup_{k \in K} \|f(k)\|$ makes V_I a Banach space and the representation I is given by right translation, $I(k)f(k') = f(k'k)$.

Lemma 7.5.18 *Let (η, V_η) be a representation of K. Then there exists a natural linear bijection*

$$\psi : \mathrm{Hom}_K\left(V_\eta, I_L^K(\gamma)\right) \to \mathrm{Hom}_L(V_\eta, V_\gamma).$$

Proof We define ψ as $\psi(\alpha)(v) = \alpha(v)(1)$. To show injectivity assume $\psi(\alpha) = 0$. Then $\alpha(v)(1) = 0$ for every $v \in V_\eta$. For $k \in K$ one gets $\alpha(v)(k) = (I(k)\alpha(v))(1) = $

$\alpha(\eta(k)v)(1) = 0$, so that $\alpha = 0$. To show surjectivity, let $\beta : V_\eta \to V_\gamma$ be an L-homomorphism. We define $\alpha : V_\eta \to I_L^K(\gamma)$ by $\alpha(v)(k) = \beta(\eta(k)v)$. Then $\psi(\alpha) = \beta$ and so ψ is surjective.   □

If (π, V_π) is an arbitrary representation of the compact group K and if (τ, V_τ) is an irreducible representation of K, then we define the τ-isotype $V_\pi(\tau)$ or the τ-isotypical component of π as the image of the canonical map

$$\mathrm{Hom}_K(V_\tau, V_\pi) \otimes V_\tau \to V_\pi$$

$$\alpha \otimes v \mapsto \alpha(v).$$

The space $V_\pi(\tau)$ is the sum of all subrepresentations of π, which are isomorphic to τ. The space V_π is a direct sum

$$V_\pi = \bigoplus_{\tau \in \hat{K}} V_\pi(\tau),$$

which is orthogonal if π is unitary.

For a given representation $\tau \in \hat{K}$ let

$$e_\tau(k) = (\dim \tau) \, \overline{\mathrm{tr}(\tau(k))}, \quad k \in K.$$

Lemma 7.5.19 *The function e_τ is an idempotent in the convolution algebra $C(K)$, i.e. one has $e_\tau * e_\tau = e_\tau$. The map*

$$P_\tau = \int_K e_\tau(k)\pi(k)\,dk$$

is a projection onto the isotype $V_\pi(\tau)$. Here the Haar measure on K is normalized such that $\mathrm{vol}(K) = 1$ holds. If π is a unitary representation, then P_τ is an orthogonal projection.

Let γ be another irreducible unitary representation of K, not isomorphic to τ. Then one has

$$e_\tau * e_\gamma = 0.$$

Proof This is a direct consequence of the Peter–Weyl theorem.   □

Definition 7.5.20 Let $F \subset \hat{K}$ be a finite subset. Then the function

$$e_F \overset{\text{def}}{=} \sum_{\tau \in F} e_\tau$$

is an idempotent in $C(K)$, since by the last lemma

$$e_F * e_F = \sum_{\tau, \gamma \in F} e_\tau * e_\gamma = \sum_{\tau \in F} e_\tau = e_F.$$

Let G be a locally compact group and K a compact subgroup. A representation (π, V_π) of G is called a *K-admissible representation* if the representation $\pi|_K$ decomposes into a direct sum with finite multiplicities, i.e. if

$$\pi|_K = \bigoplus_{\tau \in \hat{K}} [\pi : \tau]\tau,$$

where the multiplicities

$$[\pi : \tau] = \dim \mathrm{Hom}_K(V_\tau, V_\pi) = \dim \mathrm{Hom}_K(V_\pi, V_\tau)$$

are finite. The representation π is K-admissible if and only if every isotype $V_\pi(\tau)$ is finite-dimensional. By $\widehat{G}_K$ we denote the set of all isomorphy classes of irreducible unitary G-representations, which are K-admissible. A representation of G is called an *admissible representation* if there exists a compact subgroup K of G, such that the representation is K-admissible.

Example 7.5.21 Let $G = G_p$ and π_λ be the principal series representation attached to the quasi-character $\lambda \in \mathrm{Hom}(A_p, \mathbb{C}^\times)$. We show that π_λ is admissible with respect to the compact open subgroup K_p. Let (τ, V_τ) be an irreducible representation of K_p. According to Exercise 7.14, the representation τ factors over a finite quotient K_p/K, where K is an open subgroup of K_p. It therefore suffices to show that for every open subgroup K of K_p the space of K-invariants V_λ^K is finite-dimensional. Every $f \in V_\lambda$ is uniquely determined by its restriction to K_p, which lies in $L^2(K_p)$. Hence one can consider V_λ^K as a subspace of $L^2(K_p)^K = L^2(K_p/K)$. As K_p/K is finite, this space is finite-dimensional.

Lemma 7.5.22 *Let G be a locally compact group with two compact subgroups $L \subset K$. For every irreducible representation (π, V_π) we have*

$$\pi \text{ is } L\text{-admissible} \quad \Rightarrow \quad \pi \text{ is } K\text{-admissible}.$$

If L has finite index in K, the converse direction also holds.

If G possesses an open compact subgroup K, then every admissible representation of G is K-admissible.

Proof Let the representation π be L-admissible. Let (τ, V_τ) be an irreducible representation of K. Since V_τ is finite-dimensional, it decomposes as an L-representation into a finite sum of irreducible representations. Therefore the vector space $\mathrm{Hom}_L(V_\tau, V_\pi)$ is finite-dimensional and so is its subspace $\mathrm{Hom}_K(V_\tau, V_\pi)$.

Now assume that the group L has finite index in K and let π be K-admissible. Let (γ, V_γ) be an irreducible representation of L. We want to show that $\mathrm{Hom}_L(V_\pi, V_\gamma)$ is finite-dimensional. By Lemma 7.5.18 this space has the same dimension as $\mathrm{Hom}_K(V_\pi, I_L^K(\gamma))$. Since π is admissible with respect to the group K, it suffices to show that the induced representation $I_L^K(\gamma)$ is finite-dimensional. Let $K = \bigcup_{j=1}^n Lk_j$. Then every $f \in I_L^K(\gamma)$ is uniquely determined by the finitely many values $f(k_1), \ldots, f(k_n)$, so the dimension of $I_L^K(\gamma)$ is at most $\dim(V_\gamma)[K : L]$, hence finite.

We prove the last assertion of the lemma. Let K be a compact open subgroup of G, and let L be a give compact subgroup, such that the representation π is L-admissible. We have to show that π is also K-admissible. The group $K \cap L$ is open in L, so has finite index in L, so π is $K \cap L$-admissible, hence K-admissible. $\qquad\square$

Theorem 7.5.23 (Tensor product theorem) *For every set S of places, every admissible irreducible unitary representation π of the group G_S is isomorphic to a tensor product $\bigotimes_{p \in S} \pi_p$, where all π_p are admissible, irreducible, and unitary. Almost all π_p are unramified. All π_p are uniquely determined up to unitary equivalence.*

We give the proof in the rest of the section.

Let G be a locally compact group. The set $C_c(G)$ of all continuous functions of compact support is a complex algebra under the usual convolution product,

$$f * g(x) = \int_G f(y)g\left(y^{-1}x\right) dy = \int_G f(xy)g\left(y^{-1}\right) dy, \quad f, g \in C_c(G).$$

Let there be given a compact subgroup $K \subset G$. For $\alpha, \beta \in C(K)$ we also have the K-convolution,

$$\alpha * \beta(k) = \int_K \alpha(l)\beta\left(l^{-1}k\right) dl$$

which makes $C(K)$ a convolution algebra. We also want to convolve functions on K with functions on G. For $\alpha \in C(K)$ and $f \in C_c(G)$ we define

$$\alpha * f(x) = \int_K \alpha(k)f\left(k^{-1}x\right) dk \quad \text{and} \quad f * \alpha(x) = \int_K f(xk)\alpha\left(k^{-1}\right) dk.$$

We normalize the Haar measure on the compact group K by the condition $\mathrm{vol}(K) = 1$. If K is a set of positive measure with respect to the Haar measure of G, which is for instance the case if K is an open subgroup of G, like K_p in G_p, then we insist that the Haar measure of G is normalized to give K the measure 1 as well. Then the two Haar measures agree on K.

We consider a function $f \in C(K)$ as a function on G by setting $f(x) = 0$ if $x \in G \smallsetminus K$. In this sense we can consider the sum

$$C_c(G) + C(K)$$

as a subvector space of the space of all maps from G to $\mathbb{C}$. If K is open in G, then $C(K) \subset C_c(G)$ and therefore $C_c(G) + C(K) = C_c(G)$. If K is not open in G, then the sum $C_c(G) + C(K)$ is a direct sum by Exercise 6.6.

Lemma 7.5.24 *The two convolution products of G and K are compatible in the sense that*

$$f * (g * h) = (f * g) * h$$

holds for all possible combinations of f, g, h, lying in $C_c(G)$ or in $C(K)$. Further one always has

$$(f * g)^* = g^* * f^*,$$

where $f^*(x) = \Delta(x^{-1})\overline{f(x^{-1})}$ and $\Delta(x)$ is the modular function of G. We subsume these assertions by saying that

$$C_c(G) + C(K)$$

is a *-algebra.

A unitary representation (π, V_π) of G yields a *-representation of $C_c(G) + C(K)$ by $\pi(f) = \int_G f(x)\pi(x)\,dx$ for $f \in C_c(G)$ and $\pi(f) = \int_K f(k)\pi(k)\,dk$ for $f \in C(K)$.

Proof There is a deeper truth behind this assertion. These convolution products are special cases of a more general construction by which one can convolve Radon measures on G. Since we don't need the general statement, we restrict to $C_c(G) + C(K)$. The verifications are standard. As an example we show associativity in the case $f, g \in C_c(G)$ and $h \in C(K)$. In this case we have

$$
\begin{aligned}
f * (g * h)(x) &= \int_G f(y) g * h(y^{-1}x)\,dy \\
&= \int_G f(y) \int_K g(y^{-1}xk)h(k^{-1})\,dk\,dy \\
&= \int_K \int_G f(y)g(y^{-1}xk)h(k^{-1})\,dy\,dk \\
&= \int_K f * g(xk)h(k^{-1})\,dk = (f * g) * h(x). \qquad \square
\end{aligned}
$$

Definition 7.5.25 Let G be a locally compact group and let K be a compact subgroup. A function f on G is called *K-finite* if the set of all functions $x \mapsto f(k_1 x k_2)$, $k_1, k_2 \in K$, spans a finite-dimensional subspace of $\mathrm{Abb}(G, \mathbb{C})$. It is easy to see that the convolution product of two K-finite functions in $C_c(G)$ is again K-finite. Analogously we define the set of K-finite elements of $C(K)$.

Example 7.5.26 Let $f \in C_c(G)$. Consider the representation ρ of the compact group $K \times K$ on the Hilbert space $L^2(G)$ given by

$$\rho(k, l)\varphi(x) = \varphi(k^{-1}xl).$$

The Hilbert space $L^2(G)$ has the isotypical decomposition

$$L^2(G) = \bigoplus_{\tau \in \widehat{K \times K}} L^2(G)(\tau).$$

Accordingly, f can be written as

$$f = \sum_{\tau \in \widehat{K \times K}} f_\tau.$$

The function f is K-finite if and only if this sum is finite, i.e. if $f_\tau = 0$ for almost all τ. So the set of K-finite vectors is the algebraic direct sum of all isotypes $L^2(G)(\tau)$, $\tau \in \widehat{K \times K}$. It follows that the set of K-finite functions is dense in $L^2(G)$.

Definition 7.5.27 We define the *Hecke algebra* $\mathcal{H} = \mathcal{H}_{G,K}$ of the pair (G, K) as the convolution algebra of all K-finite functions in $C_c(G)$.

Lemma 7.5.28 *Let G be a locally compact group and K a compact subgroup.*

(a) *Let I_K be the set of all finite subsets of $\hat{K}$. For $F \in I_K$ set $e_F = \sum_{\tau \in F} e_\tau$ and*

$$C_F \overset{\text{def}}{=} e_F * \mathcal{H}_{G,K} * e_F.$$

Then C_F is a subalgebra of $\mathcal{H} = \mathcal{H}_{G,K}$. The Hecke algebra $\mathcal{H}$ is the union of all these subalgebras.
If $F = \{\tau\}$ is a singleton, we also write $C_F = C_\tau$.

(b) *Let (π, V_π) be a representation of G. Then the space $\pi(\mathcal{H})V_\pi$ is dense in V_π.*

(c) *The Hecke algebra is a *-algebra with the involution $f^*(x) = \Delta(x^{-1})\overline{f(x^{-1})}$. A unitary representation π of G defines by integration a *-representation of $\mathcal{H}$. For two unitary representations π, π' of G one has*

$$\pi \cong_G \pi' \quad \Leftrightarrow \quad \pi \cong_\mathcal{H} \pi',$$

which means that π and π' are unitarily isomorphic as G-representations if and only if they are unitarily isomorphic as $\mathcal{H}$-representations.

Proof The function e_F is idempotent, see Definition 7.5.20. As $\mathcal{H}$ is the set of all K-finite functions, it is the union of all C_F. This proves part (a).

For proving (b) let $h \in C(K)$. Then $\pi(h) = \int_K h(k)\pi(k)\,dk$. Let $F \in I_K$ and $P_F = \pi(e_F)$. Then $P_F^2 = \pi(e_F)\pi(e_F) = \pi(e_F * e_F) = \pi(e_F) = P_F$, so P_F is a projection. The image of this projection is $V_\pi(F) = \bigoplus_{\tau \in F} V_\pi(\tau)$, the F-isotype of π, and the kernel is $\bigoplus_{\tau \notin F} V_\pi(\tau)$. By Theorem 7.5.15 the union of all $V_\pi(F)$ with $F \in I_K$ is dense in V_π. So it suffices to show that $\pi(C_F)V_\pi$ is dense in $V_\pi(F)$. Let $v \in V_\pi(F)$ and let $\varepsilon > 0$. Since $\pi(e_F)$ is continuous, there exists $C > 0$, such that $\|\pi(e_F)w\| \le C\|w\|$ holds for every $w \in V_\pi$. As the map $G \times V_\pi \to V_\pi$; $(g, v) \mapsto \pi(g)v$ is continuous, there is a neighborhood of the unit in G, such that $x \in U \Rightarrow \|\pi(x)v - v\| < \varepsilon/C$. Let $f \in C_c(G)$ with support in U, such that $f \ge 0$ and $\int_G f(x)\,dx = 1$. Then

$$\|\pi(f)v - v\| = \left\| \int_G f(x)(\pi(x)v - v)\,dx \right\| \le \int_G f(x)\|\pi(x)v - v\|\,dx < \varepsilon/C.$$

One has $e_F * f * e_F \in C_F \subset \mathcal{H}$ and

$$\|\pi(e_F * f * e_F)v - v\| = \|\pi(e_F)(\pi(f)v - v)\| < \varepsilon.$$

We finally prove part (c). It is clear that the Hecke algebra is closed under the involution. Let π be a unitary G-representation. For $f \in \mathcal{H}$ one has

$$\pi(f^*) = \int_G \Delta(x^{-1})\overline{f(x^{-1})}\pi(x)\,dx = \int_G \overline{f(x)}\pi(x^{-1})\,dx$$

$$= \int_G \overline{f(x)}\pi(x)^*\,dx = \left(\int_G f(x)\pi(x)\,dx\right)^* = \pi(f)^*.$$

If π and π' are isomorphic as G-representations, then they are isomorphic as $\mathcal{H}$-representations. Conversely let $T : V_\pi \to V_{\pi'}$ be a unitary $\mathcal{H}$-isomorphism, so $T\pi(f) = \pi'(f)T$ for every $f \in \mathcal{H}$. First we note that this identity holds for all $f \in C_c(G)$. For this let $S = T\pi(f) - \pi'(f)T$. For every finite subset F of $\hat{K}$ it follows that

$$\pi'(e_F)S\pi(e_F) = T\pi(\underbrace{e_F f e_F}_{\in \mathcal{H}}) - \pi'(e_F f e_F)T = 0.$$

So $Sv = 0$ for every vector $v \in V_\pi(F)$. Since the $V_\pi(F)$ span a dense subspace of V_π, we get $S = 0$, so $T\pi(f) = \pi'(f)T$ holds for every $f \in C_c(G)$.

Let $\varepsilon > 0$, $x \in G$, and $v \in V$. By continuity of the representations π and π' there exists a neighborhood U of $x \in G$, such that $\|\pi(u)v - \pi(x)v\| < \varepsilon/2$ and $\|\pi'(u)Tv - \pi'(x)Tv\| < \varepsilon/2$ for every $u \in U$. Let $f \in C_c(G)$ with support inside U and such that $f \geq 0$ and $\int_G f(x)\,dx = 1$. As T is unitary, one has

$$\|T\pi(f)v - T\pi(x)v\| = \|\pi(f)v - \pi(x)v\|$$

$$= \left\|\int_G f(u)\big(\pi(u)v - \pi(x)v\big)\,du\right\|$$

$$\leq \int_G f(u)\|\pi(u)v - \pi(x)v\|\,du < \varepsilon/2$$

and analogously $\|T\pi(f)v - \pi'(x)Tv\| = \|\pi'(f)Tv - \pi'(x)Tv\| < \varepsilon/2$. This implies that the norm $\|T\pi(x)v - \pi'(x)Tv\|$ is less than ε. Since $\varepsilon > 0$ and $v \in V_\pi$ are arbitrary, we conclude $T\pi(x) = \pi'(x)T$. $\qquad\square$

Theorem 7.5.29 *Let G and H locally compact groups with compact subgroups K and L. The map $\widehat{G}_K \times \widehat{H}_L \to \widehat{G \times H}_{K \times L}$ given by $(\pi, \tau) \mapsto \pi \otimes \tau$ is a bijection.*

Proof Injectivity follows from Lemma 7.5.12. We show surjectivity. Let (π, V_π) be an irreducible unitary representation of G. Every $f \in C_c(G)$ yields a continuous operator on the Hilbert space V_π. For $0 \neq v \in V_\pi$ the space $\pi(C_c(G))v$ is a G-stable linear subspace of V_π. As π is irreducible, the space $\pi(\mathcal{H})v$ is a dense subspace.

Lemma 7.5.30 *Let G be a locally compact group and $K \subset G$ a compact subgroup.*

(a) *Let (π, V_π) be an irreducible representation of G and let F be a finite subset of $\hat{K}$. Then the F-isotype $V_\pi(F) = \bigoplus_{\tau \in F} V_\pi(\tau)$ is an irreducible C_F-module. Also, $\pi(\mathcal{H})V_\pi$ is an irreducible $\mathcal{H}$-module. If $V_\pi(F)$ is finite-dimensional, then $V_\pi(F)$ is a regular C_F-module.*

(b) *Let (η, V_η) be a unitary representation of G and let M be a finite-dimensional irreducible C_F-submodule of $V_\eta(F)$. Then the G-representation generated by M is irreducible.*

(c) *Let η, π be irreducible unitary, K admissible representations of G. Then*

$$\eta \cong \pi \quad \Leftrightarrow \quad V_\eta(F) \cong V_\pi(F) \text{ for every } F \in I_K,$$

where the isomorphy on the right-hand side is an isomorphy as C_F-modules.

Proof (a) In order to show irreducibility of $V_\pi(F)$, let $0 \neq U \subset V_\pi(F)$ be a closed submodule. Since V_π is irreducible, the space $\pi(C_c(G))U$ is dense in V_π. Let $P_F : V_\pi \to V_\pi(F)$ be the isotypical projection. For $h \in C(K)$ we write $\pi(h) = \int_K h(k)\pi(k)\,dk$. Then $P_F = \pi(e_F)$. For $f \in C_c(G)$ one has finite

$$\pi(f)\pi(e_\tau) = \pi(f * e_\tau) \quad \text{and} \quad \pi(e_\tau)\pi(f) = \pi(e_\tau * f),$$

as is easily verified.

The projection P_F being continuous, the space $P_F(C_c(G)U)$ is dense in $V_\pi(F)$, so $\overline{P_F(C_c(G)U)} = V_\pi(F)$. Now $P_F = \pi(e_F)$, so

$$V_\pi(F) = \overline{P_F(C_c(G)U)} = \overline{\pi(e_F)C_c(G)U}$$
$$= \overline{\pi(e_F * C_c(G) * e_F)U} = \overline{\pi(C_F)U} = \overline{U} = U,$$

and $V_\pi(F)$ is indeed irreducible. Next assume that $V_\pi(F)$ is finite-dimensional. We have just shown that $\pi(C_F)V_\pi(F)$ is dense in $V_\pi(F)$. As this space is finite-dimensional, the only dense subspace is the whole space, so $\pi(C_F)V_\pi(F) = V_\pi(F)$, which means that $V_\pi(F)$ is regular.

(b) We use the following principle. For a module M of a $\mathbb{C}$-algebra A and an element $m_0 \in M$, the *annihilator*

$$\text{Ann}_A(m_0) \overset{\text{def}}{=} \{a \in A : am_0 = 0\}$$

is a left ideal in A and the map $a \mapsto am_0$ induces a module isomorphism

$$A/\text{Ann}_A(m_0) \to Am_0.$$

In particular, if M is finite-dimensional and irreducible, then for $m_0 \neq 0$ one has $M = Am_0$, so $M \cong A/J$ with $J = \text{Ann}_A(m_0)$.

Let M be a finite-dimensional C_F-submodule of $V_\eta(F)$. We assume $M \neq 0$. Let U be the G-stable closed subspace of V_π generated by M. We show that $P_F(U) = M$, where $P_F = \pi(e_F)$ is the orthogonal projection onto the F-isotype. For this let $v_0 \neq 0$ be a vector in M and let $J = \text{Ann}_{C_F}(v_0)$. Then J is a left ideal and the map $a \mapsto av_0$ is a module isomorphism $C_F/J \overset{\cong}{\to} M$, since M is finite-dimensional.

Claim *Let $\overline{J}$ denote the annihilator $\text{Ann}_{C_c(G)}(v_0)$ of v_0 in $C_c(G)$. Then*

$$\overline{J} = \overline{J}e_F \oplus \text{Ann}_{C_c(G)}(e_F),$$

where $\text{Ann}_{C_c(G)}(e_F)$ is the annihilator of e_F in $C_c(G)$; this is the set of all $f \in C_c(G)$ with $f * e_F = 0$.

We prove the claim. First it is clear that $\overline{J}e_F \subset \mathrm{Ann}_{C_c(G)}(v_0) = \overline{J}$. Further, one has $v_0 = e_F v_0$, and therefore $\mathrm{Ann}_{C_c(G)}(e_F) \subset \mathrm{Ann}_{C_c(G)}(v_0) = \overline{J}$. It remains to show that $\overline{J}$ is a subset of the right-hand side. As e_F is idempotent, we get $C_c(G) = C_c(G)e_F \oplus \mathrm{Ann}_{C_c(G)}(e_F)$, since every $f \in C_c(G)$ can be written as $f = fe_F + (f - fe_F)$ and $f - fe_F$ lies in the annihilator of e_F. Finally by $\mathrm{Ann}_{C_c(G)}(e_F) \subset \mathrm{Ann}_{C_c(G)}(v_0) = \overline{J}$ the claim follows.

We infer that

$$C_c(G)v_0 \cong C_c(G)/\overline{J} \cong C_c(G)e_F/(\overline{J}e_F),$$

and so

$$P_F\big(C_c(G)v_0\big) \cong e_F C_c(G)e_F/(\overline{J}e_F) \cong C_F/J \cong M.$$

This implies $P_F(U) = M$. If U' is a subrepresentation of U, then $P_F(U') = 0$ or M. In the first case one has $M \subset (U')^{\perp}$ and so $U \subset (U')^{\perp}$, since $(U')^{\perp}$ is a subrepresentation. This, however, implies $U' = 0$. In the other case one has $M \subset U'$ and so $U' = U$. So U is indeed irreducible.

(c) If $(\eta, V_\eta), (\pi, V_\pi)$ are isomorphic representations, then the C_F-modules $V_\eta(F)$ and $V_\pi(F)$ are isomorphic. Conversely, assume for every F there is given a C_F-isomorphism $\phi_F : V_\eta(F) \to V_\pi(F)$. By Schur's lemma, for given F, two isomorphisms $V_\eta(F) \to V_\pi(F)$ will differ by a scalar only. We can, therefore, normalize the isomorphisms in such a way that they extend one another, i.e. that we have $\phi_F = \phi_{F'}|_{V_\eta(F)}$, if $F \subset F'$. Then one can compose the ϕ_F to a $\mathcal{H}$-isomorphism

$$\phi : \eta(\mathcal{H})V_\eta \xrightarrow{\cong} \pi(\mathcal{H})V_\pi.$$

Both sides are dense subspaces. If we can show that ϕ is isometric, this map will extend to an isomorphism of G-representations by Lemma 7.5.28. Let $0 \neq v_0 \in \eta(\mathcal{H})V_\eta$ and let $w_0 = \phi(v_0)$. We can assume that $\|v_0\| = \|w_0\| = 1$ and we claim that ϕ is isometric. For this we use ϕ to transport both inner products to the same side. We then have, say on $\pi(\mathcal{H})V_\pi$, two inner products $\langle .,. \rangle_1$ and $\langle .,. \rangle_2$ such that v_0 has norm one in both and the action of $\mathcal{H}$ is a *-representation in both inner products. On the finite-dimensional space $\pi(C_F)V_\pi$, the two inner products must agree by the lemma of Schur. As this holds for every $F \in I_K$, the map ϕ is isometric. $\square$

Let $G \supset K$ and $H \supset L$ be locally compact groups with compact subgroups. Let E be a finite subset of $\hat{K}$ and F a finite subset of $\hat{L}$. There is a natural homomorphism

$$\psi : C_c(G) \otimes C_c(H) \to G_c(G \times H)$$

given by

$$\psi(f \otimes g)(x, y) = f(x)g(y).$$

This induces a homomorphism

$$C_E \otimes C_F \to C_{E \times F}.$$

In general, this map won't be surjective.

Lemma 7.5.31 *Let the module M be a finite-dimensional irreducible $C_{E\times F}$-*-submodule of a unitary G-representation. Then M is irreducible and regular as $C_E \otimes C_F$-module.*

Proof We give $C_{E\times F} \subset L^1(G \times H)$ the topology of the L^1-norm. Then $C_E \otimes C_F$ is dense in $C_{E\times F}$, as one sees from the fact that $C_c(G)$ is dense in $L^1(G)$. The representation $\rho : C_{E\times F} \to \text{End}_{\mathbb{C}}(M)$ is a continuous map, since M comes from a unitary representation of $G \times H$. Hence the image of $C_E \otimes C_F$ is dense in the finite-dimensional image of $C_{E\times F}$ in $\text{End}(M)$, and the two images therefore do agree. $\square$

Accordingly, in order to finish the proof of Theorem 7.5.29, we only need the second assertion of the following lemma.

Lemma 7.5.32

(a) *Let A be a $\mathbb{C}$-algebra and M a simple regular A-module which is finite-dimensional over $\mathbb{C}$. Then the map $A \to \text{End}_{\mathbb{C}}(M)$ is surjective.*

(b) *Let A, B denote algebras over $\mathbb{C}$ and let $R = A \otimes B$. For two regular simple modules M, N of A respectively B, which are finite-dimensional over $\mathbb{C}$, the tensor product $M \otimes N$ is a regular simple R-module and every regular simple R-module is isomorphic to such a tensor product for uniquely determined modules M and N.*

Proof Part (a) is a well-known theorem of Wedderburn. One finds a proof, for instance, in Lang's book on algebra [Lan02]. However, in this book, the theorem is proven under the additional hypothesis that the algebra A possesses a unit element, a condition that we have replaced by asking M to be regular. Consequently, we have to show how to deduce (a) from the corresponding assertions for algebras with unit. This is an interesting and often used technique, called the *adjunction of a unit*: We equip the vector space $B = A \times \mathbb{C}$ with the product

$$(a, z)(b, w) = (ab + zb + wa, zw).$$

Then B is an algebra with unit $(0, 1)$, containing A as two-sided ideal. Every A-module M becomes a B-module by setting $(a, z)m = am + zm$. If $A' \subset \text{End}_{\mathbb{C}}(M)$ is the image of A and $B' \subset \text{End}_{\mathbb{C}}(M)$ the image of B, then the theorem of Wedderburn, as given in Lang's book, yields $B' = \text{End}_{\mathbb{C}}(M)$. Then A' is a two-sided ideal in B', different from the zero ideal. The algebra $\text{End}_{\mathbb{C}}(M) \cong M_n(\mathbb{C})$, however, does not possess any two-sided ideals other than zero and the whole algebra, which implies $A' = B'$ as claimed.

Now for part (b). By (a) it suffices to show the first assertion for the case $A = \text{End}_{\mathbb{C}}(M)$ and $B = \text{End}_{\mathbb{C}}(N)$. The canonical map from the algebra $\text{End}_{\mathbb{C}}(M) \otimes \text{End}_{\mathbb{C}}(N)$ to $\text{End}_{\mathbb{C}}(M \otimes N)$, however, is surjective, and therefore $M \otimes N$ is a simple module.

Finally, let there be given a $\mathbb{C}$-finite-dimensional simple $A \otimes B$-module V. Then V contains a simple A-module M, since V is finite-dimensional. Let $N =$

$\mathrm{Hom}_A(M, V)$. This vector space is a B-module in the obvious way. Consider the map $\phi : M \otimes N \to V$ given by

$$\phi(m \otimes \alpha) = \alpha(m).$$

Then ϕ is a non-zero $A \otimes B$-homomorphism, hence surjective, as V is simple. We have to show that ϕ has trivial kernel. For this let $\alpha_1, \ldots, \alpha_k$ be a basis of N and $m_1, \ldots, m_l$ a basis of M. Let $c_{i,j} \in \mathbb{C}$ be given with $\phi(\sum_{i,j} c_{i,j} m_i \otimes \alpha_j) = 0$. We have to show that all coefficients $c_{i,j}$ are zero. The condition reads

$$0 = \phi\left(\sum_{i,j} c_{i,j} m_i \otimes \alpha_j\right) = \sum_{i,j} c_{i,j} \alpha_j(m_i) = \sum_j \alpha_j \left(\sum_i c_{i,j} m_i\right).$$

Let $P : M \to M$ be a projection onto a one-dimensional subspace, say $\mathbb{C}m_0$. By part (a) there exists $a \in A$ with $am = Pm$ for every $m \in M$. Accordingly,

$$0 = \sum_j \alpha_j \left(P\left(\sum_i c_{i,j} m_i\right)\right) = \sum_j \lambda_j \alpha_j(m_0),$$

where $P(\sum_i c_{i,j} m_i) = \lambda_j m_0$. For arbitrary $a \in A$ one gets

$$0 = \sum_j \lambda_j \alpha_j(am_0).$$

As a runs through A, the element am_0 will run through the simple module M, so

$$\sum_j \lambda_j \alpha_j = 0.$$

By linear independence of the α_j, all λ_j are zero. Consequently one has $\sum_i c_{i,j} m_i = 0$ for every j and by linear independence of the m_i it follows that $c_{i,j} = 0$.

The proof also shows that all simple A-submodules of V are isomorphic. Thus the uniqueness follows.   □

We now show Theorem 7.5.29. Let η be a $K \times L$-admissible irreducible unitary representation of $G \times H$. Let $E \subset \hat{K}$ and $F \subset \hat{L}$ be finite subsets. According to Lemma 7.5.31 the space $V_\eta(E \times F)$ is a finite-dimensional irreducible regular $C_E \otimes C_F$-module and by Lemma 7.5.32 the space $V_\eta(E \times F)$ is a tensor product of modules, which we write as $V_\pi(E) \otimes V_\tau(F)$. The uniqueness of the tensor factors gives us injective homomorphisms $\varphi_E^{E'} : V_\pi(E) \to V_\pi(E')$ if $E \subset E'$ and accordingly for F. The uniqueness of inner products by the lemma of Schur allows us to scale the homomorphisms such that they are isometric. We define

$$\tilde{V}_\pi \overset{\text{def}}{=} \varinjlim V_\pi(E),$$

where the limit is taken over all finite sets $E \subset \hat{K}$. We write V_π for the completion of the pre-Hilbert space $\tilde{V}_\pi$. We construct V_τ analogously. For every finite set $E \subset \hat{K}$ the algebra C_E acts on $\tilde{V}_\pi$ via continuous operators. These extend to V_π and one gets a *-representation of the Hecke algebra $\mathcal{H}_G$ and similarly for V_τ. By construction, the isotypes of V_π are just the spaces $V_\pi(E)$ for $E \in I_K$. The isometric maps

$V_\pi(E) \otimes V_\tau(F) \hookrightarrow V_\eta$ can be composed to an isometric $\mathcal{H}_G \otimes \mathcal{H}_H$-homomorphism $\Phi : V_\pi \otimes V_\tau \to V_\eta$, which by irreducibility of η is an isomorphism. We have to install a unitary representation of the group $G \times H$ on $V_\pi \otimes V_\tau$. This means that we have to give a representation π of G on V_π. Fix a vector $w \in V_\tau$ with $\|w\| = 1$. The map $V_\pi \to V_\pi \otimes w \xrightarrow{\Phi} V_\eta$ gives an isometric embedding of V_π into V_η, which commutes with the $\mathcal{H}_G$-operation. The G-representation on V_η then defines a unitary G-representation on V_π inducing the $\mathcal{H}_G$-representation. We do the same with the group H and we get irreducible unitary representations π and τ of G and H such that Φ is a $G \times H$-isomorphism. □

We finally show Theorem 7.5.23. For a finite set of places S, Theorem 7.5.23 follows directly from Theorem 7.5.29. By the same token, one can assume that the infinite place is not in S. So assume S infinite and $\infty \notin S$. Let (π, V) be an irreducible unitary representation of G_S and let K_S be the compact open subgroup $\prod_{p \in S} K_p$. By Lemma 7.5.30 there exists $\tau \in \hat{K}_S$ such that $V_\pi(\tau)$ is a non-zero irreducible module of the Hecke algebra C_τ. By Lemma 6.2.5 the representation τ is trivial on an open subgroup of K_S. So there exists a finite set of places $T \subset S$ with $\tau(\prod_{p \in S \smallsetminus T} K_p) = 1$. By Theorem 7.5.29 we can replace G_S by $G_{S \smallsetminus T}$, which means we assume $K = \prod_{p \in S} K_p$ and $\tau = 1$. Then by Exercise 7.12 the convolution algebra $C_1 = C_c(G)^K$ is commutative and therefore every irreducible *-module is one-dimensional, hence $V_\pi(\tau) = V_\pi^K$ is one-dimensional. For every finite set of places $T \subset S$ the space V_π^K generates an irreducible representation of $G_T = \prod_{p \in T} G_p$ by Lemma 7.5.30(b). With $\pi_p = \pi_{\{p\}}$ one gets $\pi_T \cong \bigotimes_{p \in T} \pi_p$. Let $\eta = \bigotimes_{p \in S} \pi_p$. Theorem 7.5.13 implies that η is irreducible. For every finite subset $T \subset S$ one gets an isometry $\eta_T = \bigotimes_{p \in T} \pi_p \hookrightarrow \pi$. These isometries can be chosen in a compatible way, so that they yield a G_S-equivariant isometry $\eta \hookrightarrow \pi$. As π is irreducible, this map is an isomorphism. The proof of the tensor product theorem is finished. □

7.5.3 Admissibility of Automorphic Representations

An irreducible representation π of $G_\mathbb{A}$ is called a *cuspidal representation* if π is isomorphic to a subrepresentation of L^2_{cusp}. We want to apply the tensor product theorem to cuspidal representations. For this we have to ensure they are admissible.

A unitary representation of the group $G_\mathbb{A}$ is called a *compact representation* if for every $f \in C_c(G_\mathbb{A})$ the induced operator $\pi(f)$ is compact. Every subrepresentation of a compact representation is compact. Proposition 7.4.3 asserts that the space of cusp forms L^2_{cusp} defines a compact representation and by Proposition 7.3.13 every compact representation is a direct sum of irreducible representations.

Theorem 7.5.33 *Every irreducible compact representation is admissible. In particular, every cuspidal representation is admissible.*

Proof Let (π, V_π) be an irreducible compact representation and let (τ, V_τ) be an irreducible representation of the compact group $K_\mathbb{A} = \prod_{p \leq \infty} K_p$, where $K_p = GL_2(\mathbb{Z}_p)$ if $p < \infty$ and $K_\infty = SO(2)$. Then τ is a tensor product $\tau = \bigotimes_p \tau_p$. Let A be the $*$-algebra $C_c(G_\mathbb{A})$ and set $A_\tau = e_\tau A e_\tau$. Then $V_\pi(\tau)$ is an irreducible A_τ-$*$-module.

The algebra A_τ is itself a tensor product of the form $A_\infty \otimes A_{\text{fin}}$, where $A_\infty = e_{\tau_\infty} C_c^\infty(G_\infty) e_{\tau_\infty}$ and similarly for A_{fin}.

Lemma 7.5.34 *The algebra A_∞ is commutative.*

Proof The irreducible representations of the group $K_\infty = SO(2)$ are given by characters ε_ν, $\nu \in \mathbb{Z}$, where

$$\varepsilon_\nu \begin{pmatrix} a & -b \\ b & a \end{pmatrix} = (a+ib)^\nu.$$

Let ε be the character giving τ_∞. The algebra A_∞ can be viewed as the convolution algebra of all functions $f \in C_c^\infty(G_\infty)$ with $f(k_1 x k_2) = \varepsilon(k_1 k_2) f(x)$ for all $k_1, k_2 \in K_\infty$. Let D be the set of diagonal matrices in G_∞. By Exercise 3.3 we know that $G_\infty = K_\infty D K_\infty$. Consider the map $\theta : G \to G$ given by

$$\theta(x) = \begin{pmatrix} -1 & \\ & 1 \end{pmatrix} x^t \begin{pmatrix} -1 & \\ & 1 \end{pmatrix}.$$

One gets $\theta(kdl) = ldk$ for $k, l \in K_\infty$ and $d \in D$. This implies that for $f \in A_\infty$ one has $f^\theta = f$, where $f^\theta(x) = f(\theta(x))$. Since θ is an anti-automorphism of G_∞, i.e. $\theta(xy) = \theta(y)\theta(x)$, one has for arbitrary $f, g \in C_c(G)$, that $(f*g)^\theta = g^\theta * f^\theta$. This implies the claim as for $f, g \in A_\infty$ it holds that

$$f*g = (f*g)^\theta = g^\theta * f^\theta = g*f. \qquad \square$$

We finish the proof of the theorem. For $f \in A_\infty$ define $\pi(f) = \int_{G_\infty} f(x)\pi(x)\,dx$ and analogously for A_{fin}. Since A_∞ is commutative, every $f \in A_\infty$ commutes with every $a \in A_\tau$, so $\pi(f)$ commutes with every $\pi(a)$, where $a \in A_\tau$. By the lemma of Schur the operator $\pi(f)$ acts on the isotype $V_\pi(\tau)$ by a scalar. As e_τ is the idempotent attached to τ, the operator $\pi(e_\tau)$ is the orthogonal projection onto the K-isotype $V_\pi(\tau)$. Choose $f \in A_\infty$ such that $\pi(f)$ acts as identity on $V_\pi(\tau)$. The operator $\pi(e_\tau)\pi(f)$ is compact on the one hand, and on the other it acts as an orthogonal projection on $V_\pi(\tau)$. Ergo its image $V_\pi(\tau)$ is finite-dimensional, as claimed. $\qquad \square$

7.6 Exercises and Remarks

Exercise 7.1 Show that for a set S of places with $\infty \in S$ one has

$$GL_2(\mathbb{A}_S) \cong GL_2(\mathbb{A}_{S \setminus \{\infty\}}) \times GL_2(\mathbb{R}).$$

Show that if $\infty \notin S$, then $GL_2(\mathbb{A}_S)$ equals the restricted product of all $GL_2(\mathbb{Q}_p)$, $p \in S$, with respect to their compact open subgroups $GL_2(\mathbb{Z}_p)$. As a subset of $\mathbb{A}_S^4$, the set $GL_2(\mathbb{A}_S)$ inherits a topology. Show that this topology does not coincide with the restricted product topology.

Exercise 7.2 Let (X, d) be a separable metric space and let $Z \subset X$ be a subset. Show that Z is separable as well.

Exercise 7.3 Show that the modular function of the group $B_p = A_p N_p$ is given by $\Delta_{B_p}(an) = a^\delta$.

Exercise 7.4 Show that a quasi-character $a \mapsto a^\lambda$ of the group A_p is unramified if and only if it is of the form $\left(\begin{smallmatrix} a_1 & \\ & a_2 \end{smallmatrix}\right)^\lambda = |a_1|^{\lambda_1} |a_2|^{\lambda_2}$ for some $\lambda_1, \lambda_2 \in \mathbb{C}$.
(Hint: λ is unramified if and only if it factors over $A_p/K_p \cap A_p \cong \mathbb{Z}^2$.)

Exercise 7.5 Let V, W be Hilbert spaces. Show that the Hermitian extension of

$$\langle v \otimes w, v' \otimes w' \rangle \overset{\text{def}}{=} \langle v, v' \rangle \langle w, w' \rangle$$

defines an inner product on the tensor product space $V \otimes W$.
(Hint: the problem is definiteness. Let $f = \sum_i v_i \otimes w_i$ be in the tensor product. One can orthonormalize the finitely many v_i and get a new presentation of f, so one can assume the v_i to be pairwise orthogonal.)

Exercise 7.6 Show that for three Hilbert spaces V_1, V_2, V_3 the tensor products $(V_1 \otimes V_2) \otimes V_3$ and $V_1 \otimes (V_2 \otimes V_3)$ are canonically isomorphic. Further show that $V_1 \otimes V_2$ is canonically isomorphic to $V_2 \otimes V_1$.
(Hint: one needs to show that the canonical isomorphisms of the algebraic tensor products are isometries.)

Exercise 7.7 Let (π, V) and (η, W) be unitary representations of the locally compact groups G and H. Show that $\rho = \pi \otimes \eta$ is a unitary representation of $G \times H$ on the Hilbert space $V \hat{\otimes} W$.
(Hint: the problem is continuity. One has to estimate expressions of the form $\|\rho(g', h')v' - \rho(g, h)v\|$ for $g, g' \in G$, $h, h' \in H$ and $v, v' \in V \hat{\otimes} W$.)

Exercise 7.8 Show that the infinite tensor product of unitary representations, as defined before Theorem 7.5.13, is a unitary representation.
(Hint: one has to show that for convergent series $g_n \to g$ in G and $v_n \to v$ in $W = \bigotimes_p V_p$ one has $\pi(g_n)v_n \to \pi(g)v$. As in the proof of Theorem 7.1.3 one can restrict to vectors in a dense subspace.)

Exercise 7.9 Use the Peter–Weyl theorem to show

$$e_\tau * e_\gamma = \begin{cases} e_\tau & \text{if } \gamma \cong \tau, \\ 0 & \text{otherwise.} \end{cases}$$

Exercise 7.10 Let (π, V) be a representation of $G_p = \mathrm{GL}_2(\mathbb{Q}_p)$. A vector $v \in V$ is called a *smooth vector* if its stabilizer in G_p is an open subgroup. Show that the space V^∞ of all smooth vectors is dense in V.
(Hint: show that the convolution algebra of all locally constant functions with compact support contains a Dirac sequence.)

Exercise 7.11 Let $\Gamma = \mathrm{SL}_2(\mathbb{Z})$, $k \in 2\mathbb{N}$ and $f \in S_k(\Gamma)$ be a cusp form. We can view $S_k(\Gamma)$ as a subspace of $L^2(G_\mathbb{Q} Z_\mathbb{R} \backslash G_\mathbb{A})$. For distinction we write $A(f) \in L^2(G_\mathbb{Q} Z_\mathbb{R} \backslash G_\mathbb{A})$ for the element given by f.
   Let p be a prime number and let $g_p \in C_c^\infty(G_p)$ be given by

$$g_p = \mathbf{1}_{K_p\left(\begin{smallmatrix} p^{-1} & \\ & 1 \end{smallmatrix}\right)K_p}.$$

For $\varphi \in L^2(Z(\mathbb{R})G_\mathbb{Q} \backslash G_\mathbb{A})$ let

$$R(g_p)\varphi(x) = \int_{G_p} g_p(y)\varphi(xy)\,dy.$$

Show:

$$R(g_p)A(f) = p^{1-k/2}A(T_p f),$$

where T_p is the Hecke operator of Chap. 2.

Exercise 7.12 Let G be a locally compact group and let K be a compact subgroup. Show that the set $C_c(G)^K$ of all K-bi-invariant functions in $C_c(G)$ is a subalgebra of $C_c(G)$. If $C_c(G)^K$ is a commutative algebra, one says that the pair (G, K) is a *Gelfand pair*. Show: For any set S of finite places $(\mathrm{GL}_2(\mathbb{A}_S), \mathrm{GL}_2(\mathbb{Z}_S))$ is a Gelfand pair, where $\mathbb{Z}_S = \prod_{p \in S} \mathbb{Z}_p$.
(Hint: find a suitable set of representatives of $K \backslash G / K$. Show that matrix transposition defines a linear map $T : C_c(G) \to C_c(G)$ with $T(f * g) = T(g) * T(f)$. Note that the Haar measure on G is invariant under transposition and forming inverses.)

Exercise 7.13 Let G be a locally compact group and let K be a compact open subgroup. Normalize the Haar measure such that K has measure 1. Let (π, V_π) be a representation of G. Show that $\pi(\mathbf{1}_K) = \int_K \pi(x)\,dx$ is a projection with image V_π^K. Show that $\pi(\mathbf{1}_K)$ is an orthogonal projection if the representation $\pi|_K$ is unitary.

Exercise 7.14 Let K be a totally disconnected compact group. Show that every irreducible representation τ factors over a finite quotient K/N of K.
(Hint: by Theorem 7.5.15 the representation τ is finite-dimensional. Show that $\mathrm{GL}_n(\mathbb{C})$ possesses a unit neighborhood that does not contain any non-trivial subgroup.)

Remarks   In this chapter we have shown that every cuspidal representation is admissible and is actually a tensor product. It is more sharply true that every irreducible unitary representation of G_p, G_∞ or $G_\mathbb{A}$ is admissible, and so in the case $G_{\mathbb{A}}$ is a tensor product of local representations. To show this, however, one needs to use more representation theory than would be appropriate for the purpose of this book. A proof for G_∞ can be found in [Kna01], for G_p in [HC70]. The global case can be deduced from the local cases.

Chapter 8
Automorphic L-Functions

We denote the algebra of 2×2 matrices over a given ring R by $M_2(R)$. For $x \in M_2(\mathbb{A})$ we write

$$|x| = |\det(x)| = \prod_{p \leq \infty} |\det(x_p)|.$$

Let $\mathcal{S}(M_2(\mathbb{A}))$ denote the space of *Schwartz–Bruhat functions* on $M_2(\mathbb{A})$, i.e. every $f \in \mathcal{S}(M_2(\mathbb{A}))$ is a finite sum of functions of the form $f = \prod_p f_p$, where $f_p = 1_{M_2(\mathbb{Z}_p)}$ for almost all p and $f_p \in \mathcal{S}(M_2(\mathbb{Q}_p))$ for all p. Here $\mathcal{S}(M_2(\mathbb{R}))$ is the space of Schwartz functions on $M_2(\mathbb{R}) \cong \mathbb{R}^4$ and $\mathcal{S}(M_2(\mathbb{Q}_p))$ is the space of all locally constant functions with compact support on $M_2(\mathbb{Q}_p)$.

8.1 The Lattice $M_2(\mathbb{Q})$

Let e be the additive character on $\mathbb{A}$ defined in Lemma 5.4.2. Given $x \in M_2(\mathbb{A})$, we write $e(x)$ for $e(\mathrm{tr}(x))$, so

$$e\begin{pmatrix} a & b \\ c & d \end{pmatrix} = e(a + d).$$

Lemma 8.1.1 $M_2(\mathbb{Q})$ *is a discrete, cocompact subgroup of the additive group* $M_2(\mathbb{A})$, *i.e. it is a lattice. For* $x \in M_2(\mathbb{A})$ *one has*

$$e(xy) = 1 \ \forall y \in M_2(\mathbb{Q}) \quad \Leftrightarrow \quad x \in M_2(\mathbb{Q}).$$

Proof For $\mathbb{Q}$ and $\mathbb{A}$ we have seen a similar assertion before in Theorem 5.4.3. Since $\mathbb{Q}$ is discrete in $\mathbb{A}$ and $M_2(\mathbb{A}) \cong \mathbb{A}^4$ has the product topology, the set $M_2(\mathbb{Q}) \cong \mathbb{Q}^4$ is discrete in $M_2(\mathbb{A})$. By

$$M_2(\mathbb{A})/M_2(\mathbb{Q}) = \mathbb{A}^4/\mathbb{Q}^4 = (\mathbb{A}/\mathbb{Q})^4$$

the group $M_2(\mathbb{Q})$ also is cocompact in $M_2(\mathbb{A})$.

A. Deitmar, *Automorphic Forms*, Universitext,
DOI 10.1007/978-1-4471-4435-9_8, © Springer-Verlag London 2013

In the last assertion the '⇐' direction is trivial. For the '⇒' direction take $x \in M_2(\mathbb{A})$ satisfying the assumption. Write $x = \begin{pmatrix} a & b \\ c & d \end{pmatrix} \in M_2(\mathbb{A})$ and let $y = \begin{pmatrix} t & 0 \\ 0 & 0 \end{pmatrix}$, $t \in \mathbb{Q}$. One has

$$e(xy) = e\left(\mathrm{tr} \begin{pmatrix} at & 0 \\ ct & 0 \end{pmatrix} \right) = e(at).$$

We can vary t in $\mathbb{Q}$ to get $a \in \mathbb{Q}$ by Theorem 5.4.3. The same argument with $y = \begin{pmatrix} 0 & t \\ 0 & 0 \end{pmatrix}$, $\begin{pmatrix} 0 & 0 \\ t & 0 \end{pmatrix}$ and $\begin{pmatrix} 0 & 0 \\ 0 & t \end{pmatrix}$ gives the claim.   □

For $f \in \mathcal{S}(M_2(\mathbb{A}))$ we define its *Fourier transform* by

$$\hat{f}(x) = \int_{M_2(\mathbb{A})} f(y)e(-xy)\,dy.$$

Using the corresponding assertions in the one-dimensional case, one shows that the Fourier transform maps the space $\mathcal{S}(M_2(\mathbb{A}))$ into itself and that it satisfies the inversion formula $\hat{\hat{f}}(x) = f(-x)$, see also the remarks preceding Theorem 5.4.12.

8.2 Local Factors

We use the notion of a *group algebra*. For this let G be a group. The group algebra $\mathbb{C}[G]$ over $\mathbb{C}$ is defined as the convolution algebra $C_c(G)$, where we equip G with the discrete topology. Any discrete group is clearly locally compact and the counting measure is a Haar measure, which we use in the definition of the convolution.

We make this more explicit. An arbitrary function $f : G \to \mathbb{C}$ lies in $\mathbb{C}[G]$ if and only if it is zero outside a finite set. For $f, g \in \mathbb{C}[G]$ their convolution product equals

$$f * g(x) = \sum_{y \in G} f(y)g(y^{-1}x).$$

The group algebra has a canonical basis given by $(\delta_y)_{y \in G}$, where

$$\delta_y(x) = \begin{cases} 1 & \text{if } x = y, \\ 0 & \text{if } x \neq y. \end{cases}$$

We compute $\delta_x * \delta_y(z) = \sum_{r \in G} \delta_x(r)\delta_y(r^{-1}z) = \delta_y(x^{-1}z) = \delta_{xy}(z)$, i.e.

$$\delta_x * \delta_y = \delta_{xy}.$$

This identity gives rise to another definition of the group algebra, according to which the group algebra $\mathbb{C}[G]$ is the vector space with a basis $(\delta_y)_{y \in G}$ indexed by elements of G, equipped with a multiplication which on the basis elements is given by $\delta_x \delta_y = \delta_{xy}$, and which is then bilinearly extended to the entire space. We refer to this second definition as the *algebraic definition*, whereas the definition via convolution is called the *analytic definition*.

We want to understand the algebra homomorphisms from $\mathbb{C}[G]$ to $\mathbb{C}$, and we want to describe them in both definitions. We are looking at a linear map

$\phi : \mathbb{C}[G] \to \mathbb{C}$, which is multiplicative, so $\phi(ab) = \phi(a)\phi(b)$ holds for all $a, b \in \mathbb{C}[G]$ and $\phi(\delta_1) = 1$. Then the map $\varphi : G \to \mathbb{C}$ given by $\varphi(y) = \phi(\delta_y)$ is a multiplicative map. Because of $1 = \phi(\delta_1) = \varphi(1) = \varphi(yy^{-1}) = \varphi(y)\varphi(y^{-1})$, every $\varphi(y)$ is invertible, and hence lies in $\mathbb{C}^\times$. Therefore every algebra homomorphism ϕ induces a group homomorphism $\varphi : G \to \mathbb{C}^\times$. The converse is true as well, as for a given group homomorphism φ one can define an algebra homomorphism ϕ by linear extension from $\phi(\delta_y) = \varphi(y)$. We obtain a canonical bijection

$$\mathrm{Hom}_{\mathrm{alg}}(\mathbb{C}[G], \mathbb{C}) \xrightarrow{\cong} \mathrm{Hom}_{\mathrm{grp}}(G, \mathbb{C}^\times).$$

Finally we want to see how to express this in terms of the convolution algebra definition of $\mathbb{C}[G]$. So let a group homomorphism $\varphi : G \to \mathbb{C}^\times$ be given and let ϕ be the corresponding algebra homomorphism. Let $f : G \to \mathbb{C}$ be a function of finite support. Then f is a finite sum $f = \sum_{y \in G} f(y)\delta_y$. Linearity implies

$$\phi(f) = \sum_{y \in G} f(y)\varphi(y).$$

This formula will be used later.

Let $p < \infty$ be a prime number. Let (π, V) be an irreducible admissible representation of G_p on a Hilbert space. Assume π to be unramified. The algebra $\mathcal{H}_p^{K_p}$ of all K_p-bi-invariant functions of compact support on G_p acts on the space V^{K_p}. Since this space is one-dimensional, the algebra $\mathcal{H}_p^{K_p}$ acts by an algebra homomorphism $\chi_\pi \in \mathrm{Hom}_{\mathrm{alg}}(\mathcal{H}_p^{K_p}, \mathbb{C})$.

Let $A_p \subset G_p$ be the subgroup of diagonal matrices and $N_p = N_{\mathbb{Q}_p}$ the group of upper triangular matrices with ones on the diagonal. Then $B_p = A_p N_p$ is the group of upper triangular matrices. For $a = \mathrm{diag}(a_1, a_2)$ let $\delta(a) = |a_1|/|a_2|$. The map $\Delta(an) = \delta(a)$ is the modular function of the group B_p, as was shown in Exercise 7.3. Normalize the Haar measure dn on N_p in such a way that $\mathrm{vol}(N_{\mathbb{Z}_p}) = 1$.

To motivate the definition of the Satake transform, we recall the principal series representation (π_λ, V_λ) attached to a quasi-character λ of A_p.

Lemma 8.2.1 *The representation π_λ is unramified if and only if the quasi-character λ is unramified. In this case one has $V_\lambda^{K_p} = \mathbb{C}p_\lambda$, where the element p_λ of V_λ is defined by*

$$p_\lambda(ank) = a^{\lambda + \delta/2},$$

for $a \in A_p, n \in N_p, k \in K_p$.

Proof Assume that π_λ is unramified and let $0 \neq \varphi \in V_\lambda^{K_p}$. Then $\varphi|_{K_p}$ is constant outside a set of measure zero, so after altering φ we can assume it to be constant on K_p. We multiply φ by a scalar so that the constant is one. So for $a \in A_p \cap K_p$ we have

$$1 = \varphi(1) = \varphi(a) = a^{\lambda + \delta/2}\varphi(1) = a^\lambda,$$

which means that λ is unramified. Conversely, let λ be unramified. We can define p_λ by the formula in the lemma. Then $p_\lambda \in V_\lambda$, since for $a \in A_p$, $n \in N_p$ and $x \in G_p$ one has

$$p_\lambda(anx) = p_\lambda(ana_1n_1k) = p_\lambda\big(\underbrace{aa_1}_{\in A_p} \underbrace{n^{a_1}n_1}_{\in N_p} k\big)$$

$$= (aa_1)^{\lambda+\delta/2} = a^{\lambda+\delta/2}p_\lambda(a_1n_1k) = a^{\lambda+\delta/2}p_\lambda(x),$$

where we have written the Iwasawa decomposition of x as $x = a_1n_1k$ and $n^{a_1} = a_1^{-1}na_1$. The function p_λ is, up to scalars, the only element of V_λ, which is constant on K_p, so p_λ spans the space $V_\lambda^{K_p}$.   □

Let λ or equivalently π_λ be unramified. There exists an algebra homomorphism $\chi_\lambda : \mathcal{H}_p^{K_p} \to \mathbb{C}$ such that

$$\pi_\lambda(f)p_\lambda = \chi_\lambda(f)p_\lambda.$$

Since $p_\lambda(1) = 1$, one has $\chi_\lambda(f) = \pi_\lambda(f)p_\lambda(1)$. We compute

$$\chi_\lambda(f) = \pi_\lambda(f)p_\lambda(1) = \int_{G_p} f(x)\pi_\lambda(x)p_\lambda(1)\,dx$$

$$= \int_{G_p} f(x)p_\lambda(x)\,dx = \int_{A_pN_pK_p} f(an)p_\lambda(ank)\,da\,dn\,dk$$

$$= \int_{A_pN_p} f(an)a^{\lambda+\delta/2}\,da\,dn.$$

Definition 8.2.2 We define the *Satake transform* of f by

$$Sf(a) = a^{\delta/2} \int_{N_p} f(an)\,dn.$$

It follows that $\chi_\lambda(f) = \int_{A_p} Sf(a)a^\lambda\,da$. We write $\overline{A} = A_p/A_p \cap K_p \cong (\mathbb{Q}_p^\times/\mathbb{Z}_p^\times)^2 \cong \mathbb{Z}^2$. The normalizer $N(A_p)$ of A_p in G_p is the group of all monomial matrices, i.e. matrices such that in every row and every column there is exactly one non-zero entry. The *Weyl group* W of A_p is defined to be the quotient $W = N(A_p)/A_p$. The latter group is isomorphic to the permutation group Per(2) in two letters and it acts on A_p by permuting the entries. It also acts on $\overline{A} \cong \mathbb{Z}^2$ and on the group algebra $\mathbb{C}[\overline{A}]$. The set of W-invariants,

$$\mathbb{C}[\overline{A}]^W = \big\{\alpha \in \mathbb{C}[\overline{A}] : w\alpha = \alpha \ \forall w \in W\big\}$$

is a subalgebra which is closed under $\alpha \mapsto \alpha^*$, where $\alpha^*(a) = \overline{\alpha(a^{-1})}$ is the canonical involution on the group algebra.

Theorem 8.2.3 *The* Satake transformation $f \mapsto Sf$ *with* $Sf(a) = \delta(a)^{\frac{1}{2}} \times \int_{N_p} f(an)\,dn$ *is an isomorphism of *-algebras,*

$$S : \mathcal{H}_p^{K_p} \xrightarrow{\cong} \mathbb{C}[\overline{A}]^W,$$

i.e. it is an algebra isomorphism with $S(f^*) = S(f)^*$.

Proof We begin by showing that the function Sf has compact support in A. Let Ω be the support of f. Then $\Omega \cap AN$ is a compact subset of AN. By the formula

$$an = \begin{pmatrix} a_1 & \\ & a_2 \end{pmatrix} \begin{pmatrix} 1 & x \\ & 1 \end{pmatrix} = \begin{pmatrix} a_1 & a_1 x \\ & a_2 \end{pmatrix}$$

one sees that the multiplication map $A \times N \to AN$ is a homeomorphism. The support of Sf is a subset of $P(\Omega \cap AN)$, where $P : AN \to A$ is the projection map. As P is continuous, the support of Sf is compact. Later in Lemma 8.2.4 we shall give the support in explicit cases.

Next we show that the Satake transformation is an algebra homomorphism. For this we compute for $f, g \in \mathcal{H}_p^{K_p}$:

$$S(f * g)(a) = \delta(a)^{\frac{1}{2}} \int_{N_p} \int_{G_p} f(y) g(y^{-1} an)\,dy\,dn$$

$$= \delta(a)^{\frac{1}{2}} \int_{N_p} \int_{A_p} \int_{N_p} f(a'n') \underbrace{g(n'^{-1} a'^{-1} an)}_{= g(a'^{-1} an'' n)}\,da'\,dn'\,dn,$$

where $n'' = (a'^{-1} a)^{-1} n'^{-1} (a'^{-1} a) \in N$. As dn is a Haar measure, we can replace $n'' n$ by n and we get

$$S(f * g)(a) = \delta(a)^{\frac{1}{2}} \int_{N_p} \int_{A_p} \int_{N_p} f(a'n') g(a'^{-1} an)\,da'\,dn'\,dn$$

$$= \int_{A_p} \underbrace{\delta(a')^{\frac{1}{2}} \int_{N_p} f(a'n)\,dn}_{= Sf(a')} \underbrace{\delta(a'^{-1}a)^{\frac{1}{2}} \int_{N_p} g(a'^{-1}an)\,dn}_{= Sg(a'^{-1}a)}\,da'$$

$$= (Sf) * (Sg)(a).$$

We show that the image of the Satake transformation lies within the set of Weyl group invariants. For this let $w = \begin{pmatrix} & 1 \\ 1 & \end{pmatrix}$ be the canonical representative of the non-trivial element of the Weyl group. Since $w = w^{-1}$, we have $waw^{-1} = waw$ for every $a \in A_p$. The element w lies in K_p, so by K_p-invariance of $f \in \mathcal{H}_p^{K_p}$ we deduce

$$Sf(waw) = \delta(waw)^{\frac{1}{2}} \int_{N_p} f(wawn)\,dn = \delta(waw)^{\frac{1}{2}} \int_{N_p} f(awnw)\,dn.$$

Writing $a = \left(\begin{smallmatrix} a_1 & \\ & a_2 \end{smallmatrix}\right)$ and $n = \left(\begin{smallmatrix} 1 & x \\ & 1 \end{smallmatrix}\right)$, one gets

$$awnw = \begin{pmatrix} a_1 & \\ & a_2 \end{pmatrix}\begin{pmatrix} 1 & \\ x & 1 \end{pmatrix} = an^t = (na)^t.$$

In Exercise 7.12 it is shown that for every $g \in G_p$ one has $K_p g K_p = K_p g^t K_p$, so it holds that $f((na)^t) = f(na)$. Noting $\delta(waw) = \delta(a)^{-1} = \frac{|a_2|}{|a_1|}$, we arrive at

$$Sf(waw) = \left(\frac{|a_2|}{|a_1|}\right)^{\frac{1}{2}} \int_{\mathbb{Q}_p} f\begin{pmatrix} a_1 & a_2 x \\ & a_2 \end{pmatrix} dx$$

$$= \left(\frac{|a_1|}{|a_2|}\right)^{\frac{1}{2}} \int_{\mathbb{Q}_p} f\begin{pmatrix} a_1 & a_1 x \\ & a_2 \end{pmatrix} dx = Sf(a).$$

Next we prove the *-property. We have

$$S(f^*)(a) = \delta(a)^{\frac{1}{2}} \int_{N_p} \overline{f(n^{-1}a^{-1})}\, dn = \delta(a)^{\frac{1}{2}} \int_{N_p} \overline{f(a^{-1}ana^{-1})}\, dn$$

$$= \delta(a^{-1})^{\frac{1}{2}} \int_{N_p} \overline{f(a^{-1}n)}\, dn = \overline{S(f)(a^{-1})} = S(f)^*(a).$$

For *injectivity*: For $k, l \in \mathbb{Z}$ let $A(k,l) = K_p \left(\begin{smallmatrix} p^k & \\ & p^l \end{smallmatrix}\right) K_p$. By the Elementary Divisor Theorem one has for the principal ideal domain $\mathbb{Z}_p$,

$$G_p = \bigcup_{k \le l} A(k,l) \quad \text{(disjoint union)}.$$

Lemma 8.2.4 *For* $a = \left(\begin{smallmatrix} p^i & \\ & p^j \end{smallmatrix}\right)$, *the set* aN_p *meets a double coset* $A(k,l)$ *if and only if* $k \le i, j$ *and* $l = i + j - k$.

Proof Assume $aN_p \cap K_p\left(\begin{smallmatrix} p^k & \\ & p^l \end{smallmatrix}\right) K_p \neq \emptyset$. After multiplication by p^ν with $\nu \in \mathbb{N}$ we can assume $i, j, k, l \ge 0$. Then all matrices have integral entries and p^k is the first elementary divisor of an, i.e. the greatest common divisor of all entries of the matrix $an = \left(\begin{smallmatrix} p^i & p^i x \\ & p^j \end{smallmatrix}\right)$. Therefore $k \le i, j$. The determinant of an is of the form $p^{k+l} u$, where $u \in \mathbb{Z}_p^\times$. So we have $p^{i+j} = p^{k+l} u$ and we infer that $i + j = k + l$.

For the converse direction assume $k \le i, j$ and $k + l = i + j$. Again we assume all indices are positive. Then the matrix $\left(\begin{smallmatrix} p^i & p^k \\ & p^j \end{smallmatrix}\right)$ lies in aN. It has first elementary divisor p^k and therefore lies in $K_p\left(\begin{smallmatrix} p^k & \\ & p^{l'} \end{smallmatrix}\right) K_p$ for some $l' \ge k$. But as the determinant of $\left(\begin{smallmatrix} p^i & p^k \\ & p^j \end{smallmatrix}\right)$ equals $p^{i+j} = p^{k+l} = p^{k+l'} u$ for some $u \in \mathbb{Z}_p^\times$, it follows that $l' = l$. $\square$

The following picture shows the coordinates (k, l) of the double cosets $A(k, l)$, which have non-empty intersection with the set $\left(\begin{smallmatrix} p^i & \\ & p^j \end{smallmatrix}\right) N$.

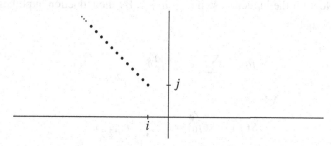

A given $0 \neq f \in \mathcal{H}_p^{K_p}$ has only finitely many double cosets in its support. So if i is very small or j is large, then the set aN has empty intersection with the support of f. Enlarging i and reducing j, one finds an $a = \begin{pmatrix} p^i & \\ & p^j \end{pmatrix}$ such that aN meets the support of f in exactly one double coset $A(k, l)$. For this a one has $Sf(a) \neq 0$. This shows injectivity of the Satake transformation.

Conversely, the following picture shows the coordinates (i, j) of those $a = \begin{pmatrix} p^i & \\ & p^j \end{pmatrix}$, such that aN meets a given double coset $A(k, l)$.

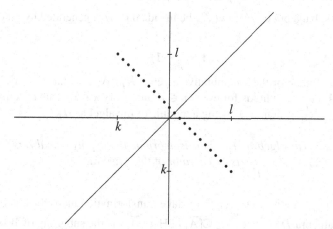

We conclude that the image of the indicator function $f = \mathbf{1}_{A(k,l)}$ under the Satake transformation is of the form

$$Sf = \sum_{v=k}^{l} c_v^{k,l} \delta_{\begin{pmatrix} p^v & \\ & p^{k+l-v} \end{pmatrix}},$$

where $c_v^{k,l} > 0$ are real numbers, satisfying the symmetry condition $c_i^{k,l} = c_{k+l-i}^{k,l}$. We use this to show that the image of the Satake transformation contains the generators

$$e_{i,j} = \delta_{\begin{pmatrix} p^i & \\ & p^j \end{pmatrix}} + \delta_{\begin{pmatrix} p^j & \\ & p^i \end{pmatrix}}, \quad i, j \in \mathbb{Z},$$

of $\mathbb{C}[\bar{A}]^W$. It suffices to consider the case $i \leq j$. We write $j = i + n$ for $n \geq 0$ and prove the claim by induction on n. For $n = 0$ the generator is a multiple of the image of $\mathbf{1}_{A(i,i)}$. Next for the induction step $n \to n+1$. By the induction hypothesis, there is $f_0 \in \mathcal{H}_p^{K_p}$ such that

$$S(f_0) = \sum_{v=i+1}^{i+n} c_v^{i,i+n+1} \delta_{\begin{pmatrix} p^v & \\ & p^{2i+n+1-v} \end{pmatrix}}.$$

Let $f = \mathbf{1}_{A(i,i+n+1)}$. We obtain

$$S(f) - S(f_0) = c_i^{i,i+n+1} e_{i,i+n+1}.$$

We have shown surjectivity of the Satake transformation. $\qquad\square$

We want to understand the Hecke algebra $\mathcal{H}_p$ a bit better. Let $\mathcal{Z}_p$ be the linear subspace

$$\mathcal{Z}_p = \mathrm{Span}\{\mathbf{1}_{p^k K_p} : k \in \mathbb{Z}\}.$$

For $k, l \in \mathbb{Z}$ one computes that

$$\mathbf{1}_{p^k K_p} * \mathbf{1}_{p^l K_p} = \mathbf{1}_{p^{k+l} K_p}.$$

So $\mathcal{Z}_p$ is a subalgebra of $\mathcal{H}_p$. Let $\mathcal{J}_p$ be the ideal of $\mathcal{H}_p$, generated by all elements of the form

$$\mathbf{1}_{p^k K_p} - \mathbf{1}_{K_p}.$$

Let V be a module of the commutative algebra $\mathcal{H}_p$. We say that V is a $\mathcal{Z}_p$-*trivial module* if $\mathbf{1}_{p^k K_p} v = v$ holds for every $v \in V$ and every $k \in \mathbb{Z}$. This is equivalent to $\mathcal{J}_p V = 0$, i.e. equivalent to V being a module of the algebra $\mathcal{H}_p/\mathcal{J}_p$.

Lemma 8.2.5 *The algebra $\mathcal{H}_p/\mathcal{J}_p$ is isomorphic to the polynomial ring $\mathbb{C}[x]$. The element $g_p = \mathbf{1}_{K_p\begin{pmatrix} p^{-1} & \\ & 1 \end{pmatrix} K_p}$ is a generator of this algebra.*

Proof A computation shows that the Satake transformation maps the algebra $\mathcal{Z}_p$ to the group algebra $D = \mathbb{C}[\overline{A}^W] \subset \mathbb{C}[\overline{A}]^W$. Here $\overline{A}^W$ is the subgroup of W-invariant elements, so

$$\overline{A}^W = \mathbb{Q}_p^\times \begin{pmatrix} 1 & \\ & 1 \end{pmatrix} / \mathbb{Z}_p^\times \begin{pmatrix} 1 & \\ & 1 \end{pmatrix}.$$

Further, $S(g_p)$ is a multiple of $\delta_{\begin{pmatrix} p^{-1} & \\ & 1 \end{pmatrix}} + \delta_{\begin{pmatrix} 1 & \\ & p^{-1} \end{pmatrix}}$. The algebra $\mathbb{C}[\overline{A}]^W$ is linearly spanned by the elements

$$a(k,l) = \delta_{\begin{pmatrix} p^k & \\ & p^l \end{pmatrix}} + \delta_{\begin{pmatrix} p^l & \\ & p^k \end{pmatrix}}, \qquad k,l \in \mathbb{Z}.$$

For $m \in \mathbb{Z}$ one has

$$a(k+m, l+m) = a(k,l)\delta_{\begin{pmatrix} p^m & \\ & p^m \end{pmatrix}},$$

and this element lies in $a(k, l) + \mathcal{J}_p$. This means that, modulo the ideal $\mathcal{J}_p$, every element $a(k, l)$ can be brought into the form $a(k, 0)$ with $k \geq 0$. Let $k \geq 1$. Then $a(1, 0)^k = a(k, 0) + c$, where c is a linear combination of elements of $\mathcal{J}_p$ and the elements $a(m, 0)$ with $0 \leq m < k$. This implies that the algebra $\mathbb{C}[g_p]$ generated by g_p maps surjectively onto $\mathcal{H}_p / \mathcal{J}_p$, and that $\mathcal{J}_p \cap \mathbb{C}[g_p] = 0$. $\square$

Let $Z_p = \mathbb{Q}_p^\times \left(\begin{smallmatrix} 1 & \\ & 1 \end{smallmatrix} \right)$ be the center of G_p. A representation (π, V_π) of G_p is called Z_p-trivial if the group Z_p acts trivially on V_π. Then Z_p-trivial G_p-modules will give $\mathcal{Z}_p$-trivial Hecke modules as follows.

Lemma 8.2.6 *Let (π, V_π) be an irreducible unramified representation of G_p. Then π is Z_p-trivial if and only if V_π^K is a $\mathcal{Z}_p$-trivial Hecke module.*

Proof Let π be a Z_p-trivial representation. For $k \in \mathbb{Z}$ one has

$$\pi(1_{p^k K_p}) = \int_{K_p} \pi(p^k x) \, dx = \underbrace{\pi(p^k)}_{=1} \int_{K_p} \pi(x) \, dx = \pi(1_{K_p}),$$

where we have written $\pi(p^k)$ for $\pi(p^k \left(\begin{smallmatrix} 1 & \\ & 1 \end{smallmatrix} \right))$. Conversely, let V_π^K be a $\mathcal{Z}_p$-trivial module and let $v_0 \in V_\pi^K \smallsetminus \{0\}$. Then $\pi(C_c(G_p))v_0$ is dense in V_π. So let $v = \pi(f)v_0$ in this subspace. For $a \in \mathbb{Q}_p^\times$ it follows that

$$\pi(a)v = \pi(a)\pi(f)v_0 = \pi(f)\pi(a)v_0 = \pi(f)\pi(1_{aK_p})v_0 = \pi(f)v_0 = v.$$

The lemma is proven. $\square$

By the Satake isomorphism we have $\mathcal{H}_p^{K_p} \cong \mathbb{C}[\overline{A}]^W$. We want to determine the algebra homomorphisms from $\mathcal{H}_p^{K_p}$ to $\mathbb{C}$. For this we need the following lemma.

Lemma 8.2.7 *Let $\mathcal{A}$ be an algebra over $\mathbb{C}$ and let $W = \{1, w\}$ be a two-element group of automorphisms of $\mathcal{A}$. The set $\mathcal{A}^W$ of W-invariants is a subalgebra of $\mathcal{A}$. The group W acts on the set $\mathrm{Hom}_{\mathrm{alg}}(\mathcal{A}, \mathbb{C})$ of algebra homomorphisms from $\mathcal{A}$ to $\mathbb{C}$ and the restriction yields a bijection*

$$\mathrm{Hom}_{\mathrm{alg}}(\mathcal{A}, \mathbb{C})/W \xrightarrow{\cong} \mathrm{Hom}_{\mathrm{alg}}(\mathcal{A}^W, \mathbb{C}).$$

Proof Since the elements of W are algebra homomorphisms, it follows that $\mathcal{A}^W$ is a subalgebra. Let $\mathcal{A}^-$ be the set of all $a \in \mathcal{A}$ such that $w(a) = -a$. Every $a \in \mathcal{A}$ can be written as

$$a = \frac{1}{2}(a + w(a)) + \frac{1}{2}(a - w(a)).$$

By $w^2 = 1$ we get $a - w(a) \in \mathcal{A}^-$ and there is a direct sum decomposition

$$\mathcal{A} = \mathcal{A}^W \oplus \mathcal{A}^-.$$

Consider the restriction map

$$\mathrm{res} : \mathrm{Hom}_{\mathrm{alg}}(\mathcal{A}, \mathbb{C})/W \to \mathrm{Hom}_{\mathrm{alg}}(\mathcal{A}^W, \mathbb{C}).$$

For injectivity, let $\phi, \psi \in \mathrm{Hom}_{\mathrm{alg}}(\mathcal{A}, \mathbb{C})$ with $\phi|_{\mathcal{A}^W} = \psi|_{\mathcal{A}^W}$. For $a \in \mathcal{A}^-$, the element a^2 lies in $\mathcal{A}^W$. Therefore, $\phi(a)^2 = \phi(a^2) = \psi(a^2) = \psi(a)^2$, so $\phi(a) = \pm\psi(a)$.

First case: $\phi(a) = \psi(a)$ for every $a \in \mathcal{A}^-$. Then one obtains $\phi = \psi$.

Second case: there is $a_0 \in \mathcal{A}^-$ with $\phi(a_0) \neq \psi(a_0)$. Then it follows that $0 \neq \phi(a_0) = -\psi(a_0)$. For any given $b \in \mathcal{A}^-$ one gets $a_0 b \in \mathcal{A}^W$ and so

$$\phi(b) = \frac{\phi(a_0 b)}{\phi(a_0)} = -\frac{\psi(a_0 b)}{\psi(a_0)} = -\psi(b) = \psi(-b) = \psi(w(b)).$$

We infer that $\phi = \psi^w$, which shows the injectivity.

To show surjectivity, let $\phi : \mathcal{A}^W \to \mathbb{C}$ be an algebra homomorphism. We have to show that ϕ extends to a homomorphism ψ from $\mathcal{A}$ to $\mathbb{C}$. We distinguish two cases:

First case: $\phi(b^2) = 0$ for every $b \in \mathcal{A}^-$. In this case it follows that $\phi(bb') = 0$ for all $b, b' \in \mathcal{A}^-$, as one sees from $2bb' = (b + b')^2 - b^2 - b'^2$. We set

$$\psi(a + b) = \phi(a),$$

for $a \in \mathcal{A}^W$ and $b \in \mathcal{A}^-$. We have to show that ψ is multiplicative. So let $a, a' \in \mathcal{A}^W$ and $b, b' \in \mathcal{A}^-$. Then

$$\psi((a + b)(a' + b')) = \psi(aa' + ab' + a'b + bb') = \phi(aa')$$
$$= \phi(a)\phi(a') = \psi(a + b)\psi(a' + b').$$

Second case: there is $b_0 \in \mathcal{A}^-$ with $\phi(b_0^2) \neq 0$. Choose a root $\psi(b_0)$ of $\phi(b_0^2)$ and set

$$\psi(a + b) = \phi(a) + \frac{\phi(bb_0)}{\psi(b_0)}.$$

A computation shows that this map is multiplicative. The lemma is proven. $\qquad\square$

The algebra homomorphisms from $\mathcal{H}_p^{K_p}$ to $\mathbb{C}$ are therefore given by

$$\mathrm{Hom}_{\mathrm{alg}}(\mathcal{H}_p^{K_p}, \mathbb{C}) \cong \mathrm{Hom}_{\mathrm{alg}}(\mathbb{C}[\overline{A}]^W, \mathbb{C})$$
$$\cong \mathrm{Hom}_{\mathrm{alg}}(\mathbb{C}[\overline{A}], \mathbb{C})/W$$
$$\cong \mathrm{Hom}_{\mathrm{grp}}(\overline{A}, \mathbb{C}^\times)/W.$$

For an irreducible unramified representation (π, V_π) the space $V_\pi^{K_p}$ is one-dimensional. Let $v \in V_\pi^{K_p} \smallsetminus \{0\}$. Then there exists an algebra homomorphism $\chi_\pi : \mathcal{H}_p^{K_p} \to \mathbb{C}$ such that

$$\pi(f)v = \chi_\pi(f)v$$

holds for every $f \in \mathcal{H}_p^{K_p}$. Hence to an irreducible unitary unramified representation π we attach a group homomorphism $\lambda = \lambda_\pi : \overline{A} \to \mathbb{C}^\times$, which is unique up to the action of the Weyl group, such that for every $f \in \mathcal{H}_p^{K_p}$ one has

$$\chi_\pi(f) = \int_{A_p} Sf(a)a^\lambda \, da.$$

The homomorphism λ is called the *Satake parameter* of the representation.

Lemma 8.2.8 *Let (π, V_π) be an irreducible unramified unitary representation of G_p.*

(a) *The Satake parameter λ_π is unitary, so $|a^{\lambda_\pi}| = 1$ for every $a \in A_p$.*
(b) *Let $f \in L^1(G_p)$ be K_p-bi-invariant. Then the integral $Sf(a) = a^{\delta/2} \times \int_{N_p} f(an)\, dn$ exists for every $a \in A_p$ and the integral $\chi_\pi(f) = \int_{A_p} Sf(a) a^{\lambda_\pi}\, da$ exists and one has $\pi(f)v = \chi_\pi(f)v$.*

Proof (a) As χ_π is a *-homomorphism, one has $\chi_\pi(f^*) = \overline{\chi_\pi(f)}$ for every $f \in \mathcal{H}_p^{K_p}$. Also, the Satake transform is a *-homomorphism, so that

$$\int_{A_p} \overline{Sf(a^{-1}) a^\lambda}\, da = \overline{\chi_\pi(f)} = \chi_\pi(f^*) = \int_{A_p} \overline{Sf(a^{-1})} a^\lambda\, da$$

$$= \int_{A_p} \overline{Sf(a)} (a^{-1})^\lambda\, da,$$

which implies the claim.

(b) The existence of the integrals is clear by part (a), integrability of f and the Iwasawa integral formula. Let $C_j \subset C_{j+1}$ be a sequence of compact, K_p-bi-invariant subsets on G_p such that $G_p = \bigcup_j C_j$. Then the sequence $s_j(x) = 1_{C_j}(x) f(x) \pi(x) v$ is an approximating sequence for the integral since

$$\int_{G_p} \| s_j(x) - f(x)\pi(x)v \|\, dx = \int_{G_p \smallsetminus C_j} \| f(x)\pi(x)v \|\, dz$$

$$\leq \int_{G_p \smallsetminus C_j} |f(x)|\, dx \|v\|,$$

and the latter tends to zero as $j \to \infty$ since f is integrable. Hence the sequence

$$\int_{G_p} s_j(x)\, dx = \int_{G_p} 1_{C_j}(x) f(x) \pi(x) v\, dx = \pi(1_{C_j} f)v = \chi_\pi(1_{C_j} f)v$$

$$= \int_{A_p} a^{\delta/2} \int_{N_p} 1_{C_j}(an) f(an)\, dn\, a^\lambda\, da\, v$$

converges to $\pi(f)v$ on the one hand, and to $\chi_\pi(f)v$ on the other. $\square$

The group $\overline{A}_p = A_p / A_p \cap K_p$ is isomorphic to $\mathbb{Z}^2$. To give an explicit isomorphism, one needs to choose two generators of the group $\overline{A}$. We choose the generators

$$\begin{pmatrix} p & \\ & 1 \end{pmatrix} \quad \text{and} \quad \begin{pmatrix} 1 & \\ & p \end{pmatrix}.$$

Then

$$\mathrm{Hom}_{\mathrm{alg}}\big(\mathcal{H}_p^{K_p}, \mathbb{C}\big) \cong \mathrm{Hom}_{\mathrm{grp}}\big(\mathbb{Z}^2, \mathbb{C}^\times\big)/W$$
$$\cong \big(\mathbb{C}^\times\big)^2/W$$
$$\cong T/W,$$

where $T \subset \mathrm{GL}_2(\mathbb{C})$ is the group of diagonal matrices and $W \cong \mathbb{Z}/2\mathbb{Z}$ acts by permuting the entries. So to the algebra homomorphism χ_π one attaches an element λ_π of T/W. With this, one defines the *local L-factor* of π as

$$L(\pi) \overset{\mathrm{def}}{=} \det(1 - \lambda_\pi)^{-1}.$$

Note that the thus defined local factor depends on the choice of the isomorphism $\overline{A} \cong \mathbb{Z}^2$. This choice is justified, however, by the next proposition. Note also that we use the letter λ_π for the homomorphism of $\overline{A}$ as well as the corresponding element of T/W. Let $\varpi_1 = \big(\begin{smallmatrix} p & \\ & 1 \end{smallmatrix}\big)$ and $\varpi_2 = \big(\begin{smallmatrix} 1 & \\ & p \end{smallmatrix}\big)$. Then

$$L(\pi)^{-1} = \big(1 - \lambda_\pi(\varpi_1)\big)\big(1 - \lambda_\pi(\varpi_2)\big).$$

We define

$$|\pi| = \max_j\big(\big|\lambda_\pi(\varpi_j)\big|\big).$$

So if π is unitary, we have $|\pi| = 1$. If π is an unramified representation and $s \in \mathbb{C}$, then the map $\pi_s : x \mapsto |x|^s \pi(x)$ also is an unramified representation on the same space as π. One has

$$|\pi_s| = p^{-\mathrm{Re}(s)}|\pi|.$$

We define

$$L(\pi, s) \overset{\mathrm{def}}{=} L(\pi_s).$$

The indicator function $\mathbf{1}_{M_2(\mathbb{Z}_p)}$, which we view as a function on $G_p = \mathrm{GL}_2(\mathbb{Q}_p)$, is not in the Hecke algebra $\mathcal{H}_p^{K_p}$, as it doesn't have compact support in G_p.

Proposition 8.2.9 Let (π, V_π) be an irreducible unramified unitary representation of G_p, and let $P_1 = \pi(\mathbf{1}_{K_p})$ be the orthogonal projection onto the one-dimensional space $V_\pi^{K_p}$ of K_p-fixed vectors. If $\mathrm{Re}(s) > \frac{1}{2}$, then the Bochner integral $\pi(\mathbf{1}_{M_v}(x)|x|^{s+\frac{1}{2}}) = \int_{G_p} \mathbf{1}_{M_v}(x)|x|^{s+\frac{1}{2}}\pi(x)\,dx$ exists in the Banach space of bounded operators on V_π and one has

$$\pi\big(\mathbf{1}_{M_v}(x)|x|^{s+\frac{1}{2}}\big) = L(\pi, s)P_1.$$

Proof We use the Iwasawa integral formula to get

$$\int_{G_p} \mathbf{1}_{M_2(\mathbb{Z}_p)}(x)|x|^{s+\frac{1}{2}}\,dx \le \int_{G_p \cap M_2(\mathbb{Z}_p)} |x|^{\mathrm{Re}(s)+\frac{1}{2}}\,dx$$
$$= \int_{A_p N_p \cap M_2(\mathbb{Z}_p)} |a|^{\mathrm{Re}(s)+\frac{1}{2}}\,da\,dn$$

$$= \int_{\mathbb{Q}_p^\times \cap \mathbb{Z}_p} \int_{\mathbb{Q}_p^\times \cap \mathbb{Z}_p} |a_1|^{-1} \left(|a_1| |a_2| \right)^{\mathrm{Re}(s)+\frac{1}{2}} da_1^\times \, da_2^\times$$

$$= \sum_{k=0}^\infty p^{-(s-\frac{1}{2})k} \sum_{j=0}^\infty p^{-(s+\frac{1}{2})} < \infty.$$

So the Bochner integral exists and by Lemma 8.2.8 we get

$$\pi \left(\mathbf{1}_{M_v}(x) |x|^{s+\frac{1}{2}} \right) = \chi_\pi \left(\mathbf{1}_{M_v}(x) |x|^{s+\frac{1}{2}} \right) P_1.$$

Let $f(x) = \mathbf{1}_{M_2(\mathbb{Z}_p)}(x) |x|^{s+\frac{1}{2}}$ and let $a \in A_p$. We have $Sf(a) = 0$ for $a \notin M_2(\mathbb{Z}_p)$ and for $a \in M_2(\mathbb{Z}_p)$ we compute

$$Sf(a) = \left(\frac{|a_1|}{|a_2|} \right)^{\frac{1}{2}} \int_{\mathbb{Q}_p} f \begin{pmatrix} a_1 & a_1 x \\ & a_2 \end{pmatrix} dx$$

$$= \left(\frac{1}{|a_1||a_2|} \right)^{\frac{1}{2}} \int_{\mathbb{Q}_p} f \begin{pmatrix} a_1 & x \\ & a_2 \end{pmatrix} dx$$

$$= \left(\frac{1}{|a_1||a_2|} \right)^{\frac{1}{2}} \int_{\mathbb{Q}_p} \left(|a_1||a_2| \right)^{s+\frac{1}{2}} \mathbf{1}_{M_2(\mathbb{Z}_p)} \begin{pmatrix} 1 & x \\ & 1 \end{pmatrix} dx = |a|^s.$$

Writing $a^{\lambda_\pi} = |a_1|^{\lambda_1} |a_2|^{\lambda_2}$ for $\lambda_1, \lambda_2 \in \mathbb{C}$ and $a = \mathrm{diag}(a_1, a_2)$, we have $Sf(a) = |a|^s \mathbf{1}_{A \cap M_2(\mathbb{Z}_p)}(a)$ and

$$\chi_\pi(f) = \int_{A_p \cap M_2(\mathbb{Z})} |a|^s a^{\lambda_\pi} \, da = \sum_{k,j=0}^\infty p^{-(s+\lambda_1)k} p^{-(s+\lambda_2)j}$$

$$= \frac{1}{1 - p^{-s-\lambda_1}} \frac{1}{1 - p^{-s-\lambda_2}} = L(\pi, s).$$

The proposition is proven. $\qquad\qquad\qquad\qquad\qquad\qquad\qquad\qquad\qquad\qquad\qquad\square$

At the end of this section we shall determine the local L-function of the trivial representation. For $\pi = 1$ and $f \in \mathcal{H}_p^{K_p}$ one has

$$\chi_\pi(f) = \int_{G_p} f(x) \, dx = \int_{A_p N_p} f(an) \, da \, dn = \int_{A_p} a^{-\delta/2} Sf(a) \, da.$$

This means that $\lambda_\pi \begin{pmatrix} p & \\ & 1 \end{pmatrix} = \begin{pmatrix} p & \\ & 1 \end{pmatrix}^{-\delta/2} = 1/\sqrt{p}$ and $\lambda_\pi \begin{pmatrix} 1 & \\ & p \end{pmatrix} = \sqrt{p}$. So the local factor is

$$L(\pi, s) = \frac{1}{(1 - p^{-s+\frac{1}{2}})(1 - p^{-s-\frac{1}{2}})}.$$

This is the Euler factor of the function $\zeta(s + \frac{1}{2}) \zeta(s - \frac{1}{2})$, where ζ is the Riemann zeta function.

8.3 Global L-Functions

In this section we introduce the global L-functions, show that they extend to meromorphic functions on the plane and prove their functional equation.

Let π be an admissible irreducible unitary representation of $G_{\mathbb{A}}$. Then π is a
tensor product of the form $\pi = \bigotimes_p \pi_p$, where π_p is unramified for almost all p.

Let F be a finite set of places including ∞ and including all places, at which π
is ramified. We define the (partial) global L-function of π as

$$L^F(\pi, s) = \prod_{p \notin F} L(\pi_p, s), \quad s \in \mathbb{C}.$$

We shall show convergence of this product for $\mathrm{Re}(s) > \frac{3}{2}$ later.

As an example, consider the trivial representation $\pi = \mathrm{triv}$. In this case we can
choose $F = \{\infty\}$ and in the last section we have seen that

$$L^{\{\infty\}}(\mathrm{triv}, s) = \zeta\left(s + \frac{1}{2}\right)\zeta\left(s - \frac{1}{2}\right).$$

Note that $L^F(\pi, s)$ depends only on those π_p with $p \notin F$, so it depends only on
the representation $\bigotimes_{p \notin F} \pi_p$ of $G_{\mathbb{A}^F}$.

From now on let (π, V_π) be a fixed cuspidal representation. Then π is admissible
and therefore a tensor product $\pi = \bigotimes_p \pi_p$ of local representations. We choose a set
of places F as above and an isometric $G_{\mathbb{A}}$-homomorphism

$$\eta : V_\pi \hookrightarrow L^2(G_{\mathbb{Q}} Z_{\mathbb{R}} \backslash G_{\mathbb{A}}).$$

Further we fix a vector $v = \bigotimes_p v_p \in V_\pi$ such that $v_p \in V_{\pi_p}^{K_p}$ if π_p is unramified. Let
$\varphi = \eta(v)$. By changing v if necessary, we assume that φ lies in the image of $R(h)$
for some $h \in C_c^\infty(G_{\mathbb{A}})$. Then the function φ is smooth and by Proposition 7.4.3 it is
rapidly decreasing, so in particular the function φ is bounded.

We can also assume $\varphi(1) = 1$. The latter can be achieved by replacing $\varphi(x)$ by
$c\varphi(xy)$ for suitable $y \in G_{\mathbb{A}}$ and $c \in \mathbb{C}$. Let $\mathbb{A}_F$ be the product of the fields $\mathbb{Q}_p$ with
$p \in F$ and $\mathbb{A}^F$ the restricted product of all $\mathbb{Q}_p$ with $p \notin F$. We consider the global
zeta integral,

$$\zeta(f, \varphi, s) = \int_{G_{\mathbb{A}}} f(x)\varphi(x)|x|^{s + \frac{1}{2}} \, dx,$$

where $f \in \mathcal{S}(M_2(\mathbb{A}))$. For the finite set of places F we also need the local zeta
integral,

$$\zeta_F(f, \varphi, s) = \int_{G_F} f(x)\varphi(x)|x|^{s + \frac{1}{2}} \, dx,$$

where we embed $G_F = \prod_{p \in F} G_p$ in $G_{\mathbb{A}}$ by sending x to $(x, 1)$, i.e. the coordinates
outside F are set to 1.

For any ring R, let $Q_2(R)$ be the set of all $x \in M_2(R)$ with $\det(x) = 0$. Let
$S_0 = S_0(M_2(\mathbb{A}))$ be the set of all $f \in \mathcal{S}(M_2(\mathbb{A}))$, such that f and its Fourier transform $\hat{f}$ both vanish on the singular set $Q_2(\mathbb{A})$, so $f(Q_2(\mathbb{A})) = 0 = \hat{f}(Q_2(\mathbb{A}))$.

Examples of such functions are easily constructed. For instance, if $f = \prod_p f_p$, then one needs only two places p, q such that supp $f_p \subset G_p$ and supp $\hat{f}_q \subset G_q$, then one has $f \in S_0$. Note that the set S_0 is stable under the Fourier transformation.

A function $f \in S(M_2(\mathbb{A}))$ is called F-simple if $f = \prod_p f_p$ with

$$p \notin F \quad \Rightarrow \quad f_p = \mathbf{1}_{M_2(\mathbb{Z}_p)}.$$

Theorem 8.3.1 *Let π be a cuspidal representation. Then the L-series $L^F(\pi, s)$ converges for $\mathrm{Re}(s) > \frac{3}{2}$ and the ensuing L-function extends to a meromorphic function on the entire plane.*

(a) *The global zeta integral for $f \in S(M_2(\mathbb{A}))$ converges locally uniformly for $\mathrm{Re}(s) > \frac{3}{2}$ and defines a holomorphic function in that region. If $f \in S_0$, then the zeta integral extends to an entire function which satisfies the functional equation*

$$\zeta(f, \varphi, s) = \zeta(\hat{f}, \varphi^\vee, 1 - s),$$

where $\varphi^\vee(x) = \varphi(x^{-1})$.

(b) *If f is an F-simple function, then for $\mathrm{Re}(s) > \frac{3}{2}$ one has*

$$\zeta(f, \varphi, s) = L^F(\pi, s)\zeta_F(f, \varphi, s).$$

(c) *If F contains at least two primes and at least one at which π is unramified, then there exists a function $f \in S_0$ which is F-simple, such that the local zeta integrals $\zeta_F(f, \varphi, s)$ and $\zeta_F(\hat{f}, \varphi^\vee, s)$ are meromorphic and one has the functional equation*

$$L^F(\pi, s) = \frac{\zeta_F(\hat{f}, \varphi^\vee, 1 - s)}{\zeta_F(f, \varphi, s)} L^F(\pi, 1 - s).$$

The proof will essentially occupy the rest of the section.

We first show locally uniform convergence of the global zeta integral for $\mathrm{Re}(s) > \frac{3}{2}$.

The function φ is bounded, so we can estimate the function $|\varphi(x)f(x)|$ by a constant times $\mathbf{1}_{qM_2(\hat{\mathbb{Z}})}(x)(1 + \|x_\infty\|^N)^{-1}$, where $q \in \mathbb{Q}$ and $N \in \mathbb{N}$ can be chosen arbitrarily large.

The norm $\|x\|$ is the Euclidean norm on $\mathbb{R}^4 \supset G_\mathbb{R}$, so

$$\left\| \begin{pmatrix} a & b \\ c & d \end{pmatrix} \right\| = \sqrt{a^2 + b^2 + c^2 + d^2}.$$

It emerges that we have to show the convergence of the integrals

$$\int_{G_{\mathrm{fin}} \cap qM_2(\hat{\mathbb{Z}})} |x|^{\mathrm{Re}(s) + \frac{1}{2}} dx \quad \text{and} \quad \int_{G_\mathbb{R}} \frac{|x|^{\mathrm{Re}(s) + \frac{1}{2}}}{1 + \|x\|^N} dx.$$

Using Example 3.1.9 we can compute the second integral. It equals

$$\int_{\mathbb{R}}\int_{\mathbb{R}}\int_{\mathbb{R}}\int_{\mathbb{R}} \frac{|xw - yz|^{\mathrm{Re}(s)-\frac{3}{2}}}{1 + (x^2 + y^2 + z^2 + w^2)^{N/2}}\, dx\, dy\, dz\, dw.$$

By the Cauchy–Schwarz inequality we get $|xw + yz| \le (x^2 + y^2)^{\frac{1}{2}}(w^2 + z^2)^{\frac{1}{2}} \le x^2 + y^2 + z^2 + w^2$, and the integral therefore has the majorant

$$\int_{\mathbb{R}}\int_{\mathbb{R}}\int_{\mathbb{R}}\int_{\mathbb{R}} \frac{(x^2 + y^2 + z^2 + w^2)^{2\mathrm{Re}(s)-3}}{1 + (x^2 + y^2 + z^2 + w^2)^{N/2}}\, dx\, dy\, dz\, dw.$$

After a change of variable to polar coordinates (see [Rud87]), this is a constant times $\int_0^\infty \frac{r^{2\mathrm{Re}(s)}}{1+r^N}\, dr$, which converges locally uniformly for $-\frac{1}{2} < \mathrm{Re}(s) < \frac{N-1}{2}$. Since N can be chosen arbitrarily large, the claimed convergence follows for the second of the above integrals.

For the first we change variables to $y = qx$ and we get a constant times the integral

$$\int_{G_{\mathrm{fin}}\cap M_2(\widehat{\mathbb{Z}})} |x|^{s+\frac{1}{2}}\, dx.$$

The set $G_{\mathrm{fin}} \cap M_2(\widehat{\mathbb{Z}})$ is the disjoint union of the sets D_n, where $n \in \mathbb{N}_0$ and

$$D_n = \left\{x \in M_2(\widehat{\mathbb{Z}}) : |x| = n^{-1}\right\}.$$

By Lemma 2.5.1 one has $|D_n/G_{\widehat{\mathbb{Z}}}| = \sum_{d|n} d$. So we can compute, formally at first,

$$\int_{G_{\mathrm{fin}}\cap M_2(\widehat{\mathbb{Z}})} |x|^{s+\frac{1}{2}}\, dx = \sum_{n=1}^{\infty}\sum_{d|n} dn^{-s-\frac{1}{2}} = \sum_{n=1}^{\infty}\sum_{ad=n} d^{-s+\frac{1}{2}} a^{-s-\frac{1}{2}}$$

$$= \sum_{a=1}^{\infty}\sum_{d=1}^{\infty} a^{-s-\frac{1}{2}} d^{-s+\frac{1}{2}} = \zeta\!\left(s + \frac{1}{2}\right)\zeta\!\left(s - \frac{1}{2}\right).$$

Therefore, the integral converges locally uniformly for $\mathrm{Re}(s) > \frac{3}{2}$ and the convergence assertion of the theorem is proven.

Lemma 8.3.2 *Let (π, V_π) be a unitary representation of $G_{\mathbb{A}}/Z_{\mathbb{R}}$. For $v \in V_\pi$ and $f \in S(M_2(\mathbb{A}))$, the Bochner integral*

$$\int_{G_{\mathbb{A}}} f(x)|x|^{s+1/2}\pi(x)v\, dx$$

converges for every complex s with $\mathrm{Re}(s) > \frac{3}{2}$.

Proof According to Theorem 7.3.7 we need to know

$$\infty > \int_{G_{\mathbb{A}}} \|f(x)|x|^{s+1/2}\pi(x)v\|\, dx = \int_{G_{\mathbb{A}}} |f(x)||x|^{\mathrm{Re}(s)+1/2}\, dx \|v\|.$$

This, however, has just been shown. $\square$

Now let $p \leq \infty$ be an arbitrary place. If $z \in Z_p$ and if π_p is an irreducible unitary representation of G_p, then $\pi_p(z)$ commutes with all $\pi_p(g)$, $g \in G_p$. By the lemma of Schur it follows that $\pi(z) \in \mathbb{C}$ Id. So there exists a character $\omega_\pi : Z_p \to \mathbb{T}$, called the *central character*, of π, such that $\pi(z) = \omega_\pi(z)$ Id holds for every $z \in Z_p$.

For $f \in S(M_2(\mathbb{A}))$ define $E(f)$, $\hat{E}(f) : G_\mathbb{A} \to \mathbb{C}$ by

$$E(f)(x) = |x| \sum_{\gamma \in G_\mathbb{Q}} f(\gamma x)$$

and

$$\hat{E}(f)(x) = |x| \sum_{\gamma \in G_\mathbb{Q}} f(x\gamma).$$

Proposition 8.3.3 *For every $f \in S(M_2(\mathbb{A}))$ the sum $E(f)(x)$ converges locally uniformly in $x \in G_\mathbb{A}$ to a continuous function. For every $\alpha \in \mathbb{R}$ with $\alpha > 1$ there exists $C(\alpha) > 0$ such that for every x one has*

$$\left| E(f)(x) \right| \leq C(\alpha)|x|^{-\alpha}.$$

Proof The set $M_2(\mathbb{Q})$ is a lattice in $M_2(\mathbb{A})$ and so is $M_2(\mathbb{Q})x$ for $x \in G_\mathbb{A}$. We can assume f to be of the form $f = f_{\mathrm{fin}} f_\infty$ with $f_{\mathrm{fin}} \in S(M_2(\mathbb{A}_{\mathrm{fin}}))$ and $f_\infty \in S(M_2(\mathbb{R}))$.

The function f_{fin} has compact support, which is contained in a set of the form $\frac{1}{m} M_2(\hat{\mathbb{Z}})$ for some $m \in \mathbb{N}$. Moving scalar factors from f_{fin} to f_∞, we can assume $|f_{\mathrm{fin}}| \leq 1$ and therefore

$$\left| E(f)(x) \right| \leq |x| \sum_{\gamma \in G_\mathbb{Q}} \left| f(\gamma x) \right|$$

$$\leq |x| \sum_{\gamma \in G_\mathbb{Q} \cap \frac{1}{m} M_2(\hat{\mathbb{Z}})x_{\mathrm{fin}}^{-1}} \left| f_\infty(\gamma x_\infty) \right|.$$

By enlarging m, we can assume $x_{\mathrm{fin}} = 1$. We have

$$G_\mathbb{Q} \cap \frac{1}{m} M_2(\hat{\mathbb{Z}}) \subset M_2(\mathbb{Q}) \cap \frac{1}{m} M_2(\hat{\mathbb{Z}}) = \frac{1}{m} M_2(\mathbb{Z}).$$

Consider the lattice $\Lambda = \frac{1}{m} M_2(\mathbb{Z})$ in $M_2(\mathbb{R})$. Since the function f_∞ is rapidly decreasing, the sum $\sum_{\gamma \in \Lambda} |f_\infty(\gamma x_\infty)|$ converges locally uniformly in the variable x_∞.

We show the growth estimate. For this we use

$$\left| E(f)(x) \right| \leq |x| \sum_{\gamma \in G_\mathbb{Q} \cap \Lambda} \left| f_\infty(\gamma x_\infty) \right|.$$

Let $\|g\| = \sqrt{\mathrm{tr}(g^t g)}$ be the Euclidean norm on $M_2(\mathbb{R})$. For every $A > 0$ there exists a $C'_A > 0$ such that $|f_\infty(x_\infty)| \leq C'_A(1 + \|x_\infty\|^A)^{-1}$. There is a unique decomposition $x_\infty = yz$, where $z \in Z_\mathbb{R}$ and $|y| = 1$. One has $\|\gamma x_\infty\| = \|\gamma yz\| =$

$\|\gamma y\|\|z\|^{\frac{1}{2}} = \|\gamma y\|\|x_\infty\|^{\frac{1}{2}}$. Further one has $|x| = |x_{\mathrm{fin}}x_\infty| = |x_{\mathrm{fin}}||x_\infty|$. Since x_{fin} stays in a fixed compact set, the value $|x_{\mathrm{fin}}|$ is bounded by a fixed constant. So there exists $C(\frac{A}{2} - 1) > 0$ such that

$$|E(f)(x)| \le C\left(\frac{A}{2} - 1\right)|x|^{1-\frac{A}{2}} \sum_{\gamma \in G_\mathbb{Q} \cap \Lambda} \frac{1}{1 + \|\gamma y\|^A}.$$

The proposition now follows from the next lemma.

Lemma 8.3.4 *For $A > 4$ the map $\mathrm{SL}_2^{\pm}(\mathbb{R}) \to \mathbb{R}$, given by*

$$y \mapsto \sum_{\gamma \in G_\mathbb{Q} \cap \Lambda} \frac{1}{1 + \|\gamma y\|^A}$$

is bounded.

Proof Replacing y with $y\left(\begin{smallmatrix} -1 \\ & 1 \end{smallmatrix}\right)$ if necessary, we may assume $y \in \mathrm{SL}_2(\mathbb{R})$. Using Iwasawa decomposition, it suffices to consider $y = na$ for $n = \left(\begin{smallmatrix} 1 & x \\ & 1 \end{smallmatrix}\right)$ and a a diagonal matrix with positive entries. We can reduce y modulo the matrices in $\mathrm{SL}_2(\mathbb{Z})$ from the left and $\mathrm{SO}(2)$ from the right, so we can assume y to be in the fundamental domain D for the modular group. So in particular $|x| \le \frac{1}{2}$ and $a = \mathrm{diag}(e^t, e^{-t})$ with $t \ge \frac{1}{2}\log(\sqrt{3}/2)$.

Write $\gamma \in \Lambda \cap G_\mathbb{Q}$ as $\gamma = \frac{1}{m}\gamma'$ with an integral matrix γ'. Then

$$\frac{1}{1 + \|\gamma y\|^A} = \frac{1}{1 + \frac{1}{m^A}\|\gamma' y\|^A} = \frac{m^A}{m^A + \|\gamma' y\|^A} \le \frac{m^A}{1 + \|\gamma' y\|^A},$$

and therefore we can assume that $m = 1$. Then $\Lambda = M_2(\mathbb{Z})$ and

$$F = \left\{\begin{pmatrix} a & b \\ c & d \end{pmatrix} : 0 \le a, b, c, d \le 1\right\}$$

is a fundamental mesh for Λ. Let Λ' be the subset of all $\lambda \in \Lambda$ with all four entries non-zero. For $\lambda \in \Lambda'$ let F_λ the unique translate of F such that $\lambda \in F_\lambda$ and $|x_j| \le |\lambda_j|$ holds for every $x \in F_\lambda$ and all $j = 1, \ldots, 4$. Then $\|x\| \le \|\lambda\|$ as well and so

$$\sum_{\lambda \in \Lambda'} \frac{1}{1 + \|\lambda\|^A} \le \int_{M_2(\mathbb{R})} \frac{1}{1 + \|x\|^A}\, dx < \infty.$$

The same argument applies to the lattice Λy as long as this lattice remains rectangular, for instance, if y is a diagonal matrix $y = \mathrm{diag}(e^t, e^{-t})$. Since y has determinant 1, the map $x \mapsto xy$ on $M_2(\mathbb{R})$ has functional determinant 1 as well, so the change of variables $xy \mapsto x$ yields

$$\sum_{\lambda \in \Lambda'} \frac{1}{1 + \|\lambda y\|^A} \le \int_{M_2(\mathbb{R})} \frac{1}{1 + \|xy\|^A}\, dx = \int_{M_2(\mathbb{R})} \frac{1}{1 + \|x\|^A}\, dx < \infty.$$

Next any $y \in D$ can be written as

$$y = na = \begin{pmatrix} 1 & x \\ & 1 \end{pmatrix}\begin{pmatrix} e^t & \\ & e^{-t} \end{pmatrix} = \begin{pmatrix} e^t & \\ & e^{-t} \end{pmatrix}\underbrace{\begin{pmatrix} 1 & xe^{-2t} \\ & 1 \end{pmatrix}}_{=z},$$

where $|x| \leq \frac{1}{2}$ and $t \geq t_0$ for some t_0. This means that the matrix z stays in a compact subset of $SL_2(\mathbb{R})$. So there is a constant $D > 0$ with $\|z^{-1}\|^A \leq D$ for all z. Therefore, for $\lambda \in \Lambda$ we have

$$\frac{1}{1+\|\lambda y\|^A} = \frac{1}{1+\|\lambda a z\|^A} \leq \frac{1}{\frac{1}{\|z^{-1}\|^A} + \|\lambda a z\|^A}$$

$$= \frac{\|z^{-1}\|^A}{1+\|\lambda a z\|^A\|z^{-1}\|^A} \leq \frac{\|z^{-1}\|^A}{1+\|\lambda a\|^A} \leq \frac{D}{1+\|\lambda a\|^A}.$$

This takes care of Λ'. Next consider the set $\Omega \subset G_{\mathbb{Q}} \cap \Lambda$ of matrices with at least one entry equal to zero. For $\sigma \in SL_2(\mathbb{Z})$ let $\Omega(\sigma)$ be the set of all $\omega \in \Omega$ with $\sigma\omega \notin \Omega$. It is easy to see that there are finitely many matrices $\sigma_1, \ldots, \sigma_n \in SL_2(\mathbb{Z})$ such that $\Omega = \bigcup_{j=1}^n \Omega(\sigma_j)$. Therefore, it suffices to show that for fixed $\sigma \in SL_2(\mathbb{Z})$, the sum $\sum_{\gamma \in \Omega(\sigma)} \frac{1}{1+\|\gamma y\|^A}$ is bounded for $y \in SL_2(\mathbb{R})$. By the above, the sum $\sum_{\gamma \in \Omega(\sigma)} \frac{1}{1+\|\sigma\gamma y\|^A}$ is bounded. The inequality $\|\sigma\gamma y\| \leq \|\sigma\|\|\gamma y\|$ and the fact that $\|\sigma\| \geq 1$, imply $\frac{1}{1+\|\sigma\gamma y\|^A} \geq \frac{1/\|\sigma\|^A}{1+\|\gamma y\|^A}$, and the lemma follows.   □

Proposition 8.3.5 *Let $f \in S_0(M_2(\mathbb{A}))$. For every $N \in \mathbb{N}$ with $N \geq 2$ there exists $C(N) > 0$, such that for every $x \in G_{\mathbb{A}}$ one has the growth estimate*

$$\left|E(f)(x)\right| \leq C(N)\min\left(|x|^N, |x|^{-N}\right),$$

and the functional equation

$$E(f)(x) = \hat{E}(\hat{f})(x^{-1}).$$

Proof Using Proposition 8.3.3, the growth estimate follows from the functional equation. To show the functional equation, note that for $\gamma \in M_2(\mathbb{Q}) \smallsetminus G_{\mathbb{Q}}$ and $x \in M_2(\mathbb{A})$ one has $\det(\gamma x) = \det(\gamma)\det(x) = 0$. Therefore, for $f \in S_0$ one has

$$E(f)(x) = |x| \sum_{\gamma \in M_2(\mathbb{Q})} f(\gamma x),$$

where the sum now runs over $M_2(\mathbb{Q})$ instead of $G_{\mathbb{Q}}$. The functional equation therefore follows from the next lemma.

Lemma 8.3.6 *For $x \in G_{\mathbb{A}}$ and $f \in S$ one has*

$$|x|^2 \sum_{\gamma \in M_2(\mathbb{Q})} f(\gamma x) = \sum_{\gamma \in M_2(\mathbb{Q})} \hat{f}(x^{-1}\gamma).$$

Proof The Poisson Summation Formula in Theorem 5.4.12 asserts

$$\sum_{\gamma \in M_2(\mathbb{Q})} f(\gamma) = \sum_{\gamma \in M_2(\mathbb{Q})} \hat{f}(\gamma).$$

For $x \in G_{\mathbb{A}}$ let $f_x(y) = f(yx)$. Then

$$\widehat{f_x}(y) = \int_{M_2(\mathbb{A})} f_x(z) e(-zy)\,dz = \int_{M_2(\mathbb{A})} f(zx) e(-zy)\,dz$$

$$= |x|^{-2} \int_{M_2(\mathbb{A})} f(z) e\left(-zx^{-1}y\right) dz = |x|^{-2} \hat{f}\left(x^{-1}y\right).$$

The lemma and the proposition are proven.   □

Proposition 8.3.7 *Let* $f \in \mathcal{S}(M_2(\mathbb{A}))$ *be of the form* $\prod_p f_p$. *Assume that for* $p \notin F$ *the local factor is* $f_p = \mathbf{1}_{M_2(\mathbb{Z}_p)}$. *For* $\mathrm{Re}(s) > \frac{3}{2}$ *the identity*

$$\int_{G_{\mathbb{Q}} \backslash G_{\mathbb{A}}} E(f)(x)\varphi(x)|x|^{s-\frac{1}{2}}dx = \zeta(f, \varphi, s)$$

$$= L^F(\pi, s) \int_{G_F} f_F(x)\varphi(x)|x|^{s+\frac{1}{2}}dx$$

holds, where $f_F = \prod_{p \in F} f_p$. *If* $f \in \mathcal{S}_0$, *then, by Proposition 8.3.5, the integral on the left-hand side converges uniformly in* $s \in \mathbb{C}$ *and thus defines an entire function in* s.

Proof With the usual unfolding trick (compare Proposition 2.7.10) we compute for $\mathrm{Re}(s) > \frac{3}{2}$

$$\int_{G_{\mathbb{Q}} \backslash G_{\mathbb{A}}} E(f)(x)\varphi(x)|x|^{s-\frac{1}{2}}\,dx = \int_{G_{\mathbb{A}}} f(x)|x|^{s+\frac{1}{2}}\varphi(x)\,dx = \zeta(f, \varphi, s).$$

Let (ψ_j) be a Dirac sequence on $G_{\mathbb{A}}$. The following interchange of limit and integration is justified by means of the dominated convergence theorem. Note that the Bochner integral $\int_{G_{\mathbb{A}}} f(x)|x|^{s+\frac{1}{2}}\pi(x)v\,dx$ exists by Lemma 8.3.2 and that

$$y \mapsto \eta\left(\int_{G_{\mathbb{A}}} f(x)|x|^{s+\frac{1}{2}}\pi(x)v\,dx\right)(y) = \int_{G_{\mathbb{A}}} f(x)|x|^{s+\frac{1}{2}}\varphi(yx)\,dx$$

is a continuous function. We compute

$$\int_{G_{\mathbb{A}}} f(x)|x|^{s+\frac{1}{2}}\varphi(x)\,dx = \int_{G_{\mathbb{A}}} \lim_j \int_{G_{\mathbb{A}}} \psi_j(y) f(y^{-1}x)|y^{-1}|^{s+\frac{1}{2}}\,dy|x|^{s+\frac{1}{2}}\varphi(x)\,dx$$

$$= \lim_j \int_{G_{\mathbb{A}}} \int_{G_{\mathbb{A}}} \psi_j(y) f(y^{-1}x)|y^{-1}|^{s+\frac{1}{2}}\,dy|x|^{s+\frac{1}{2}}\varphi(x)\,dx$$

$$= \lim_j \int_{G_{\mathbb{A}}} \int_{G_{\mathbb{A}}} f(x)|x|^{s+\frac{1}{2}}\psi_j(y)\varphi(yx)\,dx\,dy$$

$$= \lim_j \left\langle \left(\int_{G_{\mathbb{A}}} f(x)|x|^{s+\frac{1}{2}} R(x)\varphi \, dx \right), \psi_j \right\rangle$$

$$= \lim_j \left\langle \eta \left(\int_{G_{\mathbb{A}}} f(x)|x|^{s+\frac{1}{2}} \pi(x)v \, dx \right), \psi_j \right\rangle$$

$$= \eta \left(\int_{G_{\mathbb{A}}} f(x)|x|^{s+\frac{1}{2}} \pi(x)v \, dx \right)(1)$$

$$= \eta \left(\bigotimes_p \int_{G_p} f_p(x)|x|^{s+\frac{1}{2}} \pi_p(x)v_p \, dx \right)(1).$$

Linearity of η and Proposition 8.2.9 imply

$$\int_{G_{\mathbb{Q}}\backslash G_{\mathbb{A}}} E(f)(x)\varphi(x)|x|^{s-\frac{1}{2}} dx$$

$$= L^F(\pi, s) \, \eta \left(\bigotimes_{p \notin F} v_p \otimes \bigotimes_{p \in F} \int_{G_p} f_p(x)|x|^{s+\frac{1}{2}} \pi_p(x)v_p \, dx \right)(1)$$

$$= L^F(\pi, s) \int_{G_F} f_F(x)\varphi(x)|x|^{s+\frac{1}{2}} dx. \qquad \square$$

We now prove the theorem. Let $f \in \mathcal{S}_0$. The functional equation $E(f)(x) = \hat{E}(\hat{f})(x^{-1})$ yields

$$\zeta(f, \varphi, s) = \int_{G_{\mathbb{Q}}\backslash G_{\mathbb{A}}} E(f)(x)\varphi(x)|x|^{s-\frac{1}{2}} dx = \int_{G_{\mathbb{Q}}\backslash G_{\mathbb{A}}} \hat{E}(\hat{f})(x^{-1})\varphi(x)|x|^{s-\frac{1}{2}} dx$$

$$= \int_{G_{\mathbb{A}}/G_{\mathbb{Q}}} \hat{E}(\hat{f})(x)\varphi(x^{-1})|x|^{-s+\frac{1}{2}} dx = \int_{G_{\mathbb{A}}} \hat{f}(x)|x|^{\frac{3}{2}-s}\varphi(x^{-1}) dx.$$

The map $\varphi \mapsto \varphi^{\vee}$ with $\varphi^{\vee}(x) = \varphi(x^{-1})$ is a unitary map from $L^2(G_{\mathbb{Q}}Z\backslash G_{\mathbb{A}})$ to $L^2(G_{\mathbb{A}}/G_{\mathbb{Q}}Z)$, which is $G_{\mathbb{A}}$-equivariant, when we consider the $G_{\mathbb{A}}$ action on $L^2(G_{\mathbb{A}}/G_{\mathbb{Q}}Z)$ given by left translation $L(y)\varphi(x) = \varphi(y^{-1}x)$. This is seen by the following computation,

$$(R(y)\varphi)^{\vee}(x) = R(y)\varphi(x^{-1}) = \varphi(x^{-1}y) = \varphi((y^{-1}x)^{-1})$$

$$= \varphi^{\vee}(y^{-1}x) = L(y)(\varphi^{\vee})(x).$$

This means that $\varphi^{\vee}$ takes the role of φ, if we use left translations instead of right translations. One gets the functional equation,

$$\zeta(f, \varphi, s) = \zeta(\hat{f}, \varphi^{\vee}, 1-s).$$

Part (a) of the theorem is proven. By Proposition 8.3.7, we also have part (b).

Finally for part (c). Let p and q be different prime numbers in F. Choose a Schwartz–Bruhat function f_q with support in G_q and a smooth function f_p of compact support in $M_2(\mathbb{Q}_p)$ such that its Fourier transform $\hat{f}_p$ has support inside G_p. Further choose $f_{\infty} \in C_c^{\infty}(G_{\mathbb{R}})$. Then $f = \prod_p f_p$ is an F-simple function in $\mathcal{S}_0$. The functional equation follows from parts (a) and (b) if we can show that $L^F(\pi, s)$ has a meromorphic continuation. This assertion does not depend on the set of places F, for if F' is another set, then the functions L^F and $L^{F'}$ only differ by finitely many Euler factors of the form

$$\frac{1}{(1 - ap^{-s})(1 - bp^{-s})},$$

which are meromorphic.

It suffices to show that f can be chosen such that $\zeta_F(f, \varphi, s)$ is non-zero and extends meromorphically to $\mathbb{C}$. To do that, we choose F to contain at least two prime numbers, and to contain at least one prime number at which π is unramified. If p is a prime outside F, set $f_p = \mathbf{1}_{M_2(\mathbb{Z}_p)}$. Fix a prime $q \in F$, at which π is unramified. For every prime $p \neq q$ in F we choose f_p in $C_c^\infty(G_p)$. Let f_q be the Fourier transform of the function

$$q \mathbf{1}_{GL_2(\mathbb{Z}_q)} = q \sum_{g \in GL_2(\mathbb{Z}/q\mathbb{Z})} \mathbf{1}_{g + q M_2(\mathbb{Z}_q)}.$$

Note that this is bi-invariant under $K_q = GL_2(\mathbb{Z}_q)$. One computes

$$f_q(x) = \sum_{g \in GL_2(\mathbb{Z}/q\mathbb{Z})} e_q(xg) \mathbf{1}_{q^{-1} M_2(\mathbb{Z}_q)}(x).$$

Setting $f = \prod_p f_p$ we get

$$\zeta_F(f, \varphi, s) = \int_{G_F} f(x) \varphi(x) |x|^{s + \frac{1}{2}} \, dx$$

$$= \eta \bigg(\bigotimes_{p \neq q} \int_{G_p} |x|_p^{s + \frac{1}{2}} f_p(x) R(x) v_p \, dx$$

$$\otimes \int_{G_p} |x|_q^{s + \frac{1}{2}} f_q(x) R(x) v_q \, dx \bigg) (1).$$

Since the function $|.|^{s + \frac{1}{2}} f_q$ is K_q-invariant, we have $\int_{G_q} |x|_q^{s + \frac{1}{2}} f_q(x) R(x) v_q = \chi_{\pi_q}(|.|^{s + \frac{1}{2}} f_q) v_q$ and

$$\chi_{\pi_q}\big(|.|^{s + \frac{1}{2}} f_q\big) = \int_{A_q} S(f_q)(a) |a|^{s + \frac{1}{2}} a^{\lambda_\pi} \, da$$

if the integral converges absolutely, which we show below for $\mathrm{Re}(s) \gg 0$. We want to show that this is a non-zero rational function in q^{-s}. For $a = \begin{pmatrix} a_1 & \\ & a_2 \end{pmatrix} \in A_q$ we compute $Sf_q(a)$ as

$$Sf_q(a) = a^{\delta/2} \int_N f_q(an) \, dn$$

$$= \sum_{g \in GL_2(\mathbb{Z}/q\mathbb{Z})} \left(\frac{|a_1|}{|a_2|} \right)^{\frac{1}{2}} \int_{\mathbb{Q}_q} e_q\big(\mathrm{tr}\big(\begin{smallmatrix} a_1 & a_1 x \\ & a_2 \end{smallmatrix}\big)g\big) \mathbf{1}_{q^{-1} M_2(\mathbb{Z}_q)}\big(\begin{smallmatrix} a_1 & a_1 x \\ & a_2 \end{smallmatrix}\big) \, dx$$

$$= \sum_{g\in GL_2(\mathbb{Z}/q\mathbb{Z})} \left(\frac{1}{|a_1||a_2|}\right)^{\frac{1}{2}} \int_{\mathbb{Q}_q} e_q\big(\mathrm{tr}\big(\begin{smallmatrix} a_1 & x \\ & a_2 \end{smallmatrix}\big)g\big)\mathbf{1}_{q^{-1}M_2(\mathbb{Z}_q)}\big(\begin{smallmatrix} a_1 & x \\ & a_2 \end{smallmatrix}\big)\,dx.$$

If g is not upper triangular, then the function $x \mapsto e_q\big(\mathrm{tr}\big(\begin{smallmatrix} a_1 & x \\ & a_2 \end{smallmatrix}\big)g\big)$ is not constant on $q^{-1}\mathbb{Z}_q$, and therefore the integral is zero. Let $B \subset GL_2(\mathbb{Z}/q\mathbb{Z})$ be the subgroup of upper triangular matrices. It follows that

$$Sf_q(a) = q\mathbf{1}_{q^{-1}\mathbb{Z}_q}(a_1)\mathbf{1}_{q^{-1}\mathbb{Z}_q}(a_2)\left(\frac{1}{|a_1||a_2|}\right)^{\frac{1}{2}}\sum_{g\in B} e_q\big(\mathrm{tr}(ag)\big)$$

$$= q\mathbf{1}_{q^{-1}M_2(\mathbb{Z}_q)}(a)\left(\frac{1}{|a|}\right)^{\frac{1}{2}}\sum_{g\in B} e_q\big(\mathrm{tr}(ag)\big)$$

$$= q^2\mathbf{1}_{q^{-1}M_2(\mathbb{Z}_q)}(a)\left(\frac{1}{|a|}\right)^{\frac{1}{2}}\sum_{\alpha,\beta\in(\mathbb{Z}/q\mathbb{Z})^{\times}} e_q(\alpha a_1 + \beta a_2).$$

We conclude

$$\chi_{\pi_q}\big(|\cdot|^{s+\frac{1}{2}} f_q\big)$$

$$= \int_{A_q} S(f_q)(a)|a|^{s+\frac{1}{2}}a^{\lambda_\pi}\,da$$

$$= q^2(q-1)^2 \int_{A\cap q^{-1}M_2(\mathbb{Z}_q)} e_q(a_1+a_2)|a|^s a^{\lambda_\pi}\,da$$

$$= q^2(q-1)^2 \sum_{i,j=-1}^{\infty} q^{-i(\lambda_1+s)}q^{-j(\lambda_2+s)}\underbrace{\int_{A\cap GL_2(\mathbb{Z}_q)} e_q(q^i a_1 + q^j a_2)\,da}_{=c_{i,j}}.$$

We compute

$$c_{-1,-1} = \int_{A\cap GL_2(\mathbb{Z}_q)} e_q\left(\frac{a_1+a_2}{q}\right)da$$

$$= \int_{\mathbb{Z}_q^{\times}}\int_{\mathbb{Z}_q^{\times}} e_q\left(\frac{a_1+a_2}{q}\right)d^{\times}a_1\,d^{\times}a_2$$

$$= \frac{q^2}{(q-1)^2}\int_{\mathbb{Z}_q^{\times}}\int_{\mathbb{Z}_q^{\times}} e_q\left(\frac{a_1+a_2}{q}\right)da_1\,da_2$$

$$= \frac{q^2}{(q-1)^2}\left(\int_{\mathbb{Z}_q^{\times}} e_q\left(\frac{x}{q}\right)dx\right)^2$$

$$= \frac{q^2}{(q-1)^2}\left(\underbrace{\int_{\mathbb{Z}_q} e_q\left(\frac{x}{q}\right)dx}_{=0} - \underbrace{\int_{q\mathbb{Z}_q^{\times}} e_q\left(\frac{x}{q}\right)dx}_{=1/q}\right)^2 = \frac{1}{(q-1)^2}.$$

Analogously we get $c_{-1,j} = c_{i,-1} = \frac{-1}{q-1}$ and $c_{i,j} = 1$ for $i, j \geq 0$. So we end up by showing that $\chi_{\pi_q}(|.|^{s+\frac{1}{2}} f_q)$ equals

$$q^{\lambda_1+\lambda_2+2s+2} - \frac{(q-1)q^{\lambda_1+s+2}}{1-q^{-(\lambda_2+s)}} - \frac{(q-1)q^{\lambda_2+s+2}}{1-q^{-(\lambda_1+s)}}$$
$$+ \frac{1}{(1-q^{-(\lambda_1+s)})(1-q^{-(\lambda_2+s)})},$$

which is a non-zero rational function in q^{-s}. We write it as $Q(q^{-s})$. We have shown

$$\zeta_F(f, \varphi, s) = Q(q^{-s})\eta\left(\bigotimes_{p \neq q \in F} \int_{G_p} |x|_p^{s+\frac{1}{2}} f_p(x)R(x)v_p \, dx \otimes v_q\right)(1).$$

We have to show that we can choose the functions f_p for $p \neq q$, $p \in F$ in such a way that $\zeta_F(f, \varphi, s)$ is a non-zero function. For example, choose $k \in \mathbb{N}$ so large that the non-zero vector v_p is stable under the group $1 + p^k M_2(\mathbb{Z}_p)$. Let $f_p = 1_{1+p^k M_2(\mathbb{Z}_p)}$. Then

$$\int_{G_p} |x|_p^{s+\frac{1}{2}} f_p(x)R(x)v_p \, dx = cv_p,$$

where

$$c = \int_{G_p} |x|_p^{s+\frac{1}{2}} f_p(x) \, dx = \text{vol}\left(1 + p^k M_2(\mathbb{Z}_p)\right) > 0.$$

With this choice at every $p \neq q$ we get

$$\zeta_F(f, \varphi, s) = CQ(q^{-s}) \int_{G_\infty} |x|_\infty^{s+\frac{1}{2}} f_\infty(x)\varphi(x) \, dx,$$

where $C > 0$. The contribution at infinity, f_∞, is of compact support inside G_∞, so the integral converges for every $s \in \mathbb{C}$ and defines an entire function. It could be zero, though. Letting f_∞ run through a Dirac sequence, we see that there exists a function f_∞ such that this integral is non-zero since $\varphi(1) = 1$. It follows that $\zeta_F(f, \varphi, s)$ extends meromorphically and non-zero to $\mathbb{C}$ and by part (a) then so does $L^F(\pi, s)$. The theorem follows. $\qquad\qquad\square$

One can show that there are further local Euler factors $L(\pi_p, s)$ for the places $p \in F$, such that $L(\pi_p, s)^{-1}$ is an exponential polynomial at every finite place and $L(\pi_\infty, s)$ is an exponential times a Γ-factor, such that the function

$$L(\pi, s) = \prod_{p \leq \infty} L(\pi_p, s)$$

satisfies a functional equation of the form

$$L(\pi, s) = a_\pi b_\pi^s L(\pi, 1-s),$$

where $a_\pi, b_\pi \in \mathbb{C}$ with $b_\pi > 0$.

8.4 The Example of Classical Cusp Forms

In this section we show that the definition of an L-function given in the last section is compatible with the notion of L-functions for classical cusp forms as in Sect. 2.4. We show that in the case of a classical cusp form both definitions in fact give the same function.

Let $\Gamma = \mathrm{SL}_2(\mathbb{Z})$ be the modular group and let $f \in S_k(\Gamma)$ be a holomorphic cusp form of weight $k \in 2\mathbb{N}_0$. So $f : \mathbb{H} \to \mathbb{C}$ is a holomorphic function satisfying the equation $f(\gamma z) = (cz + d)^k f(z)$ for every $\gamma = \begin{pmatrix} * & * \\ c & d \end{pmatrix} \in \Gamma$ and having a Fourier expansion of the form

$$f(z) = \sum_{n=1}^{\infty} a_n e^{2\pi i n z}.$$

Its L-function is defined by

$$L(f, s) = \sum_{n=1}^{\infty} a_n n^{-s}.$$

The series converges for $\mathrm{Re}(s) > \frac{k}{2} + 1$ and the ensuing function possesses a continuation to an entire function satisfying the functional equation

$$\Lambda(f, s) = (-1)^{k/2} \Lambda(f, k - s),$$

where $\Lambda(f, s) = (2\pi)^{-s} \Gamma(s) L(f, s)$. We choose f to be a simultaneous eigenfunction of all Hecke operators and normalize it in such a way that the first Fourier coefficient a_1 equals 1. Then the nth coefficient a_n is the eigenvalue of f under the Hecke operator T_n. It follows that $L(f, s)$ has an Euler product expansion

$$L(f, s) = \prod_p \frac{1}{1 - a_p p^{-s} + p^{k-1-2s}},$$

where the product is extended over all prime numbers.

We need to extend the action of $\mathrm{SL}_2(\mathbb{R})$ on the upper half plane to the group $G_\infty = \mathrm{GL}_2(\mathbb{R})$. The latter group acts on the upper half plane by

$$gz = \begin{pmatrix} a & b \\ c & d \end{pmatrix} z \overset{\text{def}}{=} \begin{cases} \frac{az+b}{cz+d} & \text{if } \det g > 0, \\ \frac{a\bar{z}+b}{c\bar{z}+d} & \text{if } \det g < 0. \end{cases}$$

Define the function

$$\phi_f(x) = \det(x)^{k/2} (ci + d)^{-k} f(xi), \qquad x = \begin{pmatrix} a & b \\ c & d \end{pmatrix} \in G_\infty.$$

Then $\phi_f \in L^2(\mathbb{Z}_\mathbb{R} \Gamma \backslash G_\infty)$. We define $\varphi_f \in L^2(\mathbb{Z}_\mathbb{R} G_\mathbb{Q} \backslash G_\mathbb{A})$ by

$$\varphi_f(1, x) = \phi_f(x), \qquad x \in G_\infty.$$

Theorem 8.4.1 *The function φ_f lies in L^2_{cusp}. The center $Z_{\mathbb{A}} = \mathbb{A}^\times \left(\begin{smallmatrix} 1 \\ & 1 \end{smallmatrix}\right)$ of $G_{\mathbb{A}}$ acts trivially on φ_f, so $R(z)\varphi_f = \varphi_f$ holds for every $z \in Z_{\mathbb{A}}$. The element φ_f generates an irreducible representation π_f of $G_{\mathbb{A}}$. For $F = \{\infty\}$ one has*

$$L^F(\pi_f, s) = L^\infty(\pi_f, s) = L\left(f, s + \frac{k-1}{2}\right).$$

The theorem asserts that the closure of the space

$$\text{Span}\big(R(G_{\mathbb{A}})\varphi_f\big)$$

is the space of an irreducible $G_{\mathbb{A}}$-subrepresentation π_f of $L^2(Z_{\mathbb{R}}G_{\mathbb{Q}}\backslash G_{\mathbb{A}})$, where R is the representation of $G_{\mathbb{A}}$ defined by right translation on $L^2(Z_{\mathbb{R}}G_{\mathbb{Q}}\backslash G_{\mathbb{A}})$.

Proof The function φ_f lies in the cuspidal subspace as is immediate from the fact that f is a cusp form. Let $z \in Z_{\mathbb{A}}$. By $\mathbb{A}^\times = \mathbb{Q}^\times \mathbb{R}^\times \widehat{\mathbb{Z}}^\times$ every $z \in Z_{\mathbb{A}}$ is a product of an element of $G_{\mathbb{Q}}$, an element of $Z_{\mathbb{R}}$, and an element of $G_{\widehat{\mathbb{Z}}}$. It follows that φ_f is stable under the group $Z_{\mathbb{A}}$. Since L^2_{cusp} is a direct sum of irreducible representations, the function φ_f is a sum whose summands are in different irreducible representations. These are all $Z_{\mathbb{A}}$-trivial. By Lemma 8.2.6 the vector φ_f is trivial under the algebra $\mathcal{Z}_p$ for every $p < \infty$. By Exercise 7.11 and Lemma 8.2.5 the one-dimensional space $\mathbb{C}\varphi_f$ is stable under the Hecke algebra $\mathcal{H}_p^{K_p}$ for every prime number p. According to Lemma 7.5.30, the vector φ_f generates an irreducible representation of the group G_{fin}.

The same arguments work at the infinite place as well. First we note that φ_f lies in the SO(2)-isotypical space $L^2(Z_{\mathbb{R}}G_{\mathbb{Q}}\backslash G_{\mathbb{A}})(\tau)$, where $\tau = \varepsilon_{-k}$ is the character of the group SO(2), which is given by the weight k. We have to show that $R(h)\varphi_f = c(h)\varphi_f$ holds with $c(h) \in \mathbb{C}$, if $h \in C_\tau$. Since L^2_{cusp} is a direct sum of irreducibles, one has $\varphi_f = \sum_{i\in I} \varphi_i$, where each φ_i lies in one irreducible representation. For $h \in C_\tau$ one has $R(h)\varphi_f = \sum_{i\in I} c_i(h)\varphi_i$ with scalars $c_i(h) \in \mathbb{C}$. The vectors φ_i are smooth by Lemma 3.4.2. Let D be the differential operator of Exercise 3.7. Then $D\varphi_i = P_i(D\varphi_f) = 0$, where P_i is the projection onto the ith summand. It follows that $DR(h)\varphi_f = 0$. This means that $R(h)\varphi_f$ comes from a holomorphic function in the sense of Exercise 3.7. We infer that $R(h)\varphi_f = \varphi_{f'}$ holds for a cusp form f'. This has the same eigenvalues as f under the finite Hecke operators $R(g_p)$, since these operators commute with $R(h)$. By Theorem 2.5.21, the function f' has, up to a scalar factor, the same Fourier coefficients as f, and hence we get $f' = cf$ for some $c \in \mathbb{C}$. This means that $\mathbb{C}\varphi_f$ is an irreducible module of the Hecke algebra C_τ, therefore φ_f generates an irreducible representation U of the group $\tilde{H} = G_{\text{fin}} \times \text{SL}_2(\mathbb{R})$. The group $Z_{\mathbb{R}}$ acts trivially and the group $H = \tilde{H}/Z_{\mathbb{R}}$ has index 2 in $G = G_{\mathbb{A}}/Z_{\mathbb{R}}$, more precisely, $G = H \cup \omega H$ with $\omega = \left(\begin{smallmatrix} -1 \\ & 1 \end{smallmatrix}\right)$. There are two cases: either $R(\omega)U = U$, and then U is stable under $G_{\mathbb{A}}$ and therefore irreducible; or one has $R(\omega)U \perp U$, in which case $U \oplus R(\omega)U$ is irreducible. So in either case φ_f generates an irreducible representation of the group $G_{\mathbb{A}}$.

Now for the L-functions. Let a_n be the nth Fourier coefficient of f. Then

$$L(f,s) = \sum_{n=1}^{\infty} a_n n^{-s} = \prod_{p<\infty} \frac{1}{1 - a_p p^{-s} + p^{k-1-2s}},$$

and so

$$L\left(f, s + \frac{k-1}{2}\right) = \prod_p \frac{1}{1 - a_p p^{\frac{1-k}{2}} p^{-s} + p^{-2s}}.$$

By Exercise 7.11 the eigenvalue of $R(g_p)$ on $\pi_f^{K_p}$ equals $a_p p^{1-k/2}$, where $g_p = 1_{K_p\left(\begin{smallmatrix} p^{-1} & \\ & 1 \end{smallmatrix}\right)K_p}$.

Lemma 8.4.2 *We compute the Satake transform of g_p as a function on $\overline{A} = A_p/A_p \cap K_p$ to*

$$S(g_p) = p^{\frac{1}{2}}(1_{\left(\begin{smallmatrix} p^{-1} & \\ & 1 \end{smallmatrix}\right)} + 1_{\left(\begin{smallmatrix} 1 & \\ & p^{-1} \end{smallmatrix}\right)}).$$

Proof We compute

$$S(g_p)(a) = \left(\frac{|a_1|}{|a_2|}\right)^{\frac{1}{2}} \int_{\mathbb{Q}_p} 1_{K_p\left(\begin{smallmatrix} p^{-1} & \\ & 1 \end{smallmatrix}\right)K_p} \begin{pmatrix} a_1 & a_1 x \\ & a_2 \end{pmatrix} dx.$$

The double coset $K_p\left(\begin{smallmatrix} p^{-1} & \\ & 1 \end{smallmatrix}\right)K_p$ is the union of the simple cosets

$$\bigcup_{0 \le b < p} \begin{pmatrix} 1 & -b/p \\ & 1/p \end{pmatrix} K_p \cup \begin{pmatrix} p^{-1} & \\ & 1 \end{pmatrix} K_p.$$

If $\left(\begin{smallmatrix} a_1 & a_1 x \\ & a_2 \end{smallmatrix}\right)$ lies in the class $\left(\begin{smallmatrix} 1 & -b/p \\ & 1/p \end{smallmatrix}\right) K_p$, then there is a $k \in K_p$ such that $\left(\begin{smallmatrix} a_1 & a_1 x \\ & a_2 \end{smallmatrix}\right) = \left(\begin{smallmatrix} 1 & -b/p \\ & 1/p \end{smallmatrix}\right) k$. Then k is an upper triangular matrix, so $k = \left(\begin{smallmatrix} \alpha & \beta \\ & \delta \end{smallmatrix}\right)$ with $\alpha, \delta \in \mathbb{Z}_p^{\times}$ and $\beta \in \mathbb{Z}_p$. Therefore,

$$\begin{pmatrix} a_1 & a_1 x \\ & a_2 \end{pmatrix} = \begin{pmatrix} 1 & -b/p \\ & 1/p \end{pmatrix} \begin{pmatrix} \alpha & \beta \\ & \delta \end{pmatrix} = \begin{pmatrix} \alpha & \beta - \delta b/p \\ & \delta/p \end{pmatrix}.$$

We distinguish two cases:

1. If $b \ne 0$, then $a_1 \in \mathbb{Z}_p^{\times}$, $a_2 \in \frac{1}{p}\mathbb{Z}_p^{\times}$ and $x \in \frac{1}{p}\mathbb{Z}_p^{\times}$. Since the additive volume of $\frac{1}{p}\mathbb{Z}_p^{\times} = \frac{1}{p}\mathbb{Z}_p \setminus \mathbb{Z}_p$ equals $(p-1)$, the integral over $x \in \frac{1}{p}\mathbb{Z}_p^{\times}$ yields a summand of the form

$$\frac{p-1}{p^{\frac{1}{2}}} 1_{\left(\begin{smallmatrix} 1 & \\ & p^{-1} \end{smallmatrix}\right)}.$$

2. If $b = 0$, then $a_1 \in \mathbb{Z}_p^{\times}$, $a_2 \in \frac{1}{p}\mathbb{Z}_p^{\times}$ and $x \in \mathbb{Z}_p$. The integral over $x \in \mathbb{Z}_p$ in this case yields a summand

$$\frac{1}{p^{\frac{1}{2}}} 1_{\left(\begin{smallmatrix} 1 & \\ & p^{-1} \end{smallmatrix}\right)}.$$

We finally consider the coset of $\begin{pmatrix} p^{-1} \\ & 1 \end{pmatrix}$. One has

$$\begin{pmatrix} a_1 & a_1 x \\ & a_2 \end{pmatrix} = \begin{pmatrix} 1/p \\ & 1 \end{pmatrix}\begin{pmatrix} \alpha & \beta \\ & \delta \end{pmatrix} = \begin{pmatrix} \alpha/p & \beta/p \\ & \delta \end{pmatrix}.$$

This implies $a_1 \in \frac{1}{p}\mathbb{Z}_p^\times$, $a_2 \in \mathbb{Z}_p^\times$ and $x \in \mathbb{Z}_p$, so we obtain a summand

$$p^{\frac{1}{2}}\mathbf{1}_{\begin{pmatrix} p^{-1} \\ & 1 \end{pmatrix}}.$$

Putting things together, the lemma follows.   □

We give the proof of the theorem. Write $\pi_f = \bigotimes_p \pi_p$. For a prime number p the unramified representation π_p is self-dual and we have $\lambda_{\pi'} = \lambda_\pi^{-1}$, and so $\lambda_{\pi_p} = \begin{pmatrix} \lambda \\ & 1/\lambda \end{pmatrix} \in T/W$ for a $\lambda \in \mathbb{C}^\times$. The local factor then equals

$$L(\pi_p, s) = \frac{1}{(1 - \lambda p^{-s})(1 - \frac{1}{\lambda}p^{-s})} = \frac{1}{1 - (\lambda + \frac{1}{\lambda})p^{-s} + p^{-2s}}.$$

Since $p^{1-k/2}a_p$ is the eigenvalue of $R(g_p)$, and hence equal to $\chi_{\pi_p}(g_p)$, it follows that

$$p^{\frac{1}{2}}\left(\lambda + \frac{1}{\lambda}\right) = p^{1-k/2}a_p,$$

so

$$\lambda + \frac{1}{\lambda} = p^{\frac{1-k}{2}}a_p.$$

The theorem is proven.   □

8.5 Exercises and Remarks

Exercise 8.1 Let $(H_i)_{i \in I}$ be a family of Hilbert spaces. Show that the prescription

$$\left\langle \sum_{i \in I} v_i, \sum_{i \in I} w_i \right\rangle = \sum_{i \in I} \langle v_i, w_i \rangle_i$$

defines an inner product on the algebraic direct sum $\bigoplus_{i \in I} H_i$, making it a pre-Hilbert space, whose completion can be described as the set of all $v \in \prod_{i \in I} H_i$ such that $\sum_{i \in I} \|v_i\|_i^2 < \infty$, where the inner product is given by the same formula as above, only it need not be finite anymore.

Exercise 8.2 Show that $da\,dn$ is a Haar measure of the group $B = A_p N_p$ of upper triangular matrices in G_p.

Exercise 8.3 Let λ be an unramified quasi-character of A_p, so $\lambda\begin{pmatrix} a_1 \\ & a_2 \end{pmatrix} = |a_1|^{\lambda_1}|a_2|^{\lambda_2}$ for two complex numbers $\lambda_1, \lambda_2 \neq 0$. Show that the Satake parameter of the representation π_λ is given by the matrix $\begin{pmatrix} p^{-\lambda_1} \\ & p^{-\lambda_2} \end{pmatrix}$.

Exercise 8.4 Let (π, V) be a representation of the locally compact group G. Show that the map $g \mapsto \|\pi(g)\|_{\mathrm{op}}$ from G to $(0, \infty)$ is bounded on every compact set $K \subset G$.

Exercise 8.5 Let (π, V) be a Hilbert representation, i.e. the space V is a Hilbert space. Show that there is a canonical isomorphism of representations $\pi \to \pi''$. Conclude that π is irreducible if and only if π' is irreducible.

Exercise 8.6 Let V be a finite-dimensional complex vector space and let V^* be its dual space. We write $(v, \alpha) = \alpha(v)$ for $v \in V$ and $\alpha \in V^*$. Let G be a group, which we equip with the discrete topology, and let $\pi : G \to \mathrm{GL}(V)$ be a representation. Then V is a $\mathbb{C}[G]$-module. For $f \in \mathbb{C}[G]$ let $f^\vee(x) = f(x^{-1})$. Show: If $\eta : G \to \mathrm{GL}(V^*)$ is a representation with $(\pi(f)v, \alpha) = (v, \eta(f^\vee)\alpha)$ for every $f \in \mathbb{C}[G]$, then $\eta \cong \pi'$.

Exercise 8.7 Show that an irreducible unramified unitary representation η of the group $G_p = \mathrm{GL}_2(\mathbb{Q}_p)$ is isomorphic to a principal series representation π_λ.

Exercise 8.8 Show that an irreducible Hilbert representation of the group $G_p = \mathrm{GL}_2(\mathbb{Q}_p)$ is unramified if and only if its dual π' is unramified and that in this case one has

$$\lambda_{\pi'} = \lambda_\pi^{-1}.$$

Exercise 8.9 Show that the contents of Sect. 8.4 are analogously true for Maaß wave forms.
(This exercise is a bit involved.)

Remarks   We end this book with a few remarks on the literature. We give only a few hints and by no means shall we try to give a comprehensive overview. We order the books alphabetically and start accordingly with the book by Tom Apostol [Apo90]. This book describes, in more detail than we have done, classical modular forms and their number-theoretical applications. If you want to learn more about those, Apostol's book is a good point to start.

The book of Daniel Bump [Bum97] contains all I intended to say in this book, and more. I can recommend it highly for further study. It is, however, a demanding read.

If you are interested in automorphic L-functions, converse theorems and their meaning within the Langlands program, then you should read the book by James Cogdell, Henry Kim and Ram Murty [CKM04].

In order to learn about the trace formula, one should read Stephen Gelbart's book [Gel96]. It contains an elementary introduction to this most important tool in automorphic theory.

A true classic is the 1969 book by Gel'fand, Graev and Pyatetskii-Shapiro [GGPS90]. It is wonderfully written. This book marks the triumph of representation-theoretic methods in the theory of automorphic forms.

The book by Dorian Goldfeld [Gol06] mostly uses classical techniques. It is easy to read and one gets applications pretty quickly.

The books [GH11a, GH11b], which appeared after the German version of this one, are ideal for further study. They are more elaborate and contain more material along the same vein as the present work.

The book of Haruzo Hida [Hid93] is quite interesting. It does not use representation theory, but instead a lot of cohomological arguments, which in the current book have not appeared at all. It therefore takes a complementary viewpoint and thus completes the scene quite nicely.

The connection to elliptic curves has, in this book, been mentioned only briefly in Chaps. 1 and 2. If you intend to go into this, there is another book by Hida, [Hid00], which you should read. It requires a modest background in algebraic geometry.

The book by Henryk Iwaniec [Iwa02] focuses on Maaß wave forms and their analytic aspects, including the trace formula. I recommend this book as an introduction to the trace formula. However, one should not stop here, as this book only uses classical tools and in particular the trace formula only gains its full strength in the representation-theoretic context.

If you find that your knowledge about representation theory of Lie groups is insufficient, I recommend the book by Anthony Knapp [Kna01]. It is one of the best math books ever written.

Appendix
Measure and Integration

In this appendix we collect some facts from measure theory and integration which are used in the book. We recall the basic definitions of measures and integrals and give the central theorems of Lebesgue integration theory. Proofs may be found for instance in [Rud87].

A.1 Measurable Functions and Integration

Let X be a set. A *σ-algebra* on X is a set $\mathcal{A}$, whose elements are subsets of X, such that

- the empty set lies in $\mathcal{A}$ and if $A \in \mathcal{A}$, then its complement $X \smallsetminus A$ lies in $\mathcal{A}$,
- the set $\mathcal{A}$ is closed under countable unions.

It follows that a σ-algebra is closed under countable sections and that with A, B the set $A \smallsetminus B$ lies in $\mathcal{A}$.

The set $\mathcal{P}(X)$ of all subsets of X is a σ-algebra and the intersection of arbitrary many σ-algebras is a σ-algebra. This implies that for an arbitrary set $S \subset \mathcal{P}(X)$ there exists a smallest σ-algebra $\mathcal{A}$ containing S. In this case we say that S generates $\mathcal{A}$. For a topological space X, the σ-algebra $\mathcal{B} = \mathcal{B}(X)$ generated by the topology of X, is called the *Borel σ-algebra* of X. The elements of $\mathcal{B}(X)$ are called *Borel sets*. If $\mathcal{A}$ is a σ-algebra on X, then the pair $(X, \mathcal{A})$ is called a *measurable space*. The elements of $\mathcal{A}$ are called *measurable sets*.

A map $f : X \to Y$ between two measurable spaces is called a *measurable map* if the preimage $f^{-1}(A)$ is measurable for every measurable set $A \in \mathcal{A}_Y$. The composition of two measurable maps is measurable.

We equip the real line $\mathbb{R}$ and the complex plane $\mathbb{C}$ with its respective Borel σ-algebra.

Lemma A.1.1 *Let $(X, \mathcal{A})$ be a measurable space.*

(a) *A function $f : X \to \mathbb{R}$ is measurable if and only if for every $a \in \mathbb{R}$ the set $f^{-1}((a, \infty))$ is in $\mathcal{A}$.*

(b) *A function $f : X \to \mathbb{C}$ is measurable if and only if* Re f *and* Im f *are measurable.*

(c) *If $f, g : X \to \mathbb{C}$ are measurable, then so are $f + g$, $f \cdot g$, and $|f|^p$ for $p > 0$.*

(d) *If $f, g : X \to \mathbb{R}$ are measurable, then so are* $\max(f, g)$ *and* $\min(f, g)$.

(e) *If a sequence of measurable functions $f_n : X \to \mathbb{C}$ converges point-wise to a function $f : X \to \mathbb{C}$, then f is measurable as well.*

In the following it is helpful to consider functions with values in the interval $[0, \infty]$, where we equip $[0, \infty]$ with the obvious topology and the corresponding Borel Σ-algebra. A function $f : X \to [0, \infty]$ is measurable if and only if $f^{-1}((a, \infty]) \in \mathcal{A}$ holds for every $a \in \mathbb{R}$. The assertions (c), (d) and (e) of the lemma remain valid for functions $f : X \to [0, \infty]$.

A *measure* μ on a measurable space $(X, \mathcal{A})$ is a map $\mu : \mathcal{A} \to [0, \infty]$ such that $\mu(\emptyset) = 0$ and

- $\mu(\bigcup_{n=1}^{\infty} A_n) = \sum_{n=1}^{\infty} \mu(A_n)$ holds for every sequence $(A_n)_{n \in \mathbb{N}}$ of pairwise disjoint sets $A_n \in \mathcal{A}$.

It is easy to deduce the following.

- $\mu(A \cup B) = \mu(A) + \mu(B) - \mu(A \cap B)$ for all $A, B \in \mathcal{A}$.
- For a sequence $(A_n)_{n \in \mathbb{N}}$ in $\mathcal{A}$ with $A_n \subseteq A_{n+1}$ for all $n \in \mathbb{N}$, the sequence $\mu(A_n)$ converges to $\mu(A)$, where $A = \bigcup_{n=1}^{\infty} A_n$.
- For a sequence $(A_n)_{n \in \mathbb{N}}$ in $\mathcal{A}$ with $A_n \supseteq A_{n+1}$ for all $n \in \mathbb{N}$ and $\mu(A_1) < \infty$, the sequence $\mu(A_n)$ converges to $\mu(A)$, where $A = \bigcap_{n=1}^{\infty} A_n$.

Let $\mu : \mathcal{A} \to [0, \infty]$ be a measure on $(X, \mathcal{A})$. The triple $(X, \mathcal{A}, \mu)$ is called a *measure space*.

Let $(X, \mathcal{A}, \mu)$ be a measure space. A *step function* is a measurable function $s : X \to [0, \infty]$ which takes only finitely many values. Any such function is of the form $s = \sum_{i=1}^{m} a_i 1_{A_i}$ with pairwise disjoint $A_i \in \mathcal{A}$. For such a step function we define its *integral* as

$$\int_X s \, d\mu \overset{\text{def}}{=} \sum_{i=1}^{m} a_i \mu(A_i) \in [0, \infty].$$

For a measurable function $f : X \to [0, \infty]$ we define

$$\int_X f \, d\mu = \sup \left\{ \int_X s \, d\mu : 0 \le s \le f; \ s \text{ is a step function} \right\}.$$

The function is called *integrable* if $\int_X f \, d\mu < \infty$. A measurable function $f : X \to \mathbb{R}$ is called integrable if $|f|$ is integrable. In that case the functions $f^+ = \max(f, 0)$ and $f^- = -\min(f, 0)$ are both integrable and we set $\int_X f \, d\mu = \int_X f^+ \, d\mu - \int_X f^- \, d\mu$. A complex valued function $f = u + iv$ is called integrable if its real and imaginary parts u, v are. In that case one defines $\int_X f \, d\mu = \int_X u \, d\mu + i \int_X v \, d\mu$.

Proposition A.1.2 *Let $(X, \mathcal{A}, \mu)$ be a measure space. A measurable function f : $X \to \mathbb{C}$ is integrable if and only if its absolute value $|f|$ has finite integral. In that case one has*

$$\left| \int_X f \, d\mu \right| \leq \|f\|_1 \stackrel{\text{def}}{=} \int_X |f| \, d\mu.$$

The following two theorems are of central importance.

Theorem A.1.3 (Monotone convergence theorem) *Let $(f_n)_{n\in\mathbb{N}}$ be a point-wise monotonically increasing sequence of measurable functions ≥ 0. For $x \in X$ set $f(x) = \lim_n f_n(x) \in [0, \infty]$. Then one has*

$$\int_X f \, d\mu = \lim_n \int_X f_n \, d\mu.$$

Theorem A.1.4 (Dominated convergence theorem) *Let $(f_n)_{n\in\mathbb{N}}$ be a sequence of complex valued integrable functions, which converges point-wise to a function f. Suppose there exists an integrable function g such that $|f_n| \leq |g|$ holds for every $n \in \mathbb{N}$. Then f is integrable and one has*

$$\int_X f \, d\mu = \lim_n \int_X f_n \, d\mu.$$

A.2 Fubini's Theorem

A measure μ on a measurable space $(X, \mathcal{A})$ is called a *σ-finite measure* if there are countably many subsets $X_j \subset X$, $j \in \mathbb{N}$ with $X = \bigcup_{j=1}^{\infty} X_j$ and $\mu(X_j) < \infty$ for every $j \in \mathbb{N}$.

Examples A.2.1

- The Lebesgue measure on $X = \mathbb{R}$ is σ-finite, since $\mathbb{R}$ can be written as a countable union of the intervals $[k, k+1]$ with $k \in \mathbb{Z}$.
- The counting measure is not σ-finite on $X = \mathbb{R}$, since $\mathbb{R}$ is uncountable.

For two σ-finite spaces $(X, \mathcal{A}, \mu)$ and $(Y, \mathcal{C}, \nu)$ one shows that there exists a unique measure $\mu \cdot \nu$ on the σ-algebra $\mathcal{A} \otimes \mathcal{C}$, which is generated by all sets of the form $\{A \times C : A \in \mathcal{A}, \ C \in \mathcal{C}\}$, such that

$$\mu \cdot \nu(A \times C) = \mu(A)\nu(C), \quad A \in \mathcal{A}, \ C \in \mathcal{C}.$$

The measure $\mu \cdot \nu$ is called the *product measure* of μ and ν.

Theorem A.2.2 (Fubini's theorem) *Let* (X, μ) *and* (Y, ν) *be* σ-*finite measure spaces and let* f *be a measurable function on* $X \times Y$.

(a) *If* $f \geq 0$, *then the partial integrals* $\int_X f(x, y) \, d\mu(x)$ *and* $\int_Y f(x, y) \, d\nu(y)$
    define measurable functions and one has the Fubini formula,

$$\int_{X \times Y} f(x, y) \, d\mu \cdot \nu(x, y) = \int_X \int_Y f(x, y) \, d\nu(y) \, d\mu(x)$$

$$= \int_Y \int_X f(x, y) \, d\mu(x) \, d\nu(y).$$

(b) *If* f *is complex valued and if one of the iterated integrals*

$$\int_X \int_Y |f(x, y)| \, d\nu(y) \, d\mu(x) \quad or \quad \int_Y \int_X |f(x, y)| \, d\mu(x) \, d\nu(y)$$

is finite, then f *is integrable with respect to the product measure and the Fubini formula holds.*

In this book, we use the Fubini theorem for Haar measures only. All Haar measures occurring in this book are σ-finite. But as we did not mention this explicitly each time, we will also give a version of Fubini's theorem for Radon measures which works without the σ-finiteness condition (see [DE09], Appendix).

Theorem A.2.3 (Theorem of Fubini for Radon measures) *Let* μ *and* ν *be Radon measures on the Borel sets of locally compact spaces* X *and* Y, *respectively. Then there exists a unique Radon measure* $\mu \cdot \nu$ *on* $X \times Y$ *such that*

1. *If* $f : X \times Y \to \mathbb{C}$ *is* $\mu \cdot \nu$-*integrable, then the partial integrals* $\int_X f(x, y) \, dx$ *and* $\int_Y f(x, y) \, dy$ *define integrable functions such that Fubini's formula holds:*

$$\int_{X \times Y} f(x, y) \, d(x, y) = \int_X \int_Y f(x, y) \, dy \, dx = \int_Y \int_X f(x, y) \, dx \, dy.$$

2. *If* f *is measurable such that* $A = \{(x, y) \in X \times Y : f(x, y) \neq 0\}$ *is* σ-*finite, and if one of the iterated integrals*

$$\int_X \int_Y |f(x, y)| \, dy \, dx \quad or \quad \int_Y \int_X |f(x, y)| \, dx \, dy$$

is finite, then f *is integrable and the Fubini formula holds.*

A.3 L^p-Spaces

Let $(X, \mathcal{A}, \mu)$ be a measure space. For $1 \le p < \infty$ write $\mathcal{L}^p(X)$ for the set of all measurable functions $f : X \to \mathbb{C}$ such that

$$\|f\|_p \overset{\text{def}}{=} \left(\int_X |f|^p \, d\mu \right)^{\frac{1}{p}} < \infty.$$

A function in $\mathcal{L}^1(X)$ is called integrable, as we already know. A function in $\mathcal{L}^2(X)$ is called *square integrable*. Further, let $\mathcal{L}^\infty(X)$ be the set of all measurable functions $f : X \to \mathbb{C}$ for which there exists a set N of measure zero such that f is bounded on the complement $X \smallsetminus N$. Then

$$\|f\|_\infty \overset{\text{def}}{=} \inf \{0 < c \le \infty : \exists \text{ set of measure zero } N \text{ with } |f(X \smallsetminus N)| \le c\}$$

is a semi-norm on the complex vector space $\mathcal{L}^\infty(X)$.

Proposition A.3.1 (Minkowski inequality) *Let* $p \in [1, \infty]$. *For all* $f, g \in \mathcal{L}^p(X)$ *one has* $f + g \in \mathcal{L}^p(X)$ *with*

$$\|f + g\|_p \le \|f\|_p + \|g\|_p.$$

So $\| \cdot \|_p$ *is a semi-norm on* $\mathcal{L}^p(X)$.

A measurable function, being zero outside a set of measure zero, is called a *null-function*. This is equivalent to $\|f\|_p = 0$ for any $1 \le p \le \infty$. Write $\mathcal{N}$ for the vector space of null-functions and for $1 \le p \le \infty$ define

$$L^p(X) \overset{\text{def}}{=} \mathcal{L}^p(X)/\mathcal{N}.$$

Then $\| \cdot \|_p$ is a norm on $L^p(X)$. It is the content of the theorem of Riesz and Fischer, that $L^p(X)$ is actually complete, i.e. a Banach space. An important special case occurs for $p = 2$, as in this case $L^2(X)$ is a Hilbert space with inner product

$$\langle f, g \rangle = \int_X f(x) \overline{g(x)} \, d\mu(x).$$

In this book we frequently use the lemma of Urysohn.

Lemma A.3.2 (Lemma of Urysohn) *Let X be a locally compact Hausdorff space. Let $K \subset X$ be compact and let $A \subset X$ be closed with $K \cap A = \emptyset$.*

(i) *There exists an open neighborhood U of K, which has compact closure $\overline{U}$ and satisfies $K \subset U \subset \overline{U} \subset X \smallsetminus A$.*
(ii) *There is a continuous function of compact support $f : X \to [0, 1]$ with $f \equiv 1$ on K and $f \equiv 0$ on A.*

(iii) *Let $B \subset X$ be closed. Let $h : B \to [0, \infty)$ in $C_0(B)$ with $h(x) \geq 1$ for every $x \in K \cap B$. Then there exists a continuous function f as in (ii) with the additional property that $f(b) \leq h(b)$ for every $b \in B$.*

This lemma is of utmost importance and the proof is not difficult. We give it here.

Proof For the first assertion let $a \in A$. For every $k \in K$ there exists an open neighborhood U_k of k with compact closure $\overline{U_k}$, which is disjoint to a neighborhood $U_{k,a}$ of a. The family $(U_k)_{k \in K}$ is an open covering of the compact set K, so there exists a finite subcovering. Let V be the union of the sets of this subcovering and let W be the finite intersection of the corresponding $U_{k,a}$. Then V and W are open disjoint neighborhoods of K and a, respectively. Further, V has compact closure. We repeat this argument with K taking the part of $\{a\}$ and $\overline{V} \cap A$ in the role of K. We obtain disjoint open neighborhoods U' of K and W' of $\overline{V} \cap A$. The set $U = U' \cap V$ satisfies the assertion (i).

For (ii), choose U as in the first part and replace A by $A \cup (X \setminus U)$. One sees that it suffices to show the claim without the condition that f has compact support. So let U be as in the first part and call this open set $U_{\frac{1}{2}}$. There exists an open neighborhood $U_{\frac{1}{4}}$ of K, having compact closure and satisfying

$$U_{\frac{1}{2}} \subset \overline{U_{\frac{1}{2}}} \subset U_{\frac{1}{4}}.$$

Likewise, one obtains an open set $U_{\frac{3}{4}}$ with compact closure such that

$$K \subset U_{\frac{3}{4}} \subset \overline{U_{\frac{3}{4}}} \subset U_{\frac{1}{2}}.$$

Let R be the set of all numbers of the form $\frac{k}{2^n}$ in the interval $[0, 1)$. Set $U_0 = X \setminus A$. By iteration of the above construction, we obtain open sets U_r, $r \in R$, with $K \subset U_r \subset \overline{U_r} \subset U_s \subset X \setminus A$ for all $r > s$ in R. We now define f. For $x \in A$ set $f(x) = 0$. For $x \in X \setminus A$ set $f(x) = \sup\{r \in R : x \in \overline{U_r}\}$. Then $f \equiv 1$ on K. For $r > s$ in R one has

$$f^{-1}(s, r) = \bigcup_{s < s' < s'' < r} U_{s'} \setminus \overline{U_{s''}}.$$

This set is open. Likewise the sets $f^{-1}([0, s))$ and $f^{-1}((r, 1])$ are open. Since the intervals of the form (r, s), $[0, s)$, and $(r, 1]$ generate the topology on $[0, 1]$, the map f is continuous.

The proof of (iii) is a variation of the last proof, where in each step of the construction of the sets U_r one uses the set $A \cup \{b \in B : h(b) \leq r\}$ in the place of A. $\square$

References

[AS64] Abramowitz, M., Stegun, I.A.: Handbook of Mathematical Functions with Formulas, Graphs, and Mathematical Tables. National Bureau of Standards Applied Mathematics Series, vol. 55. For sale by the Superintendent of Documents, US Government Printing Office, Washington (1964)

[Apo90] Apostol, T.M.: Modular Functions and Dirichlet Series in Number Theory, 2nd edn. Graduate Texts in Mathematics, vol. 41. Springer, New York (1990)

[Bum97] Bump, D.: Automorphic Forms and Representations. Cambridge Studies in Advanced Mathematics, vol. 55. Cambridge University Press, Cambridge (1997)

[BCdS+03] Bump, D., Cogdell, J.W., de Shalit, E., Gaitsgory, D., Kowalski, E., Kudla, S.S.: An Introduction to the Langlands Program. Birkhäuser Boston, Boston (2003). Lectures presented at the Hebrew University of Jerusalem, Jerusalem, March 12–16, 2001. Edited by Joseph Bernstein and Stephen Gelbart

[CKM04] Cogdell, J.W., Kim, H.H., Ram Murty, M.: Lectures on Automorphic L-Functions. Fields Institute Monographs, vol. 20. American Mathematical Society, Providence (2004)

[Con78] Conway, J.B.: Functions of One Complex Variable, 2nd edn. Graduate Texts in Mathematics, vol. 11. Springer, New York (1978)

[Dei05] Deitmar, A.: A First Course in Harmonic Analysis, 2nd edn. Universitext. Springer, New York (2005)

[DE09] Deitmar, A., Echterhoff, S.: Principles of Harmonic Analysis. Universitext. Springer, New York (2009)

[Gel96] Gelbart, S.: Lectures on the Arthur–Selberg Trace Formula. University Lecture Series, vol. 9. American Mathematical Society, Providence (1996)

[GGPS90] Gel'fand, I.M., Graev, M.I., Pyatetskii-Shapiro, I.I.: Representation Theory and Automorphic Functions. Generalized Functions, vol. 6. Academic Press, Boston (1990). Translated from the Russian by K.A. Hirsch. Reprint of the 1969 edition

[Gol06] Goldfeld, D.: Automorphic Forms and L-Functions for the Group $GL(n, \mathbb{R})$. Cambridge Studies in Advanced Mathematics, vol. 99. Cambridge University Press, Cambridge (2006). With an appendix by Kevin A. Broughan

[GH11a] Goldfeld, D., Hundley, J.: Automorphic Representations and L-Functions for the General Linear Group, vol. I. Cambridge Studies in Advanced Mathematics, vol. 129. Cambridge University Press, Cambridge (2011). With exercises and a preface by Xander Faber

[GH11b] Goldfeld, D., Hundley, J.: Automorphic Representations and L-Functions for the General Linear Group, vol. II. Cambridge Studies in Advanced Mathematics, vol. 130. Cambridge University Press, Cambridge (2011). With exercises and a preface by Xander Faber

[HC70] Harish-Chandra: Harmonic Analysis on Reductive p-Adic Groups. Lecture Notes in Mathematics, vol. 162. Springer, Berlin (1970). Notes by G. van Dijk

[HM98] Harris, J., Morrison, I.: Moduli of Curves. Graduate Texts in Mathematics, vol. 187. Springer, New York (1998)

[HH80] Hartley, B., Hawkes, T.O.: Rings, Modules and Linear Algebra. Chapman & Hall, London (1980)

[Haz01] Hazewinkel, M. (ed.): Encyclopaedia of Mathematics. Supplement, vol. III. Kluwer Academic, Dordrecht (2001)

[Heu06] Heuser, H.: Funktionalanalysis. Theorie und Anwendung [Functional Analysis. Theory and Application], 4th edn. Mathematische Leitfäden [Mathematical Textbooks]. Teubner, Stuttgart (2006)

[Hid93] Hida, H.: Elementary Theory of L-Functions and Eisenstein Series. London Mathematical Society Student Texts, vol. 26. Cambridge University Press, Cambridge (1993)

[Hid00] Hida, H.: Geometric Modular Forms and Elliptic Curves. World Scientific, River Edge (2000)

[Iwa02] Iwaniec, H.: Spectral Methods of Automorphic Forms, 2nd edn. Graduate Studies in Mathematics, vol. 53. American Mathematical Society, Providence (2002)

[KM85] Katz, N.M., Mazur, B.: Arithmetic Moduli of Elliptic Curves. Annals of Mathematics Studies, vol. 108. Princeton University Press, Princeton (1985)

[Kat04] Katznelson, Y.: An Introduction to Harmonic Analysis, 3rd edn. Cambridge Mathematical Library. Cambridge University Press, Cambridge (2004)

[Kna01] Knapp, A.W.: Representation Theory of Semisimple Groups. An Overview Based on Examples. Princeton Landmarks in Mathematics. Princeton University Press, Princeton (2001). Reprint of the 1986 original

[Lan02] Lang, S.: Algebra, 3rd edn. Graduate Texts in Mathematics, vol. 211. Springer, New York (2002)

[Lee03] Lee, J.M.: Introduction to Smooth Manifolds. Graduate Texts in Mathematics, vol. 218. Springer, New York (2003)

[Rob10] Roberts, C.E. Jr.: Ordinary Differential Equations. Applications, Models, and Computing. Textbooks in Mathematics. CRC Press, Boca Raton (2010). With 1 CD-ROM (Windows)

[Rud87] Rudin, W.: Real and Complex Analysis, 3rd edn. McGraw-Hill, New York (1987)

[Rud91] Rudin, W.: Functional Analysis, 2nd edn. International Series in Pure and Applied Mathematics. McGraw-Hill, New York (1991)

[Sil09] Silverman, J.H.: The Arithmetic of Elliptic Curves, 2nd edn. Graduate Texts in Mathematics, vol. 106. Springer, Dordrecht (2009)

[SS03] Stein, E.M., Shakarchi, R.: Complex Analysis. Princeton Lectures in Analysis, II. Princeton University Press, Princeton (2003)

[SW71] Stein, E.M., Weiss, G.: Introduction to Fourier Analysis on Euclidean Spaces. Princeton Mathematical Series, vol. 32. Princeton University Press, Princeton (1971)

[Str06] Stroppel, M.: Locally Compact Groups. EMS Textbooks in Mathematics. European Mathematical Society (EMS), Zürich (2006)

[Wei80] Weidmann, J.: Linear Operators in Hilbert Spaces. Graduate Texts in Mathematics, vol. 68. Springer, New York (1980). Translated from the German by Joseph Szücs

[Wil95] Wiles, A.: Modular elliptic curves and Fermat's last theorem. Ann. Math. (2) **141**(3), 443–551 (1995). doi:10.2307/2118559

[Yos95] Yosida, K.: Functional Analysis. Classics in Mathematics. Springer, Berlin (1995). Reprint of the sixth (1980) edition

Index

Symbols
*-algebra, 185
*-representation, 186
a_t, 80
$\underline{a}$, 80
$\underline{a}(g)$, 80
$\mathbb{A}$, 125
$\mathbb{A}_{\text{fin}}$, 125
$\mathbb{A}_S$, 125
$\mathbb{A}^1$, 129
$\mathbb{A}^\times$, 128
$C_c(X)$, 69
C_F, 199
$\widehat{\mathbb{C}}$, 16
D_Γ, 48
$d^\times x$, 115, 130
$e(x)$, 211
F-isotype, 199
F-simple, 225
$G_\mathbb{A}$, 163
G_p, 163
G_R, 163
G_S, 163
G_∞, 163
$G_\mathbb{R}^1$, 169
G^S, 163
$\widehat{G}$, 131
$\widehat{G}_K$, 196
$\overline{\Gamma}_0$, 18
$\text{GL}_2(\mathbb{Q})^+$, 43
$\text{GL}_2(R)$, 8
$\mathcal{H}_p^{K_p}$, 213
K-admissible representation, 196
K-Bessel function, 56
K-finite, 198
k_θ, 80
$\underline{k}$, 80

$\underline{k}(g)$, 80
$\text{M}_2(R)$, 8
$\mathcal{M}_k = \mathcal{M}_k(\Gamma_0)$, 27
$\text{M}_n(\mathbb{R})$, 97
N_R, 182
n_x, 80
$\underline{n}$, 80
$\underline{n}(g)$, 80
p-adic absolute value, 106
$\mathbb{P}^1(\mathbb{C})$, 16
S_k, 27
$\mathcal{S}(\text{M}_2(\mathbb{A}))$, 211
$\mathcal{S}(\mathbb{R})$, 57
σ-algebra, 241
σ-finite measure, 243
τ-isotype, 195
τ-isotypical component, 195
$\mathbb{T}$, 88, 93
Z_p-trivial, 219
$\mathcal{Z}_p$-trivial module, 218
$\mathbb{Z}$-basis, 2
$\widehat{\mathbb{Z}}$, 129

A
Abelianization, 158
Absolute value, 105
Absolute value of an idele, 129
Absolutely convergent, 97
Adeles, 125
Adjoint, 177
Adjoint operator, 68
Adjunction of a unit, 203
Admissible representation, 196
Algebra, 42, 84
Algebra generated by E, 42
Algebra homomorphism, 43
Almost all, 107

A. Deitmar, *Automorphic Forms*, Universitext,
DOI 10.1007/978-1-4471-4435-9, © Springer-Verlag London 2013